Tools, Techniques, and Advancements in Engineering Materials Science

Ajay Kumar
Department of Mechanical Engineering, School of Engineering and Technology, JECRC University, India

Parveen Kumar
Mechanical Engineering Department, Rawal Institute of Engineering and Technology, India

Tamer Ali Sebaey
Prince Sultan University, Saudi Arabia

Victor Gambhir
JECRC University, India

IGI Global
Scientific Publishing
Publishing Tomorrow's Research Today

Vice President of Editorial	Melissa Wagner
Managing Editor of Acquisitions	Mikaela Felty
Managing Editor of Book Development	Jocelynn Hessler
Production Manager	Mike Brehm
Cover Design	Phillip Shickler

Published in the United States of America by
IGI Global Scientific Publishing
701 East Chocolate Avenue
Hershey, PA, 17033, USA
Tel: 717-533-8845
Fax: 717-533-8661
E-mail: cust@igi-global.com
Website: https://www.igi-global.com

Library of Congress Cataloging-in-Publication Data

Names: Kumar, Ajay, 1986- editor. | Kumar, Parveen (Professor of mechanical engineering), editor. | Sebaey, Tamer, 1980- editor. | Gambhir, Victor, 1958- editor.
Title: Tools, techniques, and advancements in engineering materials science / edited by Ajay Kumar, Parveen Kumar, Tamer Sebaey, Victor Gambhir.
Description: Hershey, PA : Engineering Science Reference, [2025] | Includes bibliographical references and index. | Summary: "This book offers latest tools, techniques and advancements in engineering materials science that are used throughout our consumer and industrial without depleting non-renewable resources and without disrupting the established steady-state equilibrium of the environment and key natural resource systems"-- Provided by publisher.
Identifiers: LCCN 2024039278 (print) | LCCN 2024039279 (ebook) | ISBN 9798369355138 (hardcover) | ISBN 9798369355145 (paperback) | ISBN 9798369355152 (ebook)
Subjects: LCSH: Materials science. | Materials.
Classification: LCC TA403 .T66 2024 (print) | LCC TA403 (ebook) | DDC 620.1/1--dc23/eng/20241202
LC record available at https://lccn.loc.gov/2024039278
LC ebook record available at https://lccn.loc.gov/2024039279
ISBN:979-8-3693-5513-8
eISBN:979-8-3693-5515-2

British Cataloguing in Publication Data
A Cataloguing in Publication record for this book is available from the British Library.

Table of Contents

Detailed Table of Contents

Ashish Kumar Shrivastava, Sagar Institute of Science, Technology, and Research, Bhopal, India
Rajneesh Kumar Gedam, Bhabha University, Bhopal, India
Ravindra Mohan, IES College of Technology, Bhopal, India
Manish Billore, Sagar Institute of Science, Technology, and Research, Bhopal, India
Ashish Yadav, Maharana Pratap College of Technology, Gwalior, India
Abhishek Choubey, Sagar Institute of Science, Technology, and Research, Bhopal, India
Geetanjli Khambra, Department of Computer Applications, The Bhopal School of Social Sciences, Bhopal, India
Ajay Kumar, Department of Mechanical Engineering, School of Engineering and Technology, JECRC University, India
Ashutosh Singh Raghubanshi, Department of Mechanical Engineering, Sagar Institute of Science, Technology, and Research, Bhopal, India

Friction stir welding (FSW) is a prevalent solid-state joining method. This procedure entails a revolving instrument that comprises a cylindrical pin and a shoulder. The pin infiltrates the material to produce frictional heat, while the shoulder ensures stability and exerts pressure throughout the welding procedure. For more resilient materials, the creation of economical, lasting equipment that can generate robust welds is crucial for the economic feasibility of FSW. The selection of tool material and its design profoundly influence tool efficacy, weld quality, and total expenditure. In this context, structural steel has exhibited superior thermal properties, with the maximum heat flux measured at 81.26 W/mm^2. This signifies the material's efficacy in conducting heat to the welding plates, attaining the requisite temperature for a successful welding operation. Titanium tools exhibit a heat flux of 31.87 W/mm^2.

In this work author studied the different methylammonium (MA) tri-iodide-based PSCs using a solar cell capacitance simulator (SCAPS-1D). The PSCs are formed with the best combination of their electron transport layer (ETL), perovskite absorber layer (PAL) and hole transport layer (HTL). We have analyzed lead (Pb)-based perovskite solar cells (PSCs) with different HTL of thickness (<400 nm) with the aim to achieve the optimum device performance. The CH3NH3PbI3 PSCs with various HTLs such as Cu2O, Spiro-OMeTAD, PEDOT: PSS and CuSCN have been simulated and they initially achieved PCE of 29.84%, 29.11%, 29.29% and 28.36% respectively. The prime focus is to enrich the functioning of CH3NH3PbI3-based PSC by optimizing the values of various parameters such as thickness of PAL, operating temperature and defect density (Nt) which influences the performance of the device using Taguchi method. The photovoltaic parameters show that the solar devices are stable at 310 K and achieved maximum PCE of 32.90% at 10^{14} cm-3 with 500 nm thickness of PAL.

The exceptional strength of high performance concrete is attributed to the silica fumes. FAL-G, another cement substitute offers strength with water just like hydraulic polymers. There is no need for an external heat source for the FAL-G reaction, and therefore it produces less greenhouse gas. The dead load on structures is reduced since it is lightweight. Fly ash is required as an additive to make materials with the required strength.With a 0–30% cement removal, cement largely replaces silica fume and FAL-G in this design combination. After 28 days of testing, the FAL-G concrete mix's has an extraordinary high strength. Three samples were tested with varying amounts of silica fume and FAL-G replacement for seven and twenty-eight days. A compression testing apparatus was utilized to evaluate a total of twenty-four cast cubes and twenty-four cylinders.A robust correlation has been found between the various mechanical characteristics mentioned before. The outcomes so produced will enable the utilization of waste materials in concrete in a sustainable manner.

The chapter provides an in-depth review of advanced materials used in electric vehicle (EV) applications, addressing the critical role these materials play in improving energy efficiency, safety, and performance. It explores key considerations such as lightweighting, thermal management, conductivity, safety, and durability. A detailed examination of battery materials, highlights advancements aimed at enhancing range, charging speed, and safety. The chapter also delves into the use of lightweight structural materials, such as aluminum alloys and carbon fiber composites, which significantly contribute to energy efficiency. Thermal management systems using phase change materials (PCM) and thermal interface materials (TIM) are discussed for their role in maintaining optimal temperature and safety. Additionally, the chapter examines powertrain materials, including rare earth magnets and silicon carbide (SiC), for their impact on motor efficiency and durability. Future trends in EV materials, focusing on sustainability, cost reduction, and supply chain resilience, are also explored.

The process of increasing the soil's engineering properties and making it more stable is known as soil stabilization. When the construction soil is not suitable for the intended use, soil stabilization is required. Stabilization, in its broadest sense, encompasses many similar processes, including compaction, pre-consolidation, drainage, and many others. However, the term "stabilization" is typically limited to actions that change the soil's physical makeup in order to enhance its properties. For the purpose of stabilizing a natural soil, a cementing substance or chemical is added or geosynthetic material is used. Permeability and compressibility of the soil mass in earth structures can be reduced and shear strength can also be increased by soil stabilization. It can also be used for enhancement of bearing capacity of foundation soil. Although, soil stabilization is mainly used for improvement of natural soil properties in highway and embankment construction.

Rapid urbanization, exemplified by Bhopal City in Madhya Pradesh, India, poses significant challenges in municipal solid waste (MSW) disposal. Pyrolysis, a thermal decomposition process, offers potential for transforming MSW into biochar, syngas, and bio-oil while mitigating environmental impacts. This review provides an overview of computational studies on MSW pyrolysis in Bhopal City, highlighting recent advances, limitations, and future research directions. The review begins by highlighting the strain on conventional waste disposal systems in Bhopal City due to escalating waste generation. It emphasizes the essential role of computational modeling in optimizing pyrolysis processes. Key aspects of MSW pyrolysis, including reactor design, reaction kinetics, and feedstock composition, are explored. Computational studies improve our understanding of waste behavior during pyrolysis, enabling process optimization for resource recovery and reduced environmental impact. This review paper provides a concise summary of computational research on MSW pyrolysis in Bhopal, India.

In the highway or infrastructure, the requirement of raw material is increase day by day as per traffic load and design requirement with sustainable development. In the flexible pavement the life cycles an approximately 4 to 5 years and also required maintenance, that all activity increases the cost of project. To reduce the cost of project use "Reclaimed Asphalt Pavement Material" (RAP). The RAP material can be used in bearing coarse and sub-base coarse. By using the RAP material minimize the cost of project also use of useless material or eco-friendly part of construction. Use of RAP material in BC (Bituminous Concrete) & DMB (Dense Bituminous Macadam) by replacing the virgin raw material (Aggregate & bitumen) and design the BC layer & DBM layer for road pavement. The review research "A Literature Review on RAP material use in surface, base and sub-base of flexible pavement" is presented in this document.

This study assesses the pozzolanic activity of Cappadocia volcanic waste soils (CWs) using established methodologies and explores their potential for diverse applications. Targeted research aims to reduce visual pollution, and repurpose waste into economic resources and engineering materials, thereby safeguarding the cultural and natural heritage of the region. While Frattini test results confirm pozzolanic traits in most CWs, discrepancies arise in the strength activity index that necessitate further investigation. Compressive strength and electrical conductivity tests emerge as suitable alternatives, albeit with varying durations. Despite methodological variations, all CW samples exhibit pozzolanic properties, with CW4 showing the highest pozzolanicity. This study features the versatility of CWs, emphasizing their role and economic feasibility in construction, environmental engineering, agriculture, and industrial applications. Harnessing these materials promotes sustainable waste management and heritage conservation, offering a promising solution to the region's waste challenges.

Concrete is a homogenous mixture that comprises of binder material like cement, filler material like coarse and fine aggregate and water for preparation of gel like paste to bind the ingredients. Due to growing environmental concerns, the mining of river sand is discouraged lately and is replaced by quarry dust or stone dust or manufactured sand (M-sand). Further the use of pozzolanic material such as fly ash and pond ash is also widely being accepted with the use of Portland Pozzolanic Cement (PPC). Change in behavior is obvious in concrete made by PPC and manufactured sand (m-sand). This research reviews the available literature on effects of concrete's durability due to this replacement. Researchers around the world have come up with a possible solution of mixing steel fibers with concrete to cover durability aspect. This review attempts to identify the potential of m-sand in PPC based concrete to alter the durability and further the significance of steel fibers use to alter behavior of concrete in order to minimizing the undesirable effects of this combination.

Preface

This book offers latest tools, techniques and advancements in engineering materials science that are used throughout our consumer and industrial without depleting non-renewable resources and without disrupting the established steady-state equilibrium of the environment and key natural resource systems. In today's era, there is a high demand for pollution free environment, reducing hazardous waste from industries, diverting manufacturing sectors towards green production, minimizing the use of fossil fuels, to reduce carbon contents from atmosphere and to decrease hazardous environmental impact of current engineering materials that are in use to produce customized and cultured products. Engineering materials are the best option that can meet such requirement across the globe as they have renewability, biodegradability, cost effective, higher chemical and mechanical resistance and biocompatibility for various medical, pharmaceutical, electronics equipment's and other engineering applications. Thus environmental responsible materials are designed and developed by scientists/engineers to meet the current demand of environment and industries. This book has the following key features:

- The latest research and development of engineering materials will be available to academicians, professionals and field engineers.
- Research outcomes are included with illustrations and examples so that it will be useful to anyone who wants to understand the engineering materials as well as its advanced usage for advantages.
- Various characterization, synthesis, testing and optimization techniques to explore new possibilities and to develop novel materials are included.

This edited book is for researchers, academicians, upper under graduate, post graduate, Ph.D. scholars in materials, mechanical, industrial, chemical, environmental science and healthcare personnel. The editors acknowledge the professional support received from IGI Global and express their gratitude for this opportunity. Reader's observations, suggestions and queries are welcome.

Acknowledgement

The editors are grateful to IGI Global for showing their interest to publish this book in the buzz area of Tools, Techniques, and Advancements in Engineering Materials Science. The editors express their personal adulation and gratitude to methodical staff that have helped the editors to compile and finalize the manuscript.

Editors are thankful to the researchers, experts, scientists, academicians and related professionals who generously shared their knowledge and experiences. Their contributions have enriched the content and added depth to the ideas presented in this book.

Editors would like to extend appreciation to the reviewers who diligently reviewed the manuscripts and provided valuable feedback that helped refine and improve the final version of this book.

Dr. Ajay also wishes to express his gratitude to his parents, Sh. Jagdish and Smt. Kamla, and his loving brother Sh. Parveen for their true and endless support. They have made him able to walk tall before the world regardless of sacrificing their happiness and living in a small village. He cannot close these prefatory remarks without expressing his deep sense of gratitude and reverence to his life partner Mrs. Sarita Rathee for her understanding, care, support and encouragement to keep his moral high all the time. No magnitude of words can ever quantify the love and gratitude the he feels in thanking his daughters, Sejal Rathee and Mahi Rathee, and son Kushal Rathee who are the world's best children.

Finally, the editors obligate this work to the divine creator and express their indebtedness to the "ALMIGHTY" for gifting them power to yield their ideas and concepts into substantial manifestation. The editors believe that this book would enlighten the readers about each feature of Tools, Techniques, and Advancements in Engineering Materials Science.

Ajay Kumar

Parveen Kumar

Tamer Ali Sebaey

Victor Gambhir

Chapter 1

Transient Thermal Analysis of FSW Tool With Different Material by Using Computational Testing Method

Ashish Kumar Shrivastava
Sagar Institute of Science, Technology, and Research, Bhopal, India

Rajneesh Kumar Gedam
Bhabha University, Bhopal, India

Ravindra Mohan
IES College of Technology, Bhopal, India

Manish Billore
Sagar Institute of Science, Technology, and Research, Bhopal, India

Ashish Yadav
Maharana Pratap College of Technology, Gwalior, India

Abhishek Choubey
Sagar Institute of Science, Technology, and Research, Bhopal, India

Geetanjli Khambra
Department of Computer Applications, The Bhopal School of Social Sciences, Bhopal, India

Ajay Kumar
https://orcid.org/0000-0001-7306-1902
Department of Mechanical Engineering, School of Engineering and Technology, JECRC University, India

Ashutosh Singh Raghubanshi
Department of Mechanical Engineering, Sagar Institute of Science, Technology, and Research, Bhopal, India

DOI: 10.4018/979-8-3693-5513-8.ch001

ABSTRACT

Friction stir welding (FSW) is a prevalent solid-state joining method. This procedure entails a revolving instrument that comprises a cylindrical pin and a shoulder. The pin infiltrates the material to produce frictional heat, while the shoulder ensures stability and exerts pressure throughout the welding procedure. For more resilient materials, the creation of economical, lasting equipment that can generate robust welds is crucial for the economic feasibility of FSW. The selection of tool material and its design profoundly influence tool efficacy, weld quality, and total expenditure. In this context, structural steel has exhibited superior thermal properties, with the maximum heat flux measured at 81.26 W/mm^2. This signifies the material's efficacy in conducting heat to the welding plates, attaining the requisite temperature for a successful welding operation. Titanium tools exhibit a heat flux of 31.87 W/mm^2.

1. INTRODUCTION

Friction stir welding (FSW) is an environmentally sustainable technology that eliminates the need for electrodes, rendering it a cleaner option compared to conventional welding methods. FSW is frequently employed to unite lightweight metals, including copper, aluminum, magnesium, and zinc alloys (Abnar, Kazeminezhad, & Kokabi, 2015; A. Kumar, Liu, & Kumar, 2024). A restriction, however, is that the lifespan of FSW joints often is shorter than that of the original material. Although a relatively recent concept developed in 1991, FSW has demonstrated significant cost-effectiveness (Ahmed, El-Sayed Seleman, Fydrych, & ÇAm, 2023). The efficacy of Friction Stir Welding (FSW) is significantly influenced by the employed tool, which generally comprises a spinning rounded shoulder and a threaded cylindrical pin (Almoussawi, Smith, & Faraji, 2018). This instrument produces heat predominantly via friction and manipulates the softened alloy to create the weld junction (Asl, Mirsalehi, & Dehghani, 2019). In contrast to conventional fusion welding, friction stir welding (FSW) circumvents prevalent problems such as solidification cracking, porosity, and the depletion of alloying elements, as the material remains below its melting temperature (Çam, 2011). These benefits have facilitated its commercial success in welding soft alloys such as aluminum. Welding tougher alloys, such as steel and titanium, presents complications (Anand, Satyarthi, Bharti, Kumar, & Kumar, 2024; Linton & Ripley, 2008). However, Friction stir welding

(FSW) has been considered to be the most significant development in metal joining in the past decades.

The FSW tool endures considerable stress and elevated temperatures, constraining its practical use for these metals because of the exorbitant cost and restricted durability of the tools (Chen, Shi, Wu, & Jiang, 2021). Initiatives have been undertaken to create more cost-effective and reusable tools for FSW, however, a significant portion of the research remains empirical. Further research into tool design is necessary to enhance FSW applications for tougher alloys (Christy, Mourad, Sherif, & Shivamurthy, 2021). This paper examines current advancements in critical facets of FSW tools, encompassing their geometry, material composition, microstructural integrity, load-bearing capability, failure mechanisms, and overall economic viability of the procedure (D'Orazio, Forcellese, & Simoncini, 2019), which has been shown in Figure 1. An alternating, non-consumable tool with superior high temperature characteristics compared to the workpiece materials is frequently utilized (Agarwal, Kumar, Kumar, & Ghadai, 2024; Deng et al., 2015). During Friction Stir Welding (FSW), the tool is applied to the contiguous surfaces of the materials to be welded and rotated, producing frictional heat (Steuwer, Dumont, Peel, Preuss, & Withers, 2007).

Figure 1. Tools for FSW

The heat induces softening of the material, forming a plasticized region (third body) surrounding the pin and at the contact between the tool shoulder and the workpiece (Dewangan, Banjare, Tripathi, & Manoj, 2023). The shoulder aids in containing the plasticized material and generates supplementary frictional heat (Dumpala, Chandra, & Sriram, 2022). As the temperature increases, the strength of the material in contact with the revolving tool diminishes, resulting in the softened material shifting from one side of the tool to the opposite side. The instrument is subsequently maneuvered along the joint line, yielding a continuous weld (El-Sayed, Shash, Abd-Rabou, & ElSherbiny, 2021).

2. TOOL MATERIALS

A limited number of materials have been employed in the FSW/P process. These materials include but are not limited to, apparatus preparations, high-speed steel (HSS), nickel composites, metal carbides, and ceramics (Kang, Fu, Luan, Dong, & He, 2010). However, Weld quality and tool wear are critical factors in the selection of tool material, since its qualities can impact weld quality by affecting heat generation and dissipation. The weld microstructure may be influenced by interactions with degraded tool material. Moreover, the temperatures in the workpiece are influenced by the thermal conductivity of the tool material, given certain workpieces and processing conditions. The coefficient of thermal expansion may influence the thermal stresses in the tool (Hattingh, Blignault, Van Niekerk, & James, 2008).

2.1. Tool Steels

Tool steel is one of the most commonly utilized materials in the welding and processing of aluminum, copper, and magnesium alloys, capable of welding up to 50 mm in thickness of these materials (Eyvazian, Hamouda, Tarlochan, Derazkola, & Khodabakhshi, 2020). These materials are readily available and exhibit excellent machinability and thermal fatigue resistance. When it comes to shielding points from abrasion and torsion during FSW, tool preparation is essential (Fujii, Cui, Maeda, Sato, & Nogi, 2006). These welding instruments may be used to produce butt joints and lapped joints, among other similar and dissimilar welds. Butt joint welding, on the other hand, takes more work than welding in other joint configurations (Anand, Gayatri, et al., 2024; Li, Dong, Shi, Li, & Ye, 2017).

Materials are arranged strategically in FSW, with the softer material located on the retreating side (RS) and the stronger material on the advancing side (AS). To make the procedure easier, the tool is angled slightly away from the joint interface and towards the softer material (Khodir & Shibayanagi, 2007). These tool preparations have been successfully used to combine a variety of materials, including stainless steel (SS), high carbon high chromium steel (HCHCr), high-speed steel (HSS), H13, and C40. A variety of materials, such as mild steel, copper, magnesium, and aluminum, may be welded with these instruments (Xu, Liu, Chen, Luan, & Yao, 2012).

2.2. Alloys of Nickel and Cobalt

Superalloys based on nickel and cobalt are renowned for their remarkable ductility, strength, and extraordinary resistance to erosion and wear. The presence of strengthening precipitates is responsible for their remarkable capabilities (Krishnan, Maniraj, Deepak, & Anganan, 2018). During operation, these materials usually vary between 600 and 800 degrees Celsius, thus it is important to keep them below these temperatures to retain their function. However, the manufacture of complicated components, including elaborate woodwind and pad profiles, is significantly hampered by their low machinability(M. Kumar, Das, & Ballav, 2023). These superalloys are frequently used in copper-based composite friction stir welding and processing (FSW/P). Furthermore, the 3.1 mm thick nuclear-grade material Zircaloy-4 has been successfully bonded with cobalt-based alloy tools (West, Shunmugasamy, Usman, Karaman, & Mansoor, 2021).

2.3. Polycrystalline Cubic Boron Nitride (PCBN)

It has been challenging to apply friction stir welding and processing (FSW/P) to additional high-temperature materials, partly because there aren't enough appropriate instruments that can survive the extremely high temperatures required for the procedure (Diyaley, Chakraborty, Kumar, & Kalita, 2024). The FSW technical guide emphasizes how crucial it is to make sure that no tool material stays in the weld joint by preventing tool wear. To satisfy these needs, one of the most promising tool materials to date is polycrystalline cubic boron nitride, or PCBN (Rathod, Raghubanshi, Srivastava, Kiragi, & Mondal, 2019). Because of its superior mechanical and thermal qualities, PCBN, which was first created for the purpose of machining metal alloys, is now being used in FSW/P. For friction stir welding and processing of alloys and titanium composites, PCBN is particularly attractive due to its thermal stability, high strength, and hardness at increased temperatures (Raghuwanshi, Parashar, Pathak, & Raghubanshi, 2023).

2.4. Ceramic Materials

Ceramics have also been employed as FSW/P tool materials, although they frequently fracture during the immersion stage and are exceedingly brittle (Sahu, Paul, & Ganguly, 2021). Compared to carbides, ceramic cutting tools are more heat resistant and harder, although they are often weaker. Ceramics are a good choice for machining superalloys and hard materials despite this drawback. Cutting tools for ceramics may be divided into two primary categories: silicon nitride and alumina (Zhang, Wang, Zhang, & Yang, 2015).

3. MATERIALS AND METHODS

The 3D model of tools was fabricated using CADD software using a circular probe-equipped instrument with tool dimensions diameter of the probe 20mm probe length of 40mm, shoulder length of 10mm, Pin length of 3.5mm, Tool pin diameter of 3.0mm and diameter of plunge is 8 mm, which has been depicted in Figure 2 (a & b). However, the two tool materials such as structural steel (S-460) and Titanium were selected for the present study. The physical, thermal, and mechanical properties of these materials are given in Table 1. Moreover, the 3D tools model with two different materials was simulated using ANSYS R19.2 software.

Figure 2. (a) Tool Geometry and (b) Tool's 3D model

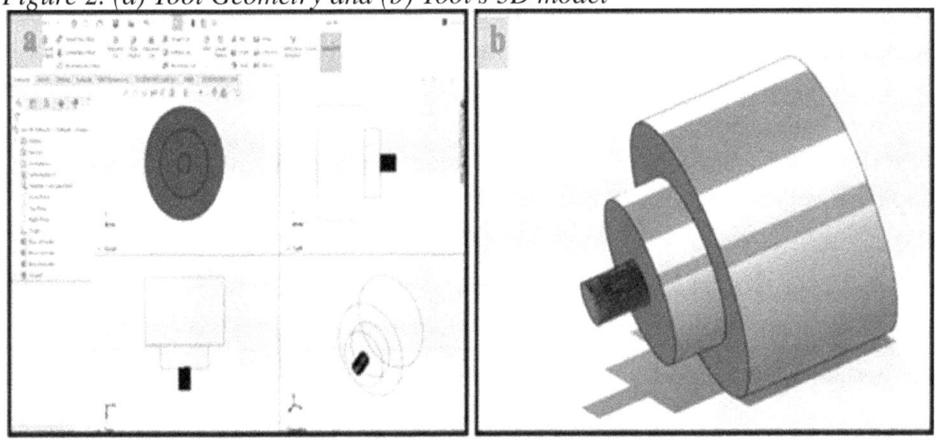

Table 1. Properties of Tool Materials

Properties	Structural steel (S-460)	Titanium
Density	7.85 g/cm^3	4.62 g/cm^3
Melting point	1410°C	1668°C
Poisson's Ratio	0.3	0.36
Hardness	60 HV	70 HV
Thermal Expansion	1.2×10^{-5} /°C	9.4×10^{-6} /°C
Modulus of elasticity	200 GPa	116 GPa
TYS	250 MPa	930 MPa
CYS	250 MPa	930 MPa
TUS	460 MPa	1070 MPa

4. RESULTS AND DISCUSSION

4.1 Modeling and Simulation

The Modeling of 3D tools of different materials has been prepared using CADD software. The developed 3D model has been imported into ANSYS R19.2 software for simulation analysis, which measured the deformation of the 3D tool at load and different boundary conditions may be applied. The force of 60 N is applied to the tool tip/pin, with an ambient temperature of 32 °C, while the temperature at the tool pin is set to 600 °C. However, tetrahedral meshing is employed using an automated method for mesh production, with an element size of 2 mm, comprising 52,464 nodes and 38,022 elements. However, the upper face of the tool is secured to the machine's chuck and delivers an acceleration of 1G in the vertical direction for frequency response. Moreover, all these parameters are measured in the deformation of 3D tools of structural steel S-460 and Titanium respectively, whose observations have been recorded and depicted in Figure 3 and Figure 4 (a & b) respectively. However, the heat flux of structural steel S-460 and Titanium 3D tools also analyzed via software and received observation has been depicted in Figure 5 (a & b).

Figure 3. Total deformation of Structural steel and Titanium Tool with temperature and their boundary conditions

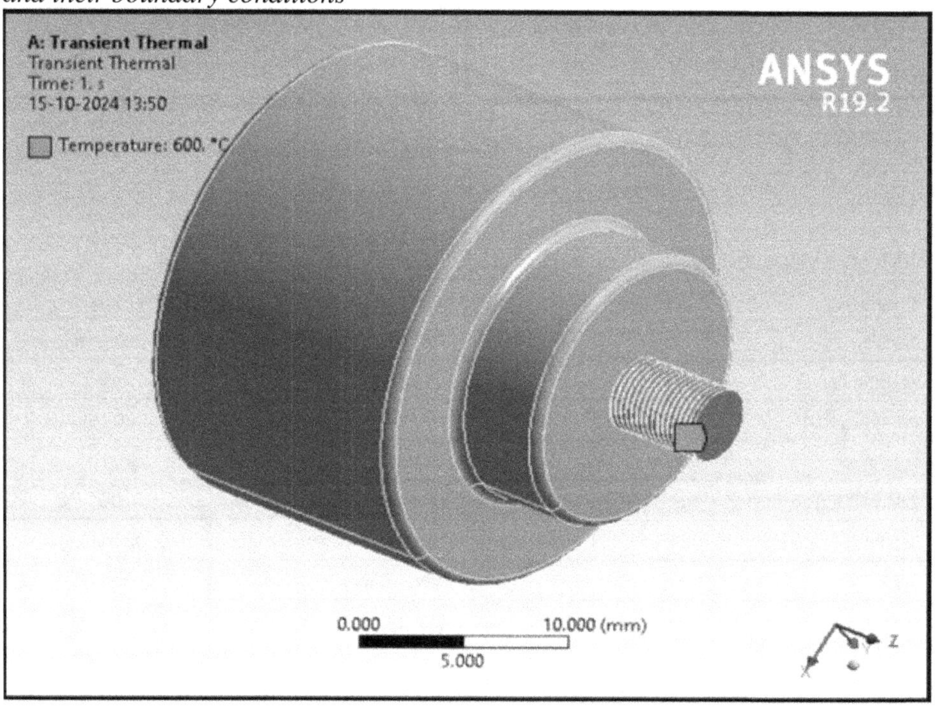

Figure 4. Total deformation 3D Tools (a) structural steel (S-460) materials and (b) Titanium materials

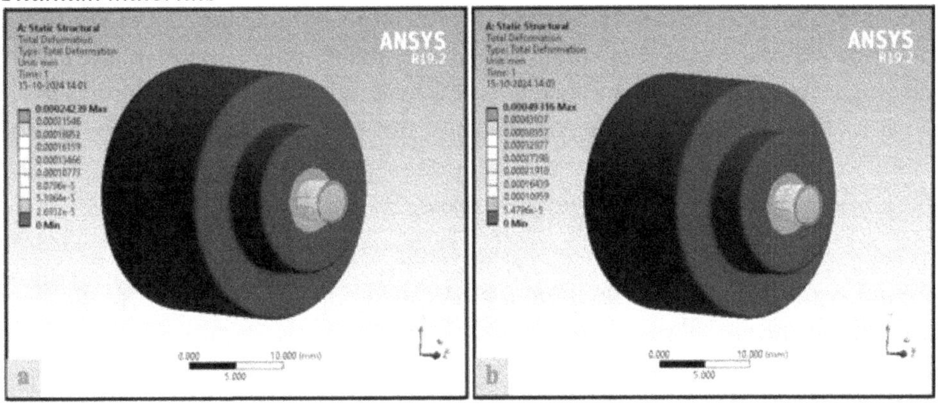

Figure 5. Total Heat Flux of 3D Tools (a) structural steel (S-460) materials and (b) Titanium materials

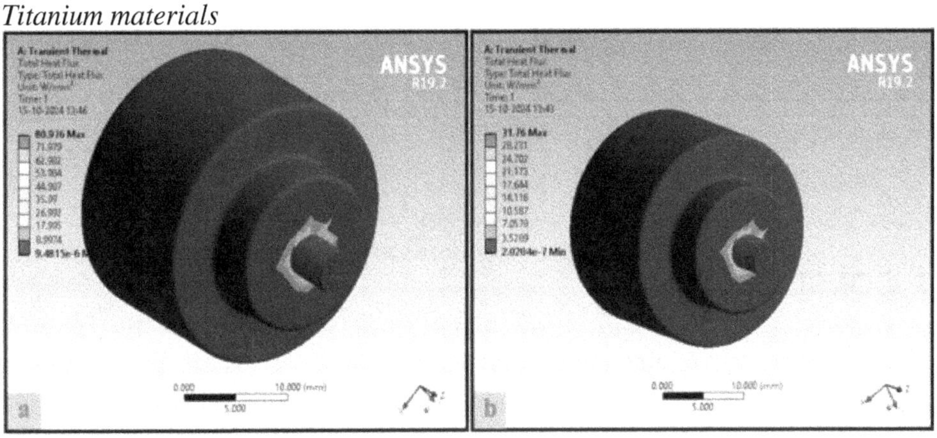

4.2 Deformation and Heat Flux

The present study is based on static structural analysis and thermal analysis to obtain total heat flux and frequency response analysis when applying 1G of acceleration to the tool's fixed geometry. This demonstrates how various material tools influence the attenuation of vibration waves transmitted from the machines to the tools, thereby affecting weld quality. Increased vibration attenuation efficiency and material properties indicate superior weld quality, which is depicted in Figure 6 (a

& b). The observation indicated that the heat flux of Titanium and structural steel S-460 3D tool are found to the 31.87 W/mm² and 81.26 W/mm² respectively. Also found that the heat flux of the steel tool is higher than the titanium tool due to the deformation of the steel tool is higher than the titanium tool. The higher heat flux also indicates that the structural steel S-460 material gives better performance for FSW tool design. However, the observed heat flux of structural steel S-460 material is sufficient for transmitting heat into the welding plates at the requisite temperature for an enhanced Friction stir welding process (FSW).

Figure 6. (a) Total Deformation of different material tools (b) Total Heat flux of different material tools

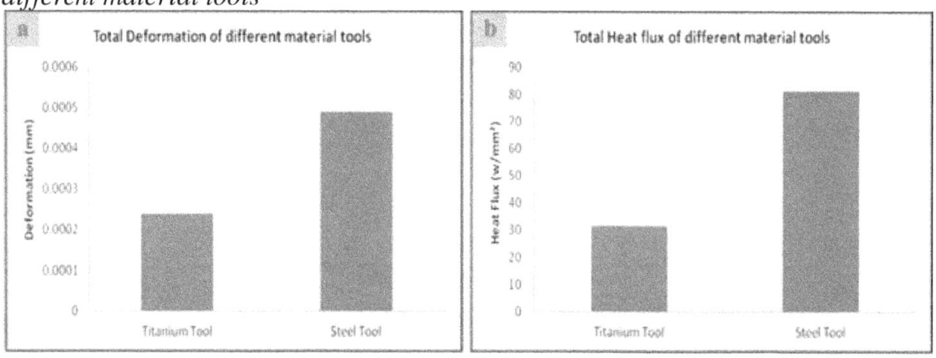

5. CONCLUSION

In Friction Stir Welding the welding temperature is crucial; if the tool fails to distribute force with the rising temperature caused by friction between the tool and the workpiece, both tool longevity and weld quality will be compromised. Therefore, we achieve optimal thermal effects for structural steel in this context. Structural steel exhibits a maximum heat flux of 81.26 W/mm², indicating its capacity to transmit heat to welding plates, hence achieving the requisite temperature for an enhanced welding process. The second titanium tool exhibits a heat flux of 31.87 W/mm². Upon comprehensive study of the results, we conclude that among the three materials for the FSW tool, structural steel demonstrates superior performance compared to the other two.

REFERENCES

Abnar, B., Kazeminezhad, M., & Kokabi, A. (2015). The effect of premixed Al-Cu powder on the stir zone in friction stir welding of AA3003-H18. *Journal of Materials Engineering and Performance*, 24(2), 1086–1093. DOI: 10.1007/s11665-014-1373-1

Agarwal, A., Kumar, P., Kumar, A., & Ghadai, R. K. (2024). *Strategic Design Optimization of Cutting Tools for Enhanced Manufacturing Efficiency Industry 4.0 Driven Manufacturing Technologies*. Springer.

Ahmed, M. M. Z., El-Sayed Seleman, M. M., Fydrych, D., & Çam, G. (2023). Review on friction stir welding of dissimilar magnesium and aluminum alloys: Scientometric analysis and strategies for achieving high-quality joints. *Journal of Magnesium and Alloys*, 11(11), 4082–4127. DOI: 10.1016/j.jma.2023.09.039

Almoussawi, M., Smith, A. J., & Faraji, M. (2018). Wear of Polycrystalline Boron Nitride Tool During the Friction Stir Welding of Steel. *Metallography, Microstructure, and Analysis*, 7(3), 252–267. DOI: 10.1007/s13632-018-0439-0

Anand, S. Gayatri, Satyarthi, M. K., Bharti, P. S., Kumar, A., & Rathee, S. (2024). Robotic Arm 3D Printing: Technological Advancements and Applications *Industry 4.0 Driven Manufacturing Technologies* (pp. 293-310): Springer.

Anand, S., Satyarthi, M. K., Bharti, P. S., Kumar, P., & Kumar, A. (2024). *Digital Twin Integration for Enhanced Control in FDM 3D Printing Industry 4.0 Driven Manufacturing Technologies*. Springer.

Asl, N. S., Mirsalehi, S. E., & Dehghani, K. (2019). Effect of TiO2 nanoparticles addition on microstructure and mechanical properties of dissimilar friction stir welded AA6063-T4 aluminum alloy and AZ31B-O magnesium alloy. *Journal of Manufacturing Processes*, 38, 338–354. DOI: 10.1016/j.jmapro.2019.01.023

Çam, G. (2011). Friction stir welded structural materials: Beyond Al-alloys. *International Materials Reviews*, 56(1), 1–48. DOI: 10.1179/095066010X12777205875750

Chen, J., Shi, L., Wu, C., & Jiang, Y. (2021). The effect of tool pin size and taper angle on the thermal process and plastic material flow in friction stir welding. *International Journal of Advanced Manufacturing Technology*, 116(9-10), 2847–2860. DOI: 10.1007/s00170-021-07650-x

Christy, J. V., Mourad, A.-H. I., Sherif, M. M., & Shivamurthy, B. (2021). Review of recent trends in friction stir welding process of aluminum alloys and aluminum metal matrix composites. *Transactions of Nonferrous Metals Society of China*, 31(11), 3281–3309. DOI: 10.1016/S1003-6326(21)65730-8

D'Orazio, A., Forcellese, A., & Simoncini, M. (2019). Prediction of the vertical force during FSW of AZ31 magnesium alloy sheets using an artificial neural network-based model. *Neural Computing & Applications*, 31(11), 7211–7226. DOI: 10.1007/s00521-018-3562-6

Deng, Y., Ye, R., Xu, G., Yang, J., Pan, Q., Peng, B., Cao, X., Duan, Y., Wang, Y., Lu, L., & Yin, Z. (2015). Corrosion behaviour and mechanism of new aerospace Al–Zn–Mg alloy friction stir welded joints and the effects of secondary Al3ScxZr1−x nanoparticles. *Corrosion Science*, 90, 359–374. DOI: 10.1016/j.corsci.2014.10.036

Dewangan, S. K., Banjare, P. N., Tripathi, M. K., & Manoj, M. K. (2023). Effect of vertical and horizontal zinc interlayer on material flow, microstructure, and mechanical properties of dissimilar FSW of Al 7075 and Mg AZ31 alloys. *International Journal of Advanced Manufacturing Technology*, 126(9), 4453–4474. DOI: 10.1007/s00170-023-11348-7

Diyaley, S., Chakraborty, S., Kumar, A., & Kalita, K. (2024). *Elephant Swarm Water Search Algorithm-Based Optimization of a Laser Beam Machining Process Industry 4.0 Driven Manufacturing Technologies*. Springer.

Dumpala, L., Chandra, P. H., & Sriram, N. (2022). Development of bobbin tool and influence of its design parameters on FSW joint. *Materials Today: Proceedings*, 56, 1582–1587. DOI: 10.1016/j.matpr.2022.02.363

El-Sayed, M. M., Shash, A. Y., Abd-Rabou, M., & ElSherbiny, M. G. (2021). Welding and processing of metallic materials by using friction stir technique: A review. *Journal of Advanced Joining Processes*, 3, 100059. DOI: 10.1016/j.jajp.2021.100059

Eyvazian, A., Hamouda, A., Tarlochan, F., Derazkola, H. A., & Khodabakhshi, F. (2020). Simulation and experimental study of underwater dissimilar friction-stir welding between aluminium and steel. *Journal of Materials Research and Technology*, 9(3), 3767–3781. DOI: 10.1016/j.jmrt.2020.02.003

Fujii, H., Cui, L., Maeda, M., Sato, Y. S., & Nogi, K. (2006). *Effect of threads on tool in friction stir welding of aluminum alloys.* Paper presented at the Materials science forum. DOI: 10.4028/0-87849-996-2.389

Hattingh, D., Blignault, C., Van Niekerk, T., & James, M. (2008). Characterization of the influences of FSW tool geometry on welding forces and weld tensile strength using an instrumented tool. *Journal of Materials Processing Technology*, 203(1-3), 46–57. DOI: 10.1016/j.jmatprotec.2007.10.028

Kang, J., Fu, R., Luan, G., Dong, C., & He, M. (2010). In-situ investigation on the pitting corrosion behavior of friction stir welded joint of AA2024-T3 aluminium alloy. *Corrosion Science*, 52(2), 620–626. DOI: 10.1016/j.corsci.2009.10.027

Khodir, S. A., & Shibayanagi, T. (2007). Dissimilar friction stir welded joints between 2024-T3 aluminum alloy and AZ31 magnesium alloy. *Materials Transactions*, 48(9), 2501–2505. DOI: 10.2320/matertrans.MRA2007093

Krishnan, M. M., Maniraj, J., Deepak, R., & Anganan, K. (2018). Prediction of optimum welding parameters for FSW of aluminium alloys AA6063 and A319 using RSM and ANN. *Materials Today: Proceedings*, 5(1), 716–723. DOI: 10.1016/j.matpr.2017.11.138

Kumar, A., Liu, Y., & Kumar, R. (2024). *Handbook of Intelligent and Sustainable Manufacturing: Tools, Principles, and Strategies*. CRC Press. DOI: 10.1201/9781003405870

Kumar, M., Das, A., & Ballav, R. (2023). Influence of the Zn interlayer on the mechanical strength, corrosion and microstructural behavior of friction stir-welded 6061-T6 aluminium alloy and AZ61 magnesium alloy dissimilar joints. *Materials Today. Communications*, 35, 105509. DOI: 10.1016/j.mtcomm.2023.105509

Li, S., Dong, H., Shi, L., Li, P., & Ye, F. (2017). Corrosion behavior and mechanical properties of Al-Zn-Mg aluminum alloy weld. *Corrosion Science*, 123, 243–255. DOI: 10.1016/j.corsci.2017.05.007

Linton, V. M., & Ripley, M. I. (2008). Influence of time on residual stresses in friction stir welds in agehardenable 7xxx aluminium alloys. *Acta Materialia*, 56(16), 4319–4327. DOI: 10.1016/j.actamat.2008.04.059

Raghuwanshi, P., Parashar, R., Pathak, T., & Raghubanshi, A. S. (2023). *Integration ofpaper-based analytical tools with smartphones and IOMT Paper-Based Diagnostic Devices for Infectious Diseases*. IOP Publishing.

Rathod, S., Raghubanshi, A. S., Srivastava, Y., Kiragi, V. R., & Mondal, D. P. (2019). Sliding wear behaviour study of alumina-silicate short fiber reinforced AA2014-Al alloy composite (SFRAC). *Materials Research Express, 6*(10), 1065a1067. DOI: 10.1088/2053-1591/ab3d60

Sahu, M., Paul, A., & Ganguly, S. (2021). Optimization of process parameters of friction stir welded joints of marine grade AA 5083. *Materials Today: Proceedings*, 44, 2957–2962. DOI: 10.1016/j.matpr.2021.01.938

Steuwer, A., Dumont, M., Peel, M., Preuss, M., & Withers, P. J. (2007). The variation of the unstrained lattice parameter in an AA7010 friction stir weld. *Acta Materialia*, 55(12), 4111–4120. DOI: 10.1016/j.actamat.2007.01.024

West, P., Shunmugasamy, V. C., Usman, C. A., Karaman, I., & Mansoor, B. (2021). Part I.: Friction stir welding of equiatomic nickel titanium shape memory alloy – microstructure, mechanical and corrosion behavior. *Journal of Advanced Joining Processes*, 4, 100071. DOI: 10.1016/j.jajp.2021.100071

Xu, W. F., Liu, J. H., Chen, D. L., Luan, G. H., & Yao, J. S. (2012). Improvements of strength and ductility in aluminum alloy joints via rapid cooling during friction stir welding. *Materials Science and Engineering A*, 548, 89–98. DOI: 10.1016/j.msea.2012.03.094

Zhang, H., Wang, M., Zhang, X., & Yang, G. (2015). Microstructural characteristics and mechanical properties of bobbin tool friction stir welded 2A14-T6 aluminum alloy. *Materials & Design (1980-2015), 65*, 559-566.

Chapter 2

Optimization of CH3NH3PbI3 Perovskite Solar Cell: A Statistical and Simulation-Based Study

Rahul Kundara
Delhi Technological University, India

Sarita Baghel
Delhi Technological University, India

ABSTRACT

In this work author studied the different methylammonium (MA) tri-iodide-based PSCs using a solar cell capacitance simulator (SCAPS-1D). The PSCs are formed with the best combination of their electron transport layer (ETL), perovskite absorber layer (PAL) and hole transport layer (HTL). We have analyzed lead (Pb)-based perovskite solar cells (PSCs) with different HTL of thickness (<400 nm) with the aim to achieve the optimum device performance. The CH3NH3PbI3 PSCs with various HTLs such as Cu2O, Spiro-OMeTAD, PEDOT: PSS and CuSCN have been simulated and they initially achieved PCE of 29.84%, 29.11%, 29.29% and 28.36% respectively. The prime focus is to enrich the functioning of CH3NH3PbI3-based PSC by optimizing the values of various parameters such as thickness of PAL, operating temperature and defect density (Nt) which influences the performance of the device using Taguchi method. The photovoltaic parameters show that the solar devices are stable at 310 K and achieved maximum PCE of 32.90% at 10^{14} cm-3 with 500 nm thickness of PAL.

DOI: 10.4018/979-8-3693-5513-8.ch002

1. INTRODUCTION

Presently, new technologies are being used to produce electricity by harvesting sunlight. These technologies are widely used around the world as non-conventional alternatives to non-renewable energy sources (Ehsanul et al., 2018). In the past few years, researchers are quite hopeful that the photovoltaic (PV) technology can replace fossil fuels for huge energy requirements in forthcoming days. However, to achieve this, there is a need to constantly overcome the limitations faced by the PV industry. Some of the challenges faced by photovoltaic technology are; to achieve high power conversion efficiencies, simplified fabrication techniques, cost effectiveness and non-toxicity towards the environment (Syed Abdul et al., 2021; Kumar et al. 2024). Perovskite Solar Cells (PSCs) have emerged as potential candidates in the field of PV technologies due to their excellent photovoltaic performance since the past studies. The organic lead halide PSCs have achieved efficiency beyond 25%. This noticeable increase in power conversion efficiency (PCE) is because of the consistent efforts made towards device configuration, stability and optimization (Michael M et al., 2012; Singh et al., 2018; Abdelkader et al., 2019; Kumar et al., 2024). Recently, Santos et al., 2020 worked upon a simulation optimization study on Pb based ($MAPbI_3$)-PSC by varying the thickness of the perovskite absorber layer (PAL), and defect density (N_t). They reported the maximum theoretical PCE of 20.26%. The TiO_2 as ETL has been achieved a maximum PCE of 16.33% with NiO_X as HTL (Sakib et al., 2023). The $FACsSnPbI_3$ based PSC obtained a PCE of 17.17 (Kenfack et al., 2023). The Poisson's equation is used in SCAPS-1D. The outcomes of model were developed and outcomes were compared with the experimental work (Behl et al., 2022). Al_2O_3 can be used ETL in PSCs (Tyagi et al., 2021; Singh et al., 2021). We need to adopt methods for creating cutting-edge smart materials based on computer, simulation, and artificial intelligence (Kumar et al., 2023). Sheet forming technique for high quality components in solar device (Mittal et al., 2020; Kumar et al., 2023; Kumar et al., 2024). Recycling and other e-waste management (Kumar et al., 2023). The additive manufacturing technologies (Kumar et al., 2022; Kumar et al., 2024). To conduct an in-depth study of the operation mechanism of PSCs, it is imperative to identify the factors which affect the PCE by a more rigorous and systematic approach, as PCE is closely related to device physics. In this work, Hole Transport Layer (HTL), PAL, N_t, and device temperature are identified as three important parameters that should be carefully optimized with the target of getting the best device efficiencies. We have used a dual approach consisting of statistical methodology and simulation program. Optimization process depends on various input parameters and seeks a huge count of experiments. Therefore, experiments are first designed by using statistical Taguchi method in order to effectively identify the impact of different process parameters and to achieve an optimum PCE. We have employed L9 Orthogonal Array (OA), which is a statistical tool given by Taguchi (Kumavat et al., 2020). Designed experiments are then simulated using SCAPS 1-D. This program uses the input parameters and gives the optimum power conversion efficiency as output. This software has been frequently applied in simulation of inorganic thin film solar cells and planar PSCs. The simulation outputs have been largely consistent with experimental results. We have performed 36 simulations by using carefully calibrated levels of concerned parameters, in order to find their optimized values. Quantum efficiency studies have also been included for better analysis. There are some existing challenges such as stability issues, toxicity and environmental concerns, manufacturing costs, lead content, scale-up challenges and encapsulation and sealing faced by the PV industry.

Figure 1. Perovskite solar cell structure

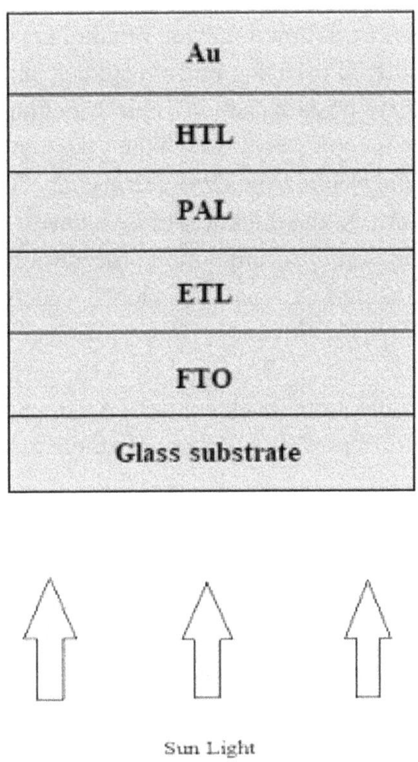

2. EXPERIMENTAL DETAILS

2.1 Numerical Simulation

The cell configuration for the simulation of PSCs is schematically shown in fig.1. Under the aegis of this study, the full SCAPS-1D software has been utilized for simulation. The Gent University of Belgium developed SCAPS software, enables the researchers to simulate a maximum of seven semiconducting layers. The primary structure of the PSC device is basically structured with three different layers such as an n-type (ETL) TiO_2, PAL and p-type (HTL) as shown in fig.1. The various HTLs are employed as Cu_2O, Spiro-OMeTAD, PEDOT: PSS and CuSCN (Mandadapu et al., 2017; Slami et al., 2019). The transparent conducting oxide (TCO) layer in the structure is fluorine doped tin oxide (FTO) where electrons are collected and holes at back contact. Au. Pt, Ag, Ni and Cu are usually employed metal back contact because of their high work function. Our simulations were done by employing material parameters gathered by using current studies analysis. The primary data set used while the simulation is attached in Table 1 including different parameters. This study of simulation is characterized with the variation in the thickness of PAL, operating temperature and N_t of various PSCs (Rai et al., 2020; Haider et al., 2019). The operating temperature, N_t and thickness of PAL have a noticeable impact on PV parameters of PSCs. The step-wise simulation procedure shown in fig.2. The study has been done at 300 K using a "air mass 1.5 global" spectrum and a light power of 1000 W/m^2. The J-V characteristics and quantum efficiency curve of Pb-based PSC with different HTLs shown in fig.3 (a) and fig.3 (b) respectively. The alignment of E_g of different layers and metal back contacts illustrated in fig.4 (Kumari et al., 2018). Initial results of different HTMs shown in table 2.

Figure 2. SCAPS-1D software simulation procedure

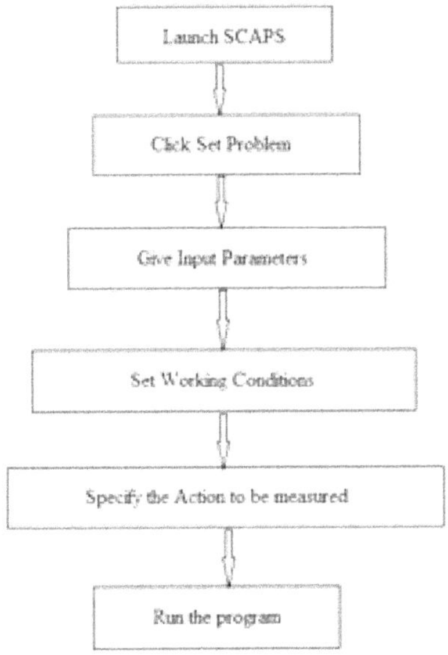

Table 1. Input parameters of distinct layers used in simulation.

Parameter	FTO (Rai et al., 2020)	TiO_2 (Rai et al., 2020)	$CH_3NH_3PbI_3$ (Husainat et al., 2019)	Spiro-OMeTAD (De Los Santos et al., 2020)	Cu_2O (Rai et al., 2020)	PEDOT: PSS (Mandadapu et al., 2017)	CuSCN (Haider et al., 2019)
Thickness (μm)	0.400	0.040	0.300	0.213	0.250	0.080	0.350
Eg (eV)	3.5	3.2	1.55	3.0	2.17	2.2	3.4
χ (eV)	4.0	3.9	3.9	2.2	3.2	2.9	1.9
ε_r	9.0	9.0	30	3.0	7.11	3.0	10
N_C (cm^{-3})	2.02×10^{18}	1×10^{21}	2.2×10^{18}	1×10^{19}	2.02×10^{17}	2.2×10^{15}	2.2×10^{19}
N_V (cm^{-3})	1.8×10^{18}	2×10^{20}	1×10^{18}	1×10^{19}	1.1×10^{19}	1.8×10^{18}	1.8×10^{18}
μ_n (cm^2/Vs)	20	20	2.2	10^{-4}	200	1×10^{-2}	1×10^{-4}
μ_p (cm^2/Vs)	10	10	2.2	10^{-4}	80	2×10^{-4}	1×10^{-2}
N_D (cm^{-3})	2×10^{19}	1×10^{19}	9×10^{20}	0	-	0	0
N_A (cm^{-3})	0	0	0	2×10^{19}	1×10^{18}	1×10^{19}	1×10^{16}
N_t (cm^{-3})	1×10^{15}	1×10^{15}	8.5×10^{13}	1×10^{14}	1×10^{14}	1×10^{14}	1×10^{17}

Table 2. Initial performances with various HTM layers.

HTM materials	**V_{OC} (V)**	**J_{SC} (mA/cm²)**	**FF (%)**	**PCE (%)**
Cu_2O	1.56	20.76	91.64	29.84
Spiro-OMeTAD	1.56	20.27	91.62	29.11
PEDOT: PSS	1.56	20.39	91.61	29.29
CuSCN	1.56	20.26	89.26	28.36

Figure 3. (a) J-V plot and (b) QE- λ curve of Pb-based PSC with different HTL

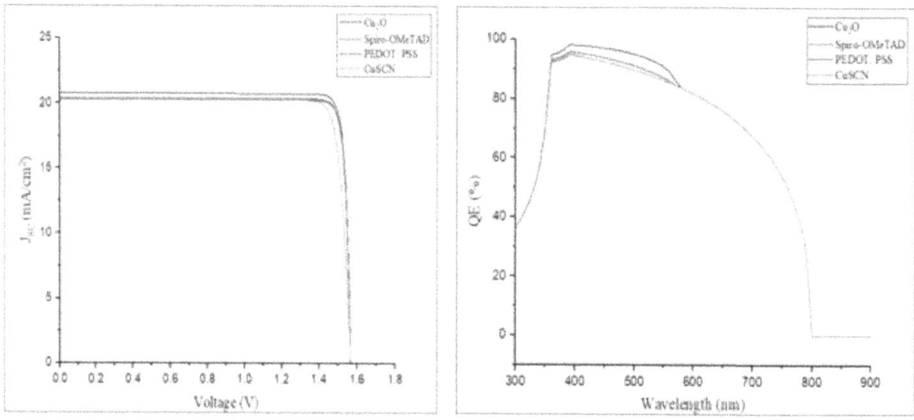

(a) (b)

Figure 4. The energy level of the device layers with gold back contact

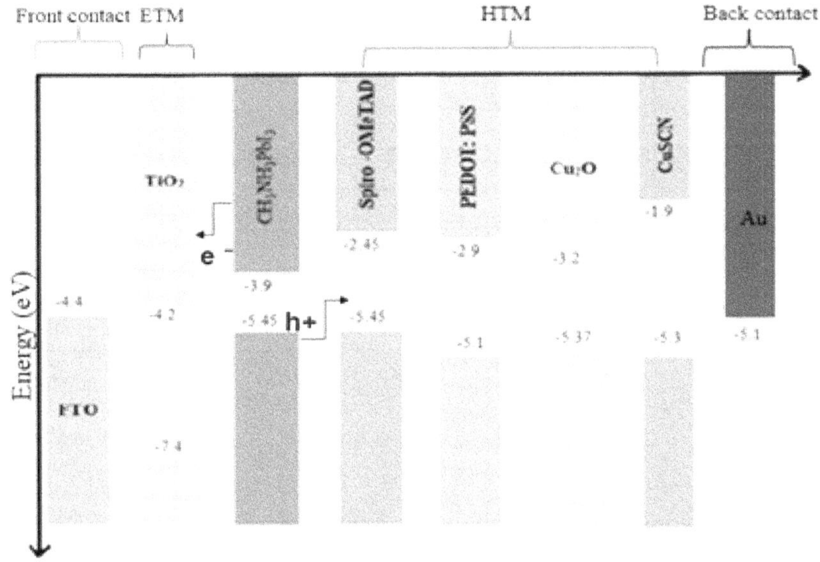

2.2. Statistical Method-(Design of Experiments)

The Taguchi method is generally employed for optimizing the different parameters involved in the functioning of a device. It includes a number of experiments by adopting specially constructed tables called an orthogonal array. The Experimental design of the statistical Taguchi method helps in finding optimum process factors or parameters (Nian et al., 1999). We have used three factors: thickness of PAL, operating temperature, and defect density in this work, as shown below in table 4. The levels of all three factors have been selected after performing test runs of the simulated experiments. The results of the test experiments are shown in fig. 5. The L9 OA was employed in designing the experiments shown below in table 3. This method allows the optimum process parameters to be found. The Taguchi method is the most vital design of the experiment (DOE); it endows an effortless and emphatic approach to determine the optimum parameters. The DOE can be quite helpful in determining how various individual parameters affect the operation of the system (Ghani et al., 2004). The parameters that have the greatest influence on the performance measure can be found by examining the effects of each individual parameter. This statistical method significantly helps in saving cost, time, and outcomes in the more detailed procedure.

Table 3. Taguchi's L9 (3^3) orthogonal array employed in simulation.

Experiment No.	Thickness	Operating Temperature	Defect Density
1	1	1	1
2	1	2	2
3	1	3	3
4	2	1	2
5	2	2	3
6	2	3	1
7	3	1	3
8	3	2	1
9	3	3	2

Table 4. Levels defined for L9 orthogonal array.

Levels	Defect Density	Thickness	Operating Temperature
1	10^{14} cm^{-3}	300 nm	300 K
2	10^{15} cm^{-3}	400 nm	310 K
3	10^{16} cm^{-3}	500 nm	320 K

Figure 5. change in efficiency with thickness of PAL, operating temperature and defect density

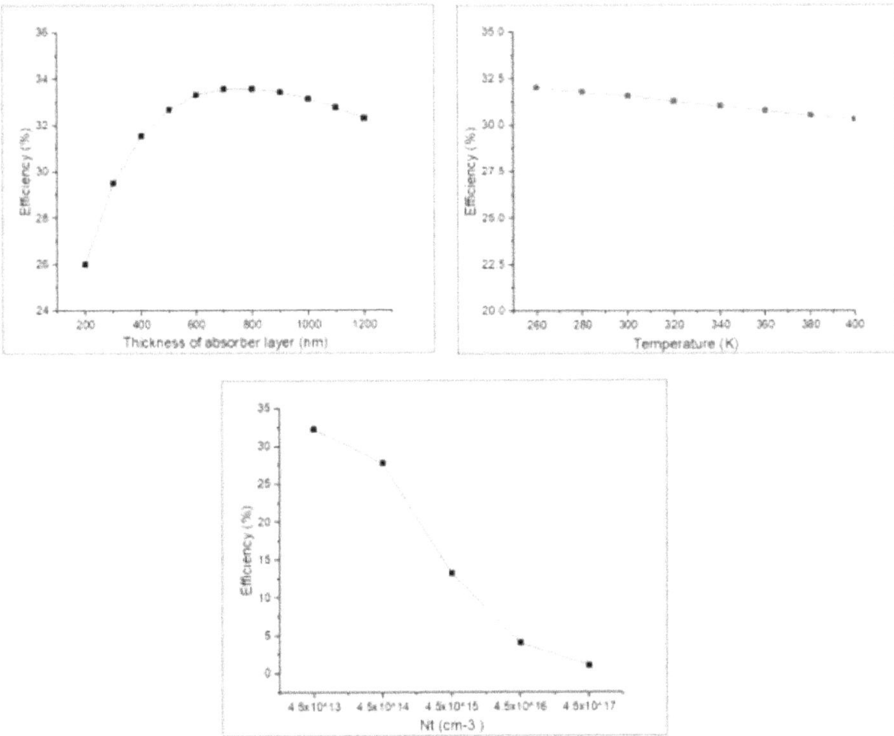

3. RESULT AND DISCUSSION

3.1 Impact of Hole Transport Layer

In the present work we inquired the influence of PAL thickness, operating temperature and N_t on cell functioning using SCAPS-1D simulator. The four PSCs with the same ETL as TiO_2 with different hole transporting layers such as Cu_2O, Spiro-OMeTAD, PEDOT: PSS and CuSCN are employed in the simulation. The impact of different values of operating temperature on the FF, V_{OC}, J_{SC}, and the PCE are also studied (Toshniwal et al., 2017; Devi et al., 2019). The thickness of PAL is a limiting and significant determinant to achieve the optimum performance of a PSC. The efficiency of a PSC can be in synced with a variation in thickness of PAL (Rozhko et al., 2021). The PAL thickness fluctuated between 300 nm to 500 nm demonstrated in table 4. The fig.4 shows the optimum value of PAL thickness, operating temperature and N_t for the simulation process. For a thin PAL of thickness 300 nm, the amount of photogenerated current is less. Despite these results, the extraction of charge is high as justified with high value of V_{OC} which is a sign of few recombination. The value of J_{SC} rises with enrichment in the value of thickness due to higher absorption of light. When the J_{SC} increases with PAL thickness, it results in the origination of large e−-hole pairs in the absorber layer leading in the direction of maximum efficiency of PSC (Kumar et al., 2022). Because it rises rapidly from 20.74 to 23.20 mA/cm^2, 20.26 to 23.10 mA/cm^2, 20.32 to 23.12 mA/cm^2 and 20.24 to 23.09 mA/cm^2 for Cu_2O, Spiro-OMeTAD, PEDOT: PSS and CuSCN respectively employed as HTM layer, for PAL thickness which varied between 300 nm to 500 nm. A steady and immediate rise in the value of J_{SC} is because of photons of energy greater than the E_g of PAL. For lower value of the thickness, long wavelength photons have lower absorption levels. Around 300–400 nm, J_{SC} is low because of less absorption and there is no significance of using the thickness of film greater than 500 nm. So, a PAL thickness of 500 nm is suitable. In this work, the best value of J_{SC} was found for PAL at a thickness of 500 nm. For the thicker absorber layer, long wavelength photons are absorbed due to the reduction in the effective E_g. For large values of diffusion lengths (greater than 500 nm), one can proceed to research upon how thickness does not impact functioning of the device (Bag et al., 2020). The utmost efficiency is achieved at an optimum thickness of 500 nm. However, a lowered efficiency for the higher thickness absorber layer is a result of the decreased electric field, which impacts the charge carriers' recombination behavior inside PAL. The saturation in PCE is a result of saturation in absorption and in J_{SC} for thicker absorber layer. Low efficiency due to fractional absorption of light and fewer photogeneration for thin perovskite layer. The optimum PAL thickness for Pb-based PSC is 500 nm. A fall in V_{OC} with PAL thickness can be correlated with the reduction the effective band gap by the thickness of perovskite layer and an increment in recombination with thickness (Kanoun et al., 2019). The drop in value of V_{OC} by thickness is attributed to enhancement in the dark saturation current results in recombination's. This can be elucidated by the reliance of V_{OC}, which is as follow (Ouslimane et al., 2021):

$$V_{oc} = \frac{KT}{q} \mathrm{Ln}\left[\frac{Jsc}{Jo} + 1\right] \tag{1}$$

The following outcomes distinctly exhibit rise in value of J_{SC} with PAL thickness and fall in V_{OC} and FF. The Fill factor diminishes with thickness because of rise in series resistance (R_s) across the cell and internal power dissipation in a coarse PAL which causes low efficiency. Identical to PAL thickness, cell PCE is also influenced with thickness of HTL. The change in FF with HTL thickness is a sign of variation in R_s (Hima et al., 2020; Raj et al., 2021). As the HTL thickness increases from 80 to 350 nm, cell PCE rises up to 32% because of increase in FF. In turn the increased thickness lowers the FF and PCE. A thin HTL (< 50 nm) does not cover PAL perfectly and opportunity of recombination is large therefore path length increased to reach the back contact thus increase in R_s (Bag et al., 2020). Therefore, FF increases with HTL thickness enhancement from 80 to 350 nm because of large coverage. Operating temperature is a noteworthy determinant for availing the optimum performance of a device. Commonly, solar panels frequently work at temperatures greater than 300 K. It was reported that rise in temperature enhances strain and stress inside structures (Mandadapu et al., 2018). The simulative study was conducted with operating temperature changed between 300 K to 320 K demonstrated below in table 4. The highest efficiency of 32.90%, 32.72%, 32.77% and 31.75% for Cu_2O, spiro-OMeTAD, PEDOT: PSS and CuSCN respectively has been achieved at a temperature of 310 K. We saw that J_{SC} is gradually increased with increment in value of the temperature, because of reduction in Eg and much **e** $^-$-hole pairs generation. The reduction in V_{OC} with respect to operating temperature can be elucidated by the erection of a great number of interfacial defects together with an increase of R_s and decrement in the carrier diffusion length. With the increase in the value of temperature, a decrease in PCE is noticed because of the downfall of PAL at high temperature (Ava et al., 2019). The impact of different aspects on the solar device performance and functioning have been examined. Such parameters are PAL thickness, N_t and operating temperature. The absorber layer's N_t is a very significant determinant that holds the capacity to highly impact the PCE of PSC. The great number of defect concentrations in the perovskite layer results in a larger recombination because of generation of pinholes, a large rate of decadence of the film, decreased stability and deficiency in total performance of device and its functioning (Ouslimane et al., 2021). The simulation was done for the different values of the total defect density such as 10^{14}, 10^{15} and 10^{16} cm^{-3} for Pb-based PSC with different HTM layers. The performance of all devices reduced when the increment in value of N_t in PAL as can be seen below in table.5. The moment of the highest and most impactful defect density as 10^{16} cm^{-3} for any combination of thickness and operating temperature, we have achieved the least value of efficiency. As the defect density increases, the fill factor decreases which results in less efficiency (Chowdhury et al., 2020). The PCE sharply declined from 32.90% to 10.18% as the value of N_t increased from 10^{14} to 10^{16} cm^{-3} shown below in table 5. We deduced that N_t influences the PCE as rise in the number of defects results in decrease in the diffusion length of charge carriers within PAL. For the larger value of N_t, downfall in the value of J_{SC} because of a rise in the recombination rate with an increment in value of N_t (Ahmed et al., 2021). In this study we have obtained the maximum efficiency for the least value of N_t which is 10^{14} cm^{-3}. Thus, we can conclude that

10^{14} cm^{-3} is the most efficient value of the N_t for maximum PCE. This all discussion deduced that the absorber layer with 500 nm thickness, at operating temperature of 310 K and the total defect density arrived the resultant value of 10^{14} cm^{-3} which is suitable for highest performance of PSC. The OA endows a collection of minimum experiments and required output which work as purposeful functions for optimization. The OA assists in data explication and prognostication of optimum outcomes (Karna et al., 2012). The Taguchi method decreases experimental expense with the help of OA to extract optimized information from a number of experiments (Wang et al., 2007). The reason for employing L9 OA as there are three determinants entailed which are known as control factors. In present simulation factors are PAL thickness, N_t and operating temperature (Bahrudin et al., 2018). The experiment no.8 of L9 (3^3) orthogonal array of Taguchi shown in table.5 provided maximum PCE (Ozcelik et al., 2011). The final optimized efficiency using different HTM layers shown below in table 6.

Table 5. Optimized value of PV parameters of different HTL layers

		Cu_2O				Spiro-OMeTAD				PEDOT: PSS				CuSCN		
Exp. No.	**V_{OC} (V)**	**J_{SC} (mA/ cm²)**	**FF (%)**	**η (%)**	**V_{OC} (V)**	**J_{SC}(mA/ cm²)**	**FF (%)**	**η (%)**	**V_{OC} (V)**	**J_{SC}(mA/ cm²)**	**FF (%)**	**η (%)**	**V_{OC} (V)**	**J_{SC}(mA/ cm²)**	**FF (%)**	**η (%)**
E1	1.56	20.74	91.64	29.73	1.56	20.26	91.62	29.03	1.56	20.32	91.62	29.19	1.56	20.24	89.24	28.25
E2	1.50	19.67	91.15	26.90	1.49	19.18	91.13	26.22	1.49	19.29	91.14	26.37	1.49	19.15	88.32	25.37
E3	1.42	13.21	90.54	17.10	1.42	12.72	90.52	16.46	1.42	12.83	90.52	16.60	1.42	12.61	86.52	15.60
E4	1.49	20.16	91.40	27.62	1.49	19.94	91.35	27.31	1.49	19.99	91.38	27.38	1.49	19.92	88.32	26.37
E5	1.42	10.50	90.78	13.62	1.42	10.28	90.75	13.33	1.42	10.33	90.76	13.39	1.42	10.21	87.19	12.71
E6	1.55	22.30	91.17	31.58	1.55	22.08	91.11	31.25	1.55	22.13	91.15	31.33	1.55	22.07	88.54	30.36
E7	1.42	7.83	91.01	10.18	1.42	7.74	90.98	10.05	1.42	7.76	90.99	10.08	1.42	7.69	88.02	9.66
E8	1.55	23.20	91.40	32.90	1.55	23.10	91.31	32.72	1.55	23.12	91.36	32.77	1.55	23.09	88.58	31.75
E9	1.48	20.02	90.90	27.01	1.48	19.93	90.83	26.86	1.48	19.95	90.87	26.90	1.48	19.89	87.57	25.86

The J-V and QE have been deduced for PSCs with various HTM layers. The QE of PSCs increased with the thickness of PAL. With increment in thickness of PAL, photons with larger wavelengths were absorbed. The QE of PSCs increased with increase in thickness of HTM layer (Karimi et al., 2017). The QE with function of wavelength of light is ratio of current flowing via the circuit ie. the amount of charge carriers drains the solar cell to the amount of incident photons (Hossain et al., 2021). The noticeable characteristics of Pb-based PSCs is less carrier recombination at interfaces by introducing PAL amid ETL and HTL, a large value of J_{SC} may be achieved with an internal QE (Shirayama et al., 2016). To find the credibility of the efficiency, we calculated the J_{SC} and PCE of various Pb-based PSCs with different bias voltages (Nie et al., 2020). The EQE demonstrated very engrossing characteristics over the spectral range of device output. The number of defects decreases cell PCE by reducing V_{OC} and J_{SC}, required to be as small as feasible (Yang et al., 2017). The optimized J-V characteristic and QE plots respectively demonstrated below in fig. 6 (a) and 6 (b). A suitable QE of 67.4-98%, 67.4-95.5%, 67.4-95.7% and 67.4-94.4% is obtained in 400 nm to 700 nm and is concerned with Jsc of 23.20, 23.10, 23.12 and 23.09 mA/cm^2 for Cu_2O, Spiro-OMeTAD. PEDOT: PSS and CuSCN respectively. The J-V curve of PSCs was obtained for different HTM layers. The impact of various HTM layers on the J-V plots are presented in Fig.6 (a). By employing various ETL and plotting J-V curves. It was apparent that every ETL Eg and χ has its own impact onto the JV curve (Khattak et al., 2020). The QE is a vital feature in PSCs to assess our simulation. The results are obtained by employing experimental and theoretical data. The QE curves of various PSCs are presented in Fig.6 (b). The graphs show the contribution of photogenerated e^--h^+ pairs to current density (Azri et al., 2019). The table.6 hereby demonstrates that there is enough space to find out various PSCs with different HTM to progress its PV parameters. The PV parameters of all PSCs are largely depending on the PAL thickness, operating temperature, HTM, ETM, N_t and the method employed for the deposition of layers (Singh et al., 2020). These outcomes explain the potential of Pb-based PSC together with the need of different dedicated materials for solar device fabrication. By optimizing the various PV parameters of Pb-based PSC, we have obtained PCE of greater than 30%. By employing PEDOT: PSS and Spiro-OMeTAD as HTL, the PCE of 31.77% and 30.35% achieved respectively (Mandadapu et al., 2017; Slami et al., 2019). The curve between optimized PCE and dopant concentration of the FTO represents that the PCE increases with increment in dopant concentration shown below in fig.7. The change in PCE with various back contact shown below in fig.8. This mechanism can be a guidance for efficient and inexpensive organic-inorganic PSCs (Pathak et al., 2020; Jeon et al., 2015). There are many factors which affect the stability of PSC such as moisture, high temperature and exposing it to sustained light. By introducing a condense buffer layer inside the structure. For thermally stable devices a rapid dissipation of internal heat in the device is expected. Instability in solar devices and materials firstly because of reaction with water or moisture, this kept PSCs from commercialization. We have to understand the microscopic level mechanism for comprehending the downfall in the performance of PSCs for reforming the longtime stability (Zhao et al., 2015; Song et al., 2016). Further analysis on factors required more detail which are the cause of unstable PSC.

Figure 6. Optimized (a) J-V plots and (b) QE- λ curve of Pb-based PSC with different HTLs

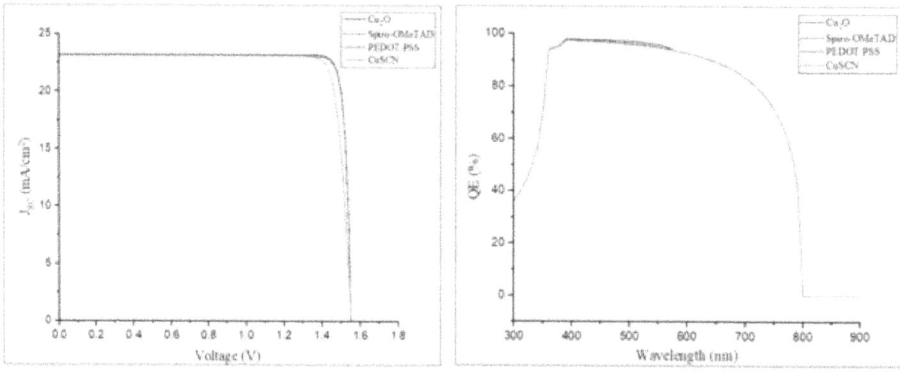

(a) (b)

Figure 7. The optimized PCE curve of Pb-based PSC with different HTLs

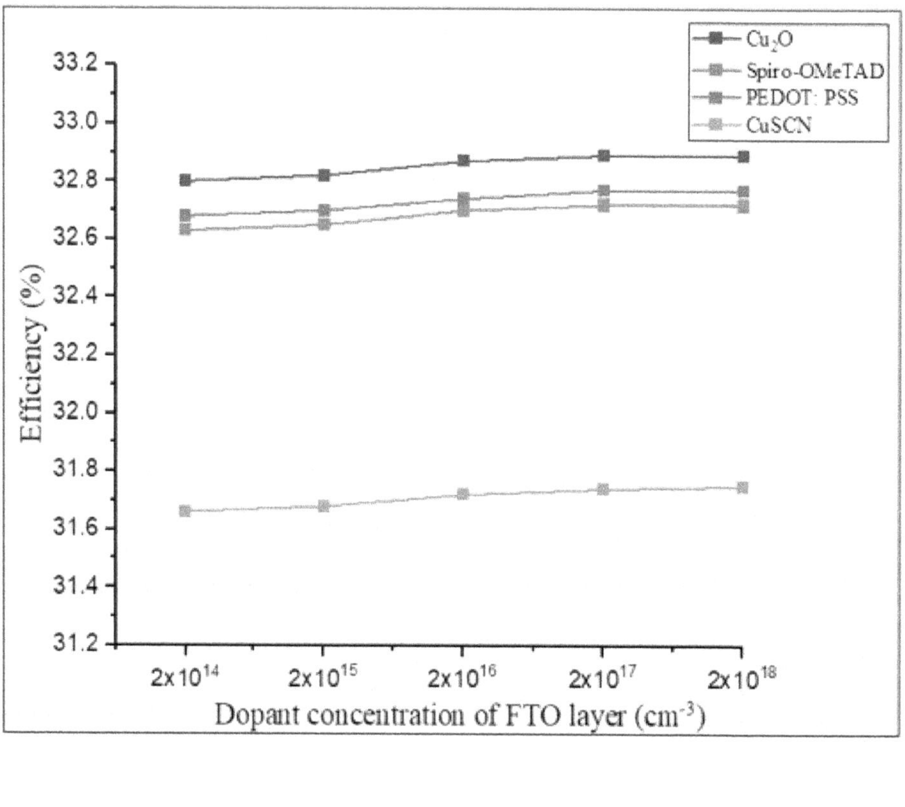

Figure 8. The optimized PCE curve of Pb-based PSCs with various back-contact

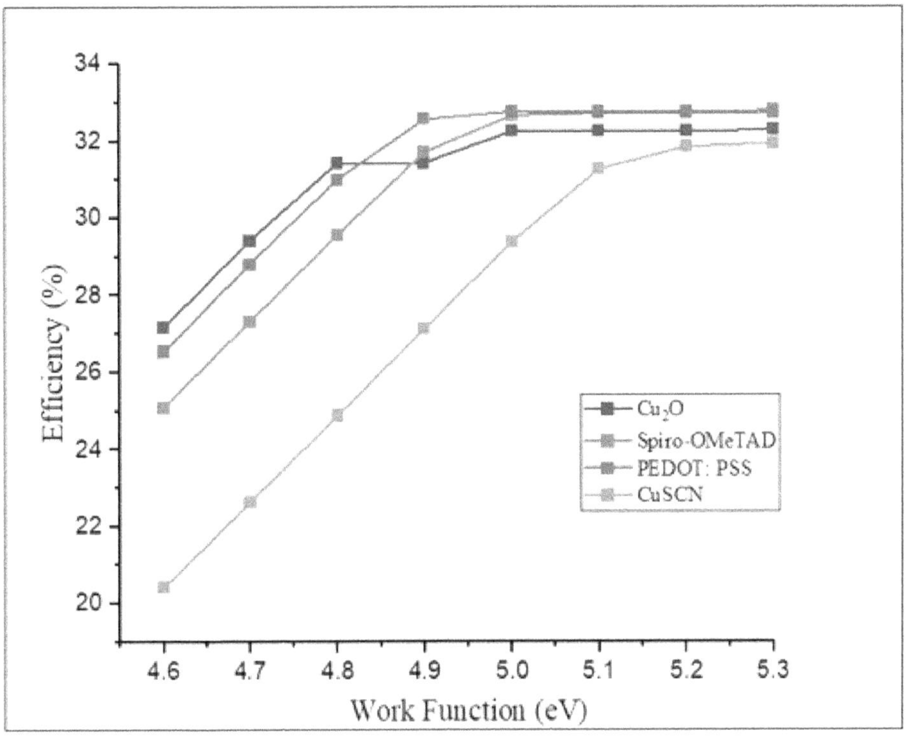

Table 6. Optimized results and its comparison with previously reported results.

HTM materials	V_{OC} (V)	J_{SC} (mA/cm^2)	FF (%)	PCE (%)
Cu_2O	1.55	23.20	91.40	32.90 (This work)
Spiro-OMeTAD	1.55	23.10	91.31	32.72 (This work)
PEDOT: PSS	1.55	23.12	91.36	32.77 (This work)
CuSCN	1.55	23.09	88.58	31.75 (This work)
Spiro-OMeTAD	1.21	22.51	74.39	20.26 (De Los Santos et al., 20207)
Cu_2O	1.15	23.34	70.31	18.92 (Rai et al., 2020)
Spiro-OMeTAD	1.47	15.15	91.00	20.34 (Husainat et al., 2019)

4. CONCLUSION

We have studied the impact of different parameters which affect the performance of PSCs. The PAL thickness, N_t in absorber layer and operating temperature of $CH_3NH_3PbI_3$ PSCs with different HTLs were scrutinized using SCAPS-1D. The band alignment of ETL, absorber layer and HTL is also very crucial factor which affects the PCE of solar devices. The band alignment diagram helps in suitable selection of ETL and HTL with the absorber layer to optimized the structure of PSC. The impact of PAL thickness on the properties of solar devices reveals that there is an optimum range of PAL thickness which is 300 to 500 nm for an efficient PSC. Besides, the optimum efficiency, exceeding 30%, has been reached for a PSC by employing Cu_2O, Spiro-OMeTAD, PEDOT: PSS and CuSCN as HTM layers. A suitable HTM layer should balance μ_p to construct a solar device with high PCE. Indeed, the gold metal back contact electrode employed for the high performance of PSCs because of its high work function. The PAL thickness was increased up to 500 nm beyond that the FF shows a monotonic decrement with the increase in PAL thickness which results in falls in PCE. The Taguchi L9 orthogonal array is applied to obtain optimized values of parameters which affects the efficiency of a PSC. The Taguchi L9 orthogonal array deduced that the optimum thickness of PAL is 500 nm and N_t is 10^{14} cm^{-3} at the operating temperature of 310 K for maximum PCE of the $CH_3NH_3PbI_3$-based PSCs. The maximum PCE of 32.90%, 32.72%, 32.77% and 31.75% has been obtained for Cu_2O, Spiro-OMeTAD, PEDOT: PSS and CuSCN respectively. This work is expected to be helpful for profound understanding of fabrication of highly efficient $CH_3NH_3PbI_3$-based PSCs. Although it has one significant limitation of stability which creates a hindrance in its commercialization.

Competing interests: The authors declare no competing interests.

Funding: The present research work has not received external funding.

Data availability statement: The authors declare that the data supporting the findings of this study are available within the paper.

REFERENCES

Ahmed, S., Jannat, F., Khan, M. A. K., & Alim, M. A. (2021). Numerical development of eco-friendly Cs2TiBr6 based perovskite solar cell with all-inorganic charge transport materials via SCAPS-1D. *Optik (Stuttgart)*, 225, 165765. DOI: 10.1016/j.ijleo.2020.165765

Ava, T. T., Al Mamun, A., Marsillac, S., & Namkoong, G. (2019). A review: Thermal stability of methylammonium lead halide based perovskite solar cells. *Applied Sciences (Basel, Switzerland)*, 9(1), 188. DOI: 10.3390/app9010188

Azri, F., Meftah, A., Sengouga, N., & Meftah, A. (2019). Electron and hole transport layers optimization by numerical simulation of a perovskite solar cell. *Solar Energy*, 181, 372–378. DOI: 10.1016/j.solener.2019.02.017

Bag, A., Radhakrishnan, R., Nekovei, R., & Jeyakumar, R. (2020). Effect of absorber layer, hole transport layer thicknesses, and its doping density on the performance of perovskite solar cells by device simulation. *Solar Energy*, 196, 177–182. DOI: 10.1016/j.solener.2019.12.014

Bahrudin, M. S., Abdullah, S. F., Ahmad, I., Zuhdi, A. W. M., & Hasani, A. H., Za'Abar, F., ... & Harif, M. N. (2018, August). J sc and V oc Optimization of Perovskite Solar Cell with Interface Defect Layer Using Taguchi Method. In 2018 IEEE International Conference on Semiconductor Electronics (ICSE) (pp. 192-196). IEEE. DOI: 10.1109/SMELEC.2018.8481203

Behl, V., Singh, V., Dahiya, V., & Kumar, A. (2022). Characterization of physicochemical and functional properties of fly ash concrete mix. *Materials Today: Proceedings*, 50, 941–945. DOI: 10.1016/j.matpr.2021.06.353

Chowdhury, M. S., Shahahmadi, S. A., Chelvanathan, P., Tiong, S. K., Amin, N., Techato, K. A., Nuthammachot, N., Chowdhury, T., & Suklueng, M. (2020). Effect of deep-level defect density of the absorber layer and n/i interface in perovskite solar cells by SCAPS-1D. *Results in Physics*, 16, 102839. DOI: 10.1016/j.rinp.2019.102839

De Los Santos, I. M., Cortina-Marrero, H. J., Ruíz-Sánchez, M. A., Hechavarría-Difur, L., Sánchez-Rodríguez, F. J., Courel, M., & Hu, H. (2020). Optimization of CH3NH3PbI3 perovskite solar cells: A theoretical and experimental study. *Solar Energy*, 199, 198–205. DOI: 10.1016/j.solener.2020.02.026

De Los Santos, I. M., Cortina-Marrero, H. J., Ruíz-Sánchez, M. A., Hechavarría-Difur, L., Sánchez-Rodríguez, F. J., Courel, M., & Hu, H. (2020). Optimization of CH3NH3PbI3 perovskite solar cells: A theoretical and experimental study. *Solar Energy*, 199, 198–205. DOI: 10.1016/j.solener.2020.02.026

Devi, C., & Mehra, R. (2019). Device simulation of lead-free MASnI 3 solar cell with CuSbS 2 (copper antimony sulfide). *Journal of Materials Science*, 54(7), 5615–5624. DOI: 10.1007/s10853-018-03265-y

Ghani, J. A., Choudhury, I. A., & Hassan, H. H. (2004). Application of Taguchi method in the optimization of end milling parameters. *Journal of Materials Processing Technology*, 145(1), 84–92. DOI: 10.1016/S0924-0136(03)00865-3

Haider, S. Z., Anwar, H., & Wang, M. (2019). Theoretical device engineering for high-performance perovskite solar cells using CuSCN as hole transport material boost the efficiency above 25%. physica status solidi (a), 216(11), 1900102.

Hima, A., & Lakhdar, N. (2020). Enhancement of efficiency and stability of CH3NH3GeI3 solar cells with CuSbS2. *Optical Materials*, 99, 109607. DOI: 10.1016/j.optmat.2019.109607

Hima, A., Lakhdar, N., Benhaoua, B., Saadoune, A., Kemerchou, I., & Rogti, F. (2019). An optimized perovskite solar cell designs for high conversion efficiency. *Superlattices and Microstructures*, 129, 240–246. DOI: 10.1016/j.spmi.2019.04.007

Hossain, J., Moon, M. M. A., Mondal, B. K., & Halim, M. A. (2021). Design guidelines for a highly efficient high-purity germanium (HPGe)-based double-heterojunction solar cell. *Optics & Laser Technology*, 143, 107306. DOI: 10.1016/j.optlastec.2021.107306

Husainat, A., Ali, W., Cofie, P., Attia, J., & Fuller, J. (2019). Simulation and analysis of methylammonium lead iodide (CH3NH3PbI3) perovskite solar cell with Au contact using SCAPS 1D simulator. *American Journal of Optics and Photonics*, 7(2), 33. DOI: 10.11648/j.ajop.20190702.12

Jeon, N. J., Noh, J. H., Yang, W. S., Kim, Y. C., Ryu, S., Seo, J., & Seok, S. I. (2015). Compositional engineering of perovskite materials for high-performance solar cells. *Nature*, 517(7535), 476–480. DOI: 10.1038/nature14133 PMID: 25561177

Kabir, E., Kumar, P., Kumar, S., Adelodun, A. A., & Kim, K. H. (2018). Solar energy: Potential and future prospects. *Renewable & Sustainable Energy Reviews*, 82, 894–900. DOI: 10.1016/j.rser.2017.09.094

Kanoun, A. A., Kanoun, M. B., Merad, A. E., & Goumri-Said, S. (2019). Toward development of high-performance perovskite solar cells based on CH3NH3GeI3 using computational approach. *Solar Energy*, 182, 237–244. DOI: 10.1016/j.solener.2019.02.041

Karimi, E., & Ghorashi, S. M. B. (2017). Simulation of perovskite solar cell with P3HT hole-transporting materials. *Journal of Nanophotonics*, 11(3), 032510–032510. DOI: 10.1117/1.JNP.11.032510

Karna, S. K., & Sahai, R. (2012). An overview on Taguchi method. International journal of engineering and mathematical sciences, 1(1), 1-7.

Kenfack, A. K., Mashamba, D. R., Thantsha, N. M., & Msimanga, M. (2023). Prediction of band gap and optimum electrical parameters of a thin homojunction perovskite solar cell based on FA1-xCsxSnyPb1-yI3 through a combination of SCAPS-1D and Machine Learning based modelling. *Materials Today. Communications*, 37, 107318. DOI: 10.1016/j.mtcomm.2023.107318

Khattak, Y. H., Baig, F., Shuja, A., Beg, S., & Soucase, B. M. (2020). Numerical analysis guidelines for the design of efficient novel nip structures for perovskite solar cell. *Solar Energy*, 207, 579–591. DOI: 10.1016/j.solener.2020.07.012

Kumar, A., Kumar, P., Srivastava, A. K., & Goyat, V. (Eds.). (2023). *Modeling, Characterization, and Processing of Smart Materials*. IGI Global., DOI: 10.4018/978-1-6684-9224-6

Kumar, A., Liu, Y., & Kumar, R. (Eds.). (2024). *Handbook of Intelligent and Sustainable Manufacturing: Tools, Principles, and Strategies*. CRC Press., DOI: 10.1201/9781003405870

Kumar, A., Mittal, R. K., & Goel, R. (Eds.). (2023). *Waste Recovery and Management: An Approach Toward Sustainable Development Goals*. CRC Press., DOI: 10.1201/9781003359784

Kumar, A., Mittal, R. K., & Haleem, A. (Eds.). (2022). *Advances in Additive Manufacturing: Artificial Intelligence, Nature-Inspired, and Biomanufacturing*. Elsevier., DOI: 10.1016/C2020-0-03877-6

Kumar, A., Rajak, D. K., Kumar, P., & Rathee, S. (Eds.). (2024). *Smart Electric and Hybrid Vehicles: Fundamentals, Strategies and Applications* (1st ed.). CRC Press., DOI: 10.1201/9781003495574

Kumar, A., Rajak, D. K., Kumar, P., & Rathee, S. (Eds.). (2024). *Smart Electric and Hybrid Vehicles: Design, Modeling, and Assessment by Industry 4.0 Approaches* (1st ed.). CRC Press., DOI: 10.1201/9781003502470

Kumar, A., Singh, H., Kumar, P., & AlMangour, B. (Eds.). (2023). *Handbook of Smart Manufacturing: Forecasting the Future of Industry 4.0*. CRC Press., DOI: 10.1201/9781003333760

Kumar, M., Kumar, A., Raj, A., Sati, P. C., Sahni, M., & Anshul, A. (2022). Organic-inorganic perovskite-based solar cell designs for high conversion efficiency: A comparative study by SCAPS simulation. *Materials Today: Proceedings*, 49, 3081–3087. DOI: 10.1016/j.matpr.2020.11.035

Kumari, N., Gohel, J. V., & Patel, S. R. (2018). Optimization of TiO2/ZnO bilayer electron transport layer to enhance efficiency of perovskite solar cell. *Materials Science in Semiconductor Processing*, 75, 149–156. DOI: 10.1016/j.mssp.2017.11.030

Kumavat, S. R., Sonvane, Y., & Gupta, S. K. (2020). Structural, optical, transport, and solar cell properties of 2D halide perovskite MAZX3 (Z= Pb, Sn, and X= Cl, Br, I). *Journal of Applied Physics*, 128(11), 114304. DOI: 10.1063/5.0016331

Lee, M. M., Teuscher, J., Miyasaka, T., Murakami, T. N., & Snaith, H. J. (2012). Efficient hybrid solar cells based on meso-superstructured organometal halide perovskites. science, 338(6107), 643-647.

Mandadapu, U., Vedanayakam, S. V., & Thyagarajan, K. (2017). Simulation and analysis of lead based perovskite solar cell using SCAPS-1D. *Indian Journal of Science and Technology*, 10(11), 65–72. DOI: 10.17485/ijst/2017/v10i11/110721

Mandadapu, U., Vedanayakam, S. V., Thyagarajan, K., & Babu, B. J. (2018). Optimisation of high efficiency tin halide perovskite solar cells using SCAPS-1D. International journal of simulation and process modelling, 13(3), 221-227.

Mittal, R. K. (2020). *Incremental sheet forming technologies: principles, merits, limitations, and applications*. CRC Press., DOI: 10.1201/9780429298905

Moiz, S. A., Alahmadi, A. N. M., & Aljohani, A. J. (2021). Design of a novel lead-free perovskite solar cell for 17.83% efficiency. *IEEE Access : Practical Innovations, Open Solutions*, 9, 54254–54263. DOI: 10.1109/ACCESS.2021.3070112

National Renewable Energy Laboratory. (2020). Best Research-Cell Efficiency Chart. Available online at: https://www.nrel.gov/pv/cell-efficiency.html (accessed April 6, 2020). [Ref list] https://www.ncbi.nlm.nih.gov/pmc/articles/PMC7399487/#B23

Nian, C. Y., Yang, W. H., & Tarng, Y. S. (1999). Optimization of turning operations with multiple performance characteristics. *Journal of Materials Processing Technology*, 95(1-3), 90–96. DOI: 10.1016/S0924-0136(99)00271-X

Nie, R., Lee, K. S., Hu, M., Paik, M. J., & Seok, S. I. (2020). Heteroleptic tin-antimony sulfoiodide for stable and lead-free solar cells. *Matter*, 3(5), 1701–1713. DOI: 10.1016/j.matt.2020.08.020

Ouslimane, T., Et-taya, L., Elmaimouni, L., & Benami, A. (2021). defect density, and operating temperature on the performance of MAPbI3 solar cells based on ZnO electron transporting material. Touria Ouslimane, Lhoussayne Et-taya, Lahoucine Elmaimouni, Abdellah Benami. *Heliyon*, 7(3), e06379. DOI: 10.1016/j.heliyon.2021.e06379 PMID: 33732928

Ozcelik, B. (2011). Optimization of injection parameters for mechanical properties of specimens with weld line of polypropylene using Taguchi method. *International Communications in Heat and Mass Transfer*, 38(8), 1067–1072. DOI: 10.1016/j.icheatmasstransfer.2011.04.025

Pathak, C., & Pandey, S. K. (2020). Design, performance, and defect density analysis of efficient eco-friendly perovskite solar cell. *IEEE Transactions on Electron Devices*, 67(7), 2837–2843. DOI: 10.1109/TED.2020.2996570

Rai, S., Pandey, B. K., & Dwivedi, D. K. (2020). Modeling of highly efficient and low cost CH3NH3Pb (I1-xClx) 3 based perovskite solar cell by numerical simulation. *Optical Materials*, 100, 109631. DOI: 10.1016/j.optmat.2019.109631

Raj, A., Kumar, M., Bherwani, H., Gupta, A., & Anshul, A. (2021). Evidence of improved power conversion efficiency in lead-free CsGeI3 based perovskite solar cell heterostructure via scaps simulation. *Journal of Vacuum Science & Technology B Microelectronics and Nanometer Structures*, 39(1).

Rozhko, A. A., Petrov, V. V., & Sayenko, A. V. (2021). Study of the effect of the thickness of the photosensitive layer of perovskite on its efficiency using SCAPS-1D software. [). IOP Publishing.]. *IOP Conference Series. Materials Science and Engineering*, 1035(1), 012032. DOI: 10.1088/1757-899X/1035/1/012032

Sakib, S., Noor, M. Y. M., Salim, M. R., Abdullah, A. S., Azmi, A. I., & Ibrahim, M. H. (2023). Effect of transport layer thickness in lead-based perovskite solar cell: A numerical simulation. *Materials Today: Proceedings*, 80, 1022–1026. DOI: 10.1016/j.matpr.2022.11.456

Shirayama, M., Kadowaki, H., Miyadera, T., Sugita, T., Tamakoshi, M., Kato, M., Fujiseki, T., Murata, D., Hara, S., Murakami, T. N., Fujimoto, S., Chikamatsu, M., & Fujiwara, H. (2016). Optical transitions in hybrid perovskite solar cells: Ellipsometry, density functional theory, and quantum efficiency analyses for CH 3 NH 3 PbI 3. *Physical Review Applied*, 5(1), 014012. DOI: 10.1103/PhysRevApplied.5.014012

Singh, N., Agarwal, A., & Agarwal, M. (2020). Numerical simulation of highly efficient lead-free all-perovskite tandem solar cell. *Solar Energy*, 208, 399–410. DOI: 10.1016/j.solener.2020.08.003

Singh, T., & Miyasaka, T. (2018). Stabilizing the efficiency beyond 20% with a mixed cation perovskite solar cell fabricated in ambient air under controlled humidity. *Advanced Energy Materials*, 8(3), 1700677. DOI: 10.1002/aenm.201700677

Singh, V., Kumar, A., Kumar, K., & Gupta, M. (2021). Synthesis and Stability of Al 2 O 3/Water Nanofluids. In Advances in Materials and Mechanical Engineering: Select Proceedings of ICFTMME 2020 (pp. 3-11). Springer Singapore.

Slami, A., Bouchaour, M., & Merad, L. (2019). Numerical study of based perovskite solar cells by SCAPS-1D. *International Journal of Energy and Environment*, 3, 17–21.

Song, Z., Abate, A., Watthage, S. C., Liyanage, G. K., Phillips, A. B., Steiner, U., Graetzel, M., & Heben, M. J. (2016). Perovskite solar cell stability in humid air: Partially reversible phase transitions in the PbI2-CH3NH3I-H2O system. *Advanced Energy Materials*, 6(19), 1600846. DOI: 10.1002/aenm.201600846

Toshniwal, A., Jariwala, A., Kheraj, V., Opanasyuk, A. S., & Panchal, C. J. (2017). Numerical simulation of tin based perovskite solar cell: Effects of absorber parameters and hole transport materials. *Journal of Nano-and Electronic Physics*, 9(3), 03038-1, 03038-4. DOI: 10.21272/jnep.9(3).03038

Tyagi, A., Pandey, S. M., Walia, R. S., Murtaza, Q., & Kumar, A. (2021). Effect of Temperature on the Sliding Wear Composite Behavior Coating of HVOF Sprayed Al2O3. Advances in Materials and Mechanical Engineering: Select Proceedings of ICFTMME 2020, 23. DOI: 10.1007/978-981-16-0673-1_3

Wang, T. Y., & Huang, C. Y. (2007). Improving forecasting performance by employing the Taguchi method. *European Journal of Operational Research*, 176(2), 1052–1065. DOI: 10.1016/j.ejor.2005.08.020

Yang, W. S., Park, B. W., Jung, E. H., Jeon, N. J., Kim, Y. C., Lee, D. U., Shin, S. S., Seo, J., Kim, E. K., Noh, J. H., & Seok, S. I. (2017). Iodide management in formamidinium-lead-halide–based perovskite layers for efficient solar cells. *Science*, 356(6345), 1376–1379. DOI: 10.1126/science.aan2301 PMID: 28663498

Zhao, X., & Park, N. G. (2015, November). Stability issues on perovskite solar cells. In Photonics (Vol. 2, No. 4, pp. 1139-1151). MDPI. DOI: 10.3390/photonics2041139

Chapter 3 Silica Fumes, Fal–G as Cement Substitutes in Concrete: An Experimental Investigation

Ravpreet Kaur
SGT University, Gurugram, India

Ranjodh Singh
DAV University, India

Ajay Kumar
https://orcid.org/0000-0001-7306-1902
Department of Mechanical Engineering, School of Engineering and Technology, JECRC University, India

ABSTRACT

The exceptional strength of high performance concrete is attributed to the silica fumes. FAL-G, another cement substitute offers strength with water just like hydraulic polymers. There is no need for an external heat source for the FAL-G reaction, and therefore it produces less greenhouse gas. The dead load on structures is reduced since it is lightweight. Fly ash is required as an additive to make materials with the required strength. With a 0–30% cement removal, cement largely replaces silica fume and FAL-G in this design combination. After 28 days of testing, the FAL-G concrete mix's has an extraordinary high strength. Three samples were tested with varying amounts of silica fume and FAL-G replacement for seven and twenty-eight days. A compression testing apparatus was utilized to evaluate a total of twenty-four cast cubes and twenty-four cylinders. A robust correlation has been found between the various mechanical characteristics mentioned before. The outcomes so produced

DOI: 10.4018/979-8-3693-5513-8.ch003

will enable the utilization of waste materials in concrete in a sustainable manner.

1 INTRODUCTION TO FAL– G CONCRETE

The modern pozzolanic binder is called FAL-G. FAL-G provides strength like other hydraulic materials when there is water present. The cementitious mixture known as FAL-G is made up of fly ash, lime, and gypsum (Baikerikar 2014). It is a cheap and eco-friendly material that is very helpful in the construction business (Kumar et al., 2024). The FAL-G reaction does not require outside heat and produces less greenhouse gas. Compared to regular Portland cement, it is a more environmentally friendly binding substance. Gypsum and fly ash are by-products of industry, and the quantity of gypsum and lime depends on the characteristics of fly ash. By using the FAL-G as a cementing material, cement usage can be reduced while maintaining strength and durability. Because FAL-G materials are lightweight (Mittal et al., 2005; Kumar et al., 2024), they reduce the dead load on structures and also enable the production of fire-resistant bricks. The amount of lime and gypsum in FAL-G affects how chemically the ingredients behave in concrete. All of these materials are widely available in the construction areas in the form of industrial wastes and byproducts. Environmentally and financially safe, FAL-G. Even in the rural housing business, it is a low-cost, environmentally friendly material that is highly helpful. The project's objectives include using FAL-G bricks and blocks as an alternative building material in a few national clusters. However, using cement as a means of overcoming the delayed strength development at young ages is not the ideal strategy for improving FAL-G's overall strength (Bagaria et al., 2023; Kumar et al. 2023). The biggest issue is replacing the cement since it emits more carbon dioxide, which worsens the consequences of climate change on the environment. Additionally, adding cement is not particularly cost-effective. Therefore, it becomes a challenge that experimenting should be economical and environmentally friendly (Bhanumathidas et al., 1993; Kumar et al., 2024).

2 APPLICATIONS OF THE FAL-G CONCRETE

Uses of FAL-G concrete is as follows:-

a. FAL-G components perform well as a very good thermal insulator. Many Bricks constructed with FAL-G material have find application in boundary walls, partitioning walls, and other structures.

b. Because FAL-G bricks are lightweight, the overall dead load on a building is decreased substantially (Bhanumathidas et al., 1999). Energy to produce cement is significantly saved as it is the major component of concrete's binder.
c. Substantially reducing the volume of landfill area needed to dispose of the fly ash. Utilize the garbage to redirect it from the waste stream.
d. Cutting down both, the demand for Portland cement and greenhouse gas emissions.
e. Fly Ash pumping is made easier, which requires less energy and allows for greater pumping lengths.
f. Additional advantages include improved workability, reduced water requirement, greater sulphate resistance (Class F) (Chhinnaraju et al., 2011), improved durability, and increased strength.
g. Also Segregation and bleeding are reduced, as well as the heat of hydration, permeability, and corrosion. The Alkali-Silica Reaction (ASR) is primarily reduced (mostly Class F). FAL-G renders concrete very permeable by nature, which results in reduced water seepage, which causes slump loss and improves workability.
h. FAL-G is much more durable than OPC when combined with Micro silica. Although the lifetime of cement is defined Dey 2016, it is rather challenging to do so for the FAL-G

Figure 1. Flow chart of using silica fume

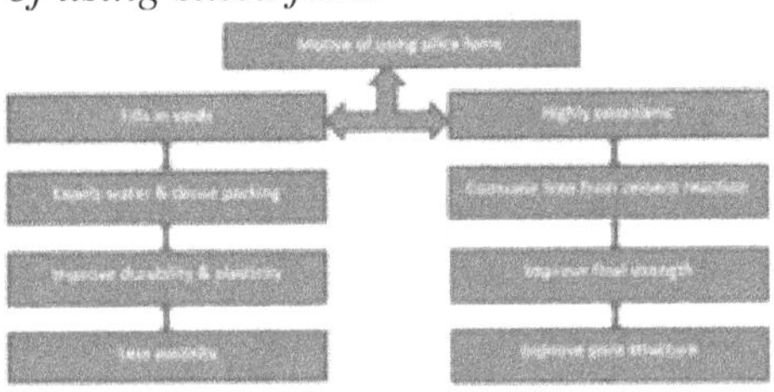

Necessity & Importance of the Research: FAL-G is a low-cost, environmentally friendly material that is excellent for use in building. Industrial byproducts and trash are used in the production of FAL-G. As a result of the light weight of the FAL-G bricks and blocks, dead loads on structures are reduced, and fire insulation bricks are also produced. Bricks are not damaged during shipping or usage because of its great strength. For 24 hours, FAL-G bricks do not require curing (Ghais et al., 2014). It is sufficient to sprinkle water over bricks before usage. Because FAL-G bricks have a consistent structure, the amount of joints and plaster is reduced by roughly 60%.Due to their high permeability and low water seepage, these FAL-G bricks lose slump and have good workability. CO_2 emissions into the environment are rising daily as a result of industry. This emission causes harm to the atmosphere's heat-trapping layer. Additionally, a rise in thermal power generation capacity leads to the production of hundreds of tons of fly ash. Large-scale fly ash disposal is not environmentally friendly. On the other hand, the air is also polluted by CO2 from the massive cement industry. One million clay bricks require about 300 tons of coal to produce, and this process generates approximately 350 tons of carbon dioxide (CO2). It is not the best technique to develop the strength of FAL-G in all aspects to use cement as a means of overcoming the delayed strength gain at young ages. The main problem is that cement needs to be changed since using it is not especially cost-effective and because it emits more carbon dioxide, which contributes to global warming (Hariharan et al., 2011). Calcium sulpho-aluminates hydrates, which are produced in the presence of an essential input (the quantity of gypsum addition that can act as its threshold limit), can be used to remedy the issue. Super plasticizers can also be used to control the ratio of cement to water, which will increase the strength of FAL-G. Super plasticizers can raise the compressive strength value of FAL-G to a range of 30 to 45 MPa, IS 456:2000.

3 LITERATURE REVIEW

In this study, *Ambalavanan and Roja et al.* found that using waste lime in some places did not produce the desired results, and that more development is required to boost the strength of FAL-G bricks.

N. Kalidas and Dr. N. Bhanumathidas (2002): The results of this well-known study provide evidence that the strength and durability of concrete are unrelated. When compared to reference concrete, it has been repeatedly shown that utilizing fly ash of the appropriate quality improves the workability of fly ash blended mortars and concrete for the designated water cementitious material ratio. Stated differently, less water is needed to attain the same level of workability. When choosing fly ash concretes to build the Hungry Horse dam, this phenomenon was one of the requirements. According to one of the authors' works, there is a decrease in water demand for mortars with the same flow capacity. Perumal et al., 2004 stated that fly ash was divided into two groups by *ASTM* using CaO as the differentiating factor: The ASTM classifies fly ash made from burning lignite and sub-bituminous coal as Class C fly ash if it contains more than 10% CaO. Due to the presence of reactive calcium-related minerals including tetra calcium sulphoaluminate and dicalcium silicate, these fly ashes self-harden. Reactive aspects make Class C fly ashes less resistant to chemical attacks and more prone to concrete deterioration, according to numerous studies. Fly ash meeting ASTM Class F standards is produced when coal is burned and has less than 10% CaO. These fly ashes often have too little calcium oxide (CaO) for significant calcium-associated mineralogy to form. As a result, Class F ashes are just pozzolanic and less reactive or dormant than Class C ashes. Class F fly ashes provide greater concrete durability and longer structural service lives due to the lack of reactive phases. Mittal et al., 2005, fly ash LT: LT ash is the result of burning fly ash between 750 and 8500 degrees Celsius. This is perfect for precast items like bricks and blocks since it is more reactive early in the hydration process. LT ashes respond more readily in the beginning and perform better in the end. Fly ash formed by burning at temperatures between 900°C and 1400°C is known as HT fly ash. As mentioned earlier, the HT fly ash's silica and alumina components form as extremely stable phases or spheroids with comparatively moderate reactivity.

Early on, HT ashes are less reactive, but later on, they perform better. The low LOI of less than 1.5% along with this quality makes HT fly ashes more preferred in cement and concrete. *2005 INSWAREB*: The construction of between 150 and 350 micro industrial facilities in various locations throughout India is the goal of this initiative. In this investigation, fly ash is used to cast FAL-G bricks. Fly Ash, Lime, and Gypsum is referred to as FAL-G. Dhoka et al., 2013 emphasized on fly ash and lime (FAL) bricks and blocks are of lower grade than FAL-G bricks and blocks. The C-S-H hydrates in calcium silicate mixtures give them strength. Calcium alumino-sulphate hydrates (CASH) are a supplement given to FAL-G combinations for ultimate strengths and late-age. FAL-G blends offer greater strength than FAL mixes. Due to the high final strengths, the one-day strengths are adequate to handle the product for stacking.

Visakhapatnam-based Eco Carbon Private Limited (2006): Fly ash, gypsum, silica fume, lime, and fly ash are used to make FAL-G bricks. An indigenous technology created in India that supports sustainable development is FAL-G bricks and blocks. A traditional and common material is cement. Due to a lack of availability, rising prices, and energy requirements, employing waste and pozzolanic materials like fly ash, silica fume, GGBS, etc., has been considered as a novel building material and technology, Vijayan et al., 2022. Adopting affordable and environmentally friendly technologies is necessary. Concrete that incorporates waste and pozzolanic elements can use less cement without sacrificing strength or durability, Bouzoubaa et al.,2001. A traditional and common material is cement. *(2010) Priyanka Pasupuleti* India is a developing country with a strong technical foundation and the capacity to create innovative technologies for sustainable growth. Buildings that contribute to the effective development of a technology frequently face challenges since they fall within the category of emerging nations. Therefore, the project's main objective is to investigate the many influences on technology created for sustainable growth in a nation like India. Thus, the case study FAL-G bricks and blocks project, an indigenous technology created in India and contributing to sustainable development, has been discovered. The purpose of the study is to identify some of the aspects that affect this project as well as to address the many factors that are involved in it. The results of the study are recorded in a variety of ways after completion. The big picture indicates that the actors with ambiguous roles are crucial to the project and that further investigation is necessary to determine if they will have a positive or negative impact. FAL-G provides strength like other hydraulic materials when there is water present. In thermal power plants, gypsum and fly ash are both readily available in large quantities. *PB K Rao and P Rattish Kumar (2010):* Silica fume is a by-product that is inexpensive and can be used in cement and other pozzolanic materials. In concrete, it serves as a partial replacement. The goal of the study is to build concrete mixes utilizing silica fume in a manner similar to that of conventional concrete, with the exception of adding a "Reduction factor" based on the results of the basic compressive strength test of cement-silica fume mortar mixes Maage 1984. The study's objective was to design silica fume concrete mixtures for normal strength grade concretes with different percentages of silica fume contents from 0 to 70% (as a partial replacement for cement) in increments of 30%. Considerable savings have been achieved by partially substituting silica fume for cement in the mix proportions (using the Reduction Factor approach). Savings of 15% have been achieved by replacing cement with silica fume up to 30–40%, Revathi et al., 2023. The ideal silica fume content in cement silica fume mix is determined to be between 30 and 40% for both classes of concrete based on material savings and cost analysis. The theoretical target strength has been validated by the design mix values for 28-day compressive strength, Silica fume user's manual 2015. According to the results of this experiment, silica fume mixes are suitable for use in regular structural concrete. When combined with Micro silica, FAL-G is far more durable than OPC. Although the lifetime of cement is defined, the lifetime of the FAL-G is rather difficult to define.

4 MATERIALS AND METHODOLOGY

The local markets in Ludhiana are the source of fly ash, silica fume, lime, and gypsum, all of which are readily available for usage in this project:-

KGR Agro Fusion (p) Limited (Hambran) Ludhiana provides fly ash. Fly ash is utilized as powder. KGR Agro Fusion (p) Limited (Hambran) provides silica fume. Harmeet Exports Chet Singh Nagar offers blend. Liquid mixture is employed. Gypsum with Lime: -Prime Enterprises Macchiwara Ludhiana provides gypsum and lime. Table 1 and Table 2 incorporates the physical properties of Cement and fine aggregates used in the study.

Table 1. Physical properties of Binder/cement

S No.	Characteristics	Units	Results obtained	Value specified
1.	Fineness (Specific Surface)	cm^2/g	3000	3500
3.	Specific Gravity	gm/cm^3	3.15	
4.	Normal Consistency of Cement (% of cement by weight)	%	29	30
5.	Setting Time			
	(i) Initial	minutes	135	30(min.)
	(ii) Final	minutes	225	600 (max.)

Table 2. Physical Properties of Fine aggregates

Characteristics	Value
Fineness modulus	2.95
Specific Gravity	2.64
Zone (fine aggregates)	Zone II
Moisture content	1.522

4.1 Gypsum

Gypsum is crucial in regulating the speed at which cement hardens. Gypsum is added to concrete to manage the cement's setting time; if it isn't, the cement will set up right away after mixing. High performance materials are produced using gypsum stated by Akarsh et al.,2021 and Thomas et al., 2007. The use of cement as a means of overcoming the sluggish strength development at young ages is not the best method for producing FAL-G strength in all respects. The primary issue arises from the need to replace cement due to increased carbon dioxide emissions that cause global warming and because cement use is not particularly cost-effective. Calcium Sulpho-aluminates hydrates, which are created when gypsum is present (the quantity of gypsum addition that can be its threshold limit), can be used to fix the problem.

4.2 Superplasticizer Made of Polycarboxylate

The super plasticizer polycarboxylate is a modern additive. It is a superplasticizer that is environmentally friendly. A concentrated substance with a high water reduction is called super plasticizer. Additionally, super plasticizer has a low alkali concentration and a high slump retention capacity. The polycarboxylate super plasticizer's characteristics and advantages:

Super plasticizer saves labor costs while reducing vibrations during concrete placement.

- The early and ultimate strengths of concrete increase with additive addition. Greater E modulus is provided by super plasticizer.
- Super plasticizer increased the adherence to steel that is being stressed and reinforced.
- It increases concrete's durability by reducing permeability.

5 TESTING

Various proportions of cement, fly ash, lime, and gypsum, fine and coarse aggregate, silica fume, water, and chemical additive were combined to produce a concrete mix that met M50 grade requirements. With a cement removal of 0–30%, cement is employed in this design combination to partially replace silica fume and FAL-G. 150 mm X 150 mm X 150 mm cubes and 150 mm X 300 mm cylinders made up the test subject. Testing was done on the cube and cylinder samples seven and twenty-eight days after cure Nath et al., 2011. The graph for compressive strength and split tensile strength is created after that. Figure 2 illustrates the compression testing machine used to test the cubes' and cylinders' strength after 7 and 28 days.

Figure 2. Laboratory testing of strength of cylindrical specimen in Compression testing machine

6 RESULTS AND DISCUSSIONS

The compressive and split tensile strengths of each sample are shown in Tables 3 and 4, respectively. The graphs and related tables' plotting is displayed in Figures 3 and 4. According to the guidelines of the IS code, strengths in compression and tension were measured in the study. The compressive strength of concrete was tested using 24 cubes that were 150 mm x 150 mm x 150 mm in size. A total of 24 cylinders measuring 150 mm by 300 mm were casted and hence split tensile strength test was performed.

Table 3. Strength in Compression (N/mm$^{2)}$ for M50 concrete at 7 and 28 days

S.No.	FAL-G (In %)	Silica – fumes (in %)	Compression Strength at 7 days (N/mm^2)	Compression Strength at 28 days (N/mm^2)
Mix-1.	0	0	35.3	53.2
Mix-2.	10.0	10.0	36.2	54.4
Mix-3.	20.0	20.0	35.7	56.2
Mix-4.	30.0	30.0	38.4	61.1

Table 4. Strength in tension (N/mm^2) for M50 concrete at 7 and 28 days

S.No.	FAL-G (In %)	Silica – fumes (in %)	Strength in Tension at 7 days (N/mm^2)	Strength in Tension at 28 days (N/mm^2)
Mix- 1	0	0	4.21	4.57
Mix- 2	10.0	10.0	3.82	5.18
Mix- 3	20.0	20.0	3.93	5.86
Mix- 4	30.0	30.0	3.87	6.24

Figure 3. Comparison of compression strength (N/mm²) -Substitution of Cement by %weight (with same %age of Micro silica and Fal-G)

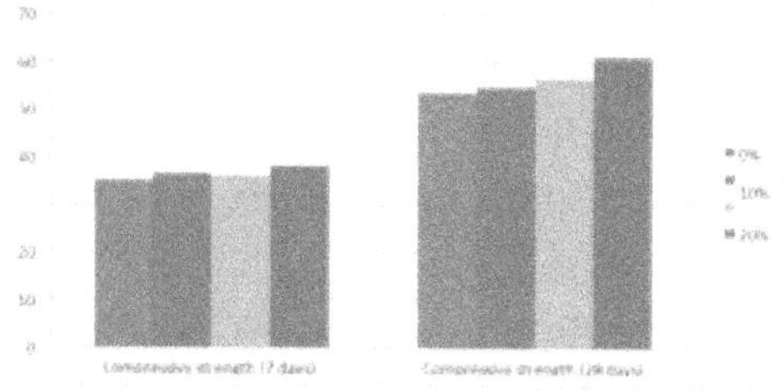

Different Percentages of replacement of cement used were 0%, 10%, 20% and 30% with micro silica and Fal-G material each.

Figure 4. Comparison of tensile Strengths-Substitution of Cement by weight % (with same %age of Micro silica and FAL-G)

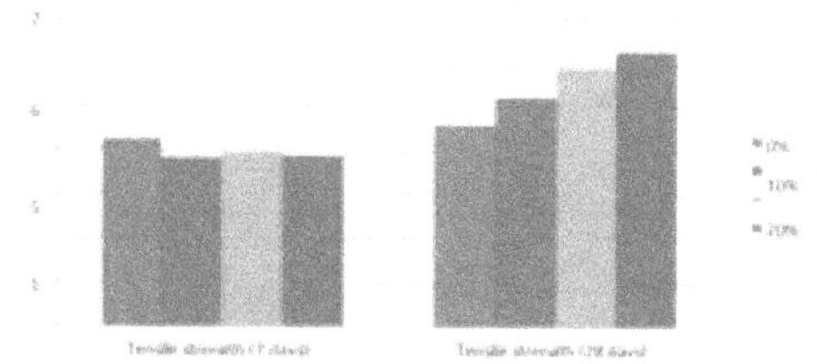

As can be seen in figure 3, it came to the conclusion that concrete had a greater tensile strength at a 60% cement replacement after 7 days. The tensile strength [6.25N/mm²] is at its highest point when 60% of cement [30% FAL-G + 30% Micro silica] is replaced. 24 cylinders measuring 150 mm by 300 mm were cast and put to the test to determine their tensile strength. High tensile strength is achieved by substituting 30% silica fume and FAL-G for Portland cements. Both fly ash and silica fume are used to strengthen cement concrete's tensile strength. Silica fumes are used in the production of FAL-G concrete, where they act as a super plasticizer, strength enhancer, and set accelerator. Gypsum intake in this should reach a threshold level where it can effectively boost or accelerate the sulfoaluminates hydrates. Research on FAL-G and silica fume has found that it performs better than Portland cement in terms of high strength, durability, outstanding resistance to chloride attack, and minimal environmental impact.

6 CONCLUSION

When there is 30% replacement by weight of cement, a compressive strength of 38.1 N/mm^2 has been recorded at 7 days. Micro-silica and FAL-G together provide a high compressive strength, as indicated by the ratio of 40% cement to 30% micro-silica. As an alternative building material, FAL-G and silica fume could be employed. Due to the presence of Super plasticizer, the ultimate setting time of Silica Fume Concrete is significantly shorter than that of Regular Portland Cement Concrete. So, this can be utilized to speed up building as well. When both silica fume and FAL-G are added to cement concrete, the tensile strength of the material is increased to 6.25 N/mm^2.

- The project will be more affordable if some of the cement is replaced with silica fume and FAL-G material. By-products of industrial waste like FAL-G composition and Micro-silica can be employed in concrete technology to create environmentally friendly buildings and structures. The compressive strength of the FAL-G and micro silica concrete blend effect is 61.06 N/mm^2, which is significantly higher than the compressive strength of the control mix.
- The ideal values for the compressive strength [61.06 N/mm2] are 60% cement replacement (30% FAL-G + 30% micro silica). For good M50 grade concrete, the activated mix effect of fly ash and micro silica is nonexistent. The most important factor in enhancing the pozzolanic characteristics of fly ash is alkali activation.
- With time, silica fume mixture gains strength. We discovered that while the FAL-G blend did not initially give strength, it did become stronger over time. But FAL-G is more durable than OPC when combined with Micro silica. Although the lifetime of cement is defined, the lifetime of the FAL-G is rather difficult to define.

REFERENCES

Akarsh, P. K., Marathe, S., & Bhat, A. K. (2021). Influence of graphene oxide on properties of concrete in the presence of silica fumes and M-sand. *Construction & Building Materials*, 268, 121093. DOI: 10.1016/j.conbuildmat.2020.121093

Ambalavanan, R., & Roja, A. (1996). Feasibility studies on utilisation of waste lime and gypsum with flyash. *Indian Concrete Journal*, 70(11), 611–615.

Bagaria, A., & Juneja, D. (2023). Experimental research on influence of marble powder, silica fume and polypropylene fiber on the porous concrete. *Materials Today: Proceedings*. Advance online publication. DOI: 10.1016/j.matpr.2023.03.240

Baikerikar, A. (2014). A review on green concrete. *Journal of Emerging Technologies and Innovative Research*, 1(6), 472–474.

Bhanumathidas, N., & Kalidas, N. (1993). The Renaissance of Mediaeval Age Cement: FaL-G. *Civil Engineering & Construction Review*, 31-33.

Bhanumathidas, N., & Kalidas, N. (1999) The Rationale for Portland Pozzolan Cement Compositions. *Proceedings of International Conference on Waste and By-products as Secondary Resources for Building Materials*.

Bouzoubaa, N., Zhang, M. H., & Malhotra, V. M. (2001). Mechanical properties and durability of concrete made with high-volume fly ash blended cements using a coarse fly ash. *Cement and Concrete Research*, 31(10), 1393–1402. DOI: 10.1016/S0008-8846(01)00592-0

Chinnaraju, K., Subramanian, K., & Senthil Kumar, S. R. R. (2011). Role of Fly Ash and Silica Fume on Durability Characteristics of High Performance Concrete. *International Journal of Advance in Civil Engineering*, 17-27.

Dey, S. (2016). Study of compressive strength of high volume fly ash concrete with varying proportion of fly ash and silica fume. *International Journal of Civil Engineering*, 3(3), 3–6. DOI: 10.14445/23488352/IJCE-V3I3P102

Dhoka, M. C. (2013). Green concrete: Using industrial waste of marble powder, quarry dust and paper pulp. *International Journal of Engineering and Science Invention*, 2(10), 67–70.

Ghais, A., Ahmed, D., Siddig, E., Elsadig, I., & Albager, S. (2014). Performance of concrete with fly ash and kaolin inclusion. *International Journal of Geosciences*, 5(12), 1445–1450. DOI: 10.4236/ijg.2014.512118

Hariharan, A. R., Santhi, A. S., & Mohan Ganesh, G. (2011). Study on Strength Development of High Strength Concrete Containing Fly Ash and Silica Fume *[IJEST]*. *International Journal of Engineering Science and Technology*, 3, 2955–2961.

Kumar, A., Kumar, P., & Liu, Y. (2024). *Industry 4.0 Driven Manufacturing Technologies*. Springer Nature., DOI: 10.1007/978-3-031-68271-1

Kumar, A., Kumar, P., Srivastava, A. K., & Goyat, V. (Eds.). (2023). *Modeling, Characterization, and Processing of Smart Materials*. IGI Global., DOI: 10.4018/978-1-6684-9224-6

Kumar, A., Liu, Y., & Kumar, R. (Eds.). (2024). *Handbook of Intelligent and Sustainable Manufacturing: Tools, Principles, and Strategies*. CRC Press., DOI: 10.1201/9781003405870

Kumar, A., Rajak, D. K., Kumar, P., & Kumar, A. (Eds.). (2024). *Advances in Sustainable Biomaterials: Bioprocessing 4.0, Characterizations, and Applications* (1st ed.). CRC Press., DOI: 10.1201/9781003434313

Maage, M. (1984). Effect of micro silica on the durability of concrete structures, Report STF65 A 84019, FCB/SINTEF. *The Norwegian Institute of technology, Tronddheim, Norway*.

Mittal, A., Kaisare, M. B., & Shetti, R. (2005). Experimental Study on use of fly ash in concrete. *Tarapur Atomic Power Project, 3*.

Mittal, A., Kaisare, M. B., & Shetti, R. (2005). Experimental Study on use of fly ash in concrete. *Tarapur Atomic Power Project, 3*.

Nath, P., & Sarker, P. (2011). Effect of fly ash on the durability properties of high strength concrete. *Procedia Engineering*, 14, 1149–1156. DOI: 10.1016/j.proeng.2011.07.144

Perumal, K., & Sundararajan, R. (2004, August). Effect of partial replacement of cement with silica fume on the strength and durability characteristics of high performance concrete. In *29th Conference on our world in Concrete & Structures* (pp. 25-26).

Revathi, S., Suresh, D., & Anwar, S. T. (2023). Behaviour of concrete with PET bottles as fibers & silica fume as partial replacement of cement. *Materials Today: Proceedings*. Advance online publication. DOI: 10.1016/j.matpr.2023.03.580

Silica Fume User's Manual. (2015). Silica Fume Association.

Thomas, M. D. A. (2007). *Optimizing the use of fly ash in concrete* (Vol. 5420). Portland Cement Association.

Vijayan, V., Jayakesh, K., & Anand, K. B. (2022). Mechanical properties of recycled aggregates concrete with sisal fiber and silica fume. *Materials Today: Proceedings*, 65, 1887–1894. DOI: 10.1016/j.matpr.2022.05.055

Chapter 4

Review of Advanced Materials Used in Electric Vehicle Applications

Amol S. Shinde
https://orcid.org/0000-0001-6711-8941
A.P. Shah Institute of Technology, Thane, India

Sameer S. Nanivadekar
https://orcid.org/0000-0002-7498-8336
A.P. Shah Institute of Technology, Thane, India

Raju Kumar Swami
https://orcid.org/0000-0002-8359-542X
Pacific Academy of Higher Education and Research University, India

Ajay Kumar
https://orcid.org/0000-0001-7306-1902
Department of Mechanical Engineering, School of Engineering and Technology, JECRC University, Jaipur, India

ABSTRACT

The chapter provides an in-depth review of advanced materials used in electric vehicle (EV) applications, addressing the critical role these materials play in improving energy efficiency, safety, and performance. It explores key considerations such as lightweighting, thermal management, conductivity, safety, and durability. A detailed examination of battery materials, highlights advancements aimed at enhancing range, charging speed, and safety. The chapter also delves into the use of lightweight structural materials, such as aluminum alloys and carbon fiber composites, which significantly contribute to energy efficiency. Thermal management

DOI: 10.4018/979-8-3693-5513-8.ch004

systems using phase change materials (PCM) and thermal interface materials (TIM) are discussed for their role in maintaining optimal temperature and safety. Additionally, the chapter examines powertrain materials, including rare earth magnets and silicon carbide (SiC), for their impact on motor efficiency and durability. Future trends in EV materials, focusing on sustainability, cost reduction, and supply chain resilience, are also explored.

1. INTRODUCTION TO MATERIAL REQUIREMENTS FOR ELECTRIC VEHICLES

The electric vehicle (EV) industry has experienced rapid growth due to the increasing demand for sustainable and energy-efficient transportation. As the market continues to expand, the materials used in EV design and production have evolved significantly to enhance performance, reduce weight, improve energy efficiency, and increase safety. Electric vehicles differ from traditional internal combustion engine (ICE) vehicles in their need to meet higher standards in terms of energy efficiency, weight reduction, safety, and durability. These requirements have driven innovation in the materials used across different components of the vehicle (Wenting et al., 2023).

1.1 Key Material Considerations for EVs

a) *Light-weighting*: Reducing the weight of EVs is a critical aspect of improving battery range and overall energy efficiency. Lightweight materials are essential to this process, allowing EVs to maximize the distance travelled per charge without sacrificing structural integrity.
b) *Thermal Management*: In electric vehicles, effective heat dissipation is necessary to maintain the performance and longevity of batteries, particularly in high-power applications where temperature control is critical (Manzetti & Mariasiu, 2015). Advanced thermal management materials are being developed to address these challenges.
c) *Conductivity*: EVs require materials with high electrical and thermal conductivity, especially for key components such as batteries and power electronics. Conductive materials ensure efficient energy transfer and heat dissipation, contributing to improved vehicle performance and reliability.
d) *Safety and Durability*: Materials used in EVs must be able to withstand extreme conditions, including high temperatures and mechanical stress while ensuring passenger safety and vehicle durability (Cheng et al., 2015). This includes the

need for impact-resistant materials that can provide protection in the event of a crash, as well as materials that extend the lifespan of key components.

This chapter will delve into the material technologies that address these essential requirements for electric vehicles and explore potential advancements that could further enhance EV performance and sustainability.

2. ADVANCED BATTERY MATERIALS

The battery is often regarded as the heart of an electric vehicle, and the development of high-performance, lightweight, and safe batteries is crucial to the success of electric vehicles (EVs). Battery technology directly affects the range, charging speed, safety, and overall efficiency of EVs, making advancements in this area a critical focus for the industry. Several materials are key to improving the energy density, longevity, and safety of EV batteries.

Figure 1. Electric Vehicle Battery Components

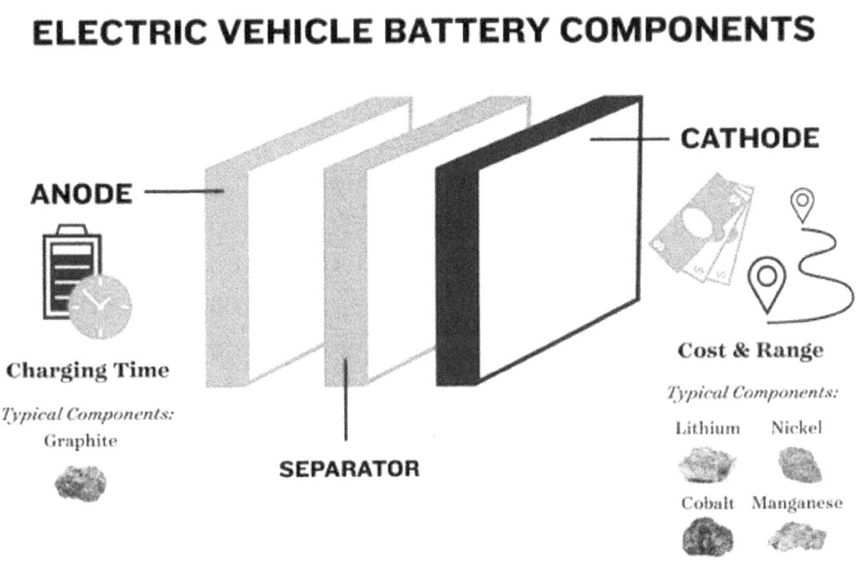

2.1 Lithium-Ion Batteries

Lithium-ion (Li-ion) batteries currently dominate the EV market due to their high energy density, long cycle life, and relatively reasonable cost. However, despite their wide adoption, ongoing research and development efforts aim to improve the performance and safety of Li-ion batteries by exploring advanced materials for their core components: the cathode, anode, and electrolyte.

a) *Cathode Materials*: Traditionally, Li-ion batteries use lithium cobalt oxide ($LiCoO_2$) as the cathode material, but this is being phased out in favour of alternatives that offer better safety, energy density, and thermal stability. Notable alternatives include lithium iron phosphate ($LiFePO_4$), which provides enhanced safety and thermal stability, and nickel-cobalt-manganese (NCM) and nickel-cobalt-aluminium (NCA) cathodes, which offer higher energy density (Wenting et al., 2023). These newer cathode materials are integral to increasing the range and safety of EVs.
b) *Anode Materials*: Graphite has long been the standard anode material in Li-ion batteries, but silicon-based anodes are gaining increasing attention. Silicon offers a much higher capacity for energy storage compared to graphite, potentially leading to significant improvements in battery capacity. However, silicon anodes expand during charging cycles, leading to mechanical stress and degradation of the material, which can affect the battery's longevity. Current research focuses on developing silicon nanowires or silicon-grapheme composites to address these stability issues (Mittal et al., 2023).
c) *Electrolytes*: Traditional Li-ion batteries use liquid electrolytes, which pose risks related to leakage and flammability. To enhance safety, solid-state electrolytes, such as ceramic or polymer-based electrolytes, are being developed. These materials offer superior thermal management and eliminate the risk of leakage, though challenges remain in terms of manufacturing scalability.

2.2 Next-Generation Battery Technologies

As the demand for better performance and longer range in EVs continues to grow, next-generation battery technologies are being developed to go beyond the limits of conventional Li-ion designs.

a) *Solid-State Batteries:* Solid-state batteries represent one of the most promising advancements in battery technology for EVs. These batteries use solid electrolytes, which provide several advantages over liquid electrolytes, such as reduced flammability risk and the potential to use high-energy-density materials like lithium metal anodes. Lithium metal has a much higher theoretical capacity than conventional anode materials, which could significantly extend the range of EVs (Mittal et al., 2023). However, manufacturing solid-state batteries at scale remains a challenge, with ongoing research aimed at overcoming these barriers.

b) *Lithium-Sulphur (Li-S) Batteries:* Lithium-sulphur batteries offer a higher theoretical energy density than Li-ion batteries, making them a promising option for future EVs, particularly for long-range applications. However, Li-S batteries face significant challenges, such as the "shuttle effect," where lithium polysulfide's migrate through the electrolyte, reducing the battery's cycle life and stability. Research into materials like sulphur-carbon composites is ongoing to address these issues and improve the commercial viability of Li-S batteries (Wenting et al., 2023).

Advancements in battery materials are pivotal to the future of electric vehicles, with each new development promising to improve energy storage, safety, and vehicle range. As research continues, these materials will likely pave the way for more efficient and reliable EVs.

3. LIGHTWEIGHT STRUCTURAL MATERIALS

Reducing vehicle weight is one of the most effective ways to increase the range of electric vehicles (EVs). A lighter vehicle requires less energy to move, allowing the battery to last longer and improving overall energy efficiency. As a result, the choice of structural materials plays a crucial role in the design of EVs, particularly in areas such as body structures, frames, and chassis components. The following materials are key in reducing vehicle weight without sacrificing performance or safety.

Figure 2. Lightweight Materials and its Challenges

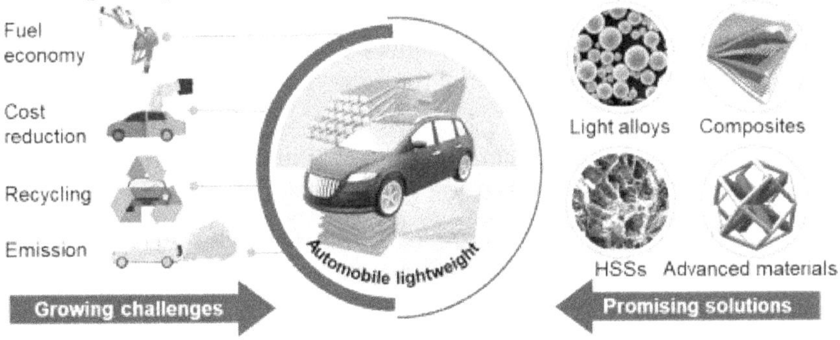

3.1 Aluminium Alloys

Aluminium is one of the most commonly used lightweight materials in EV construction due to its favourable properties, including a high strength-to-weight ratio, corrosion resistance, and recyclability. Modern aluminium alloys, such as those in the 6000 and 7000 series, offer enhanced strength and formability, making them ideal for a variety of automotive applications. **6000 Series Aluminium Alloys** are highly formable and corrosion-resistant, making them suitable for use in body panels, frames, and battery enclosures. **7000 Series Aluminium Alloys** offer higher strength than the 6000 series, making it suitable for applications where strength is a critical factor, such as suspension systems and crash structures. Aluminium's lightweight properties contribute directly to energy efficiency in electric vehicles. However, aluminium can be more expensive than traditional steel, and this cost factor must be considered in mass-market EVs.

3.2 Carbon Fibre Composites

Carbon fibre-reinforced polymers (CFRP) are known for their superior strength-to-weight ratio, making them highly effective for use in reducing the weight of electric vehicles. Carbon fibre materials provide exceptional rigidity and strength while being significantly lighter than metals like steel and aluminium. This makes CFRP particularly advantageous for high-performance EVs, where weight reduction is paramount to maximizing speed and range.

Carbon fibre composites have the highest strength-to-weight ratio among the materials currently used in automotive applications. This makes them particularly valuable in areas where weight reduction and strength are both critical, such as in structural components and body panels. Despite their advantages, the high cost of producing carbon fibre and the complexity of manufacturing CFRP components limit their use to luxury or performance-focused electric vehicles (Manzetti & Mariasiu, 2015). The intricate production process and difficulties in recycling carbon fibre also pose challenges for wider adoption. However, advancements in automated manufacturing processes and recycling techniques could help reduce costs and make carbon fibre a more accessible material for mass-market EVs in the future.

Reducing vehicle weight is a key factor in increasing the efficiency and range of electric vehicles. Lightweight materials such as aluminium alloys and carbon fibre composites are critical in this endeavour, offering significant performance benefits while contributing to the sustainability and practicality of EV designs. As research continues to improve the manufacturing processes and cost-effectiveness of these materials, they will likely play an even greater role in the future of electric vehicle construction.

Figure 3. Weight Reduction Strategies and Materials Used in Electric Vehicles

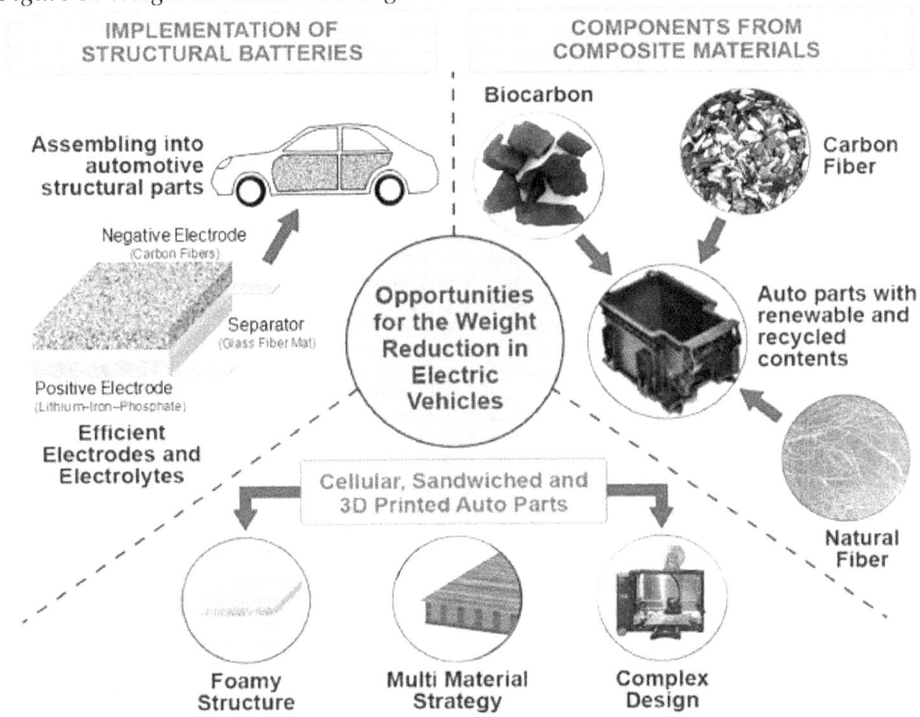

4. THERMAL MANAGEMENT MATERIALS

Thermal management is a critical factor in the design and operation of electric vehicles (EVs), particularly for systems like the battery and power electronics. Effective thermal management ensures that components operate within optimal temperature ranges, which is essential for maintaining performance, efficiency, and safety. Advanced materials used in thermal management systems help dissipate heat, prevent overheating, and extend the lifespan of critical components. The following materials are commonly used to address these challenges in EVs:

Figure 4. Materials Used in Thermal Management of Electric Vehicles

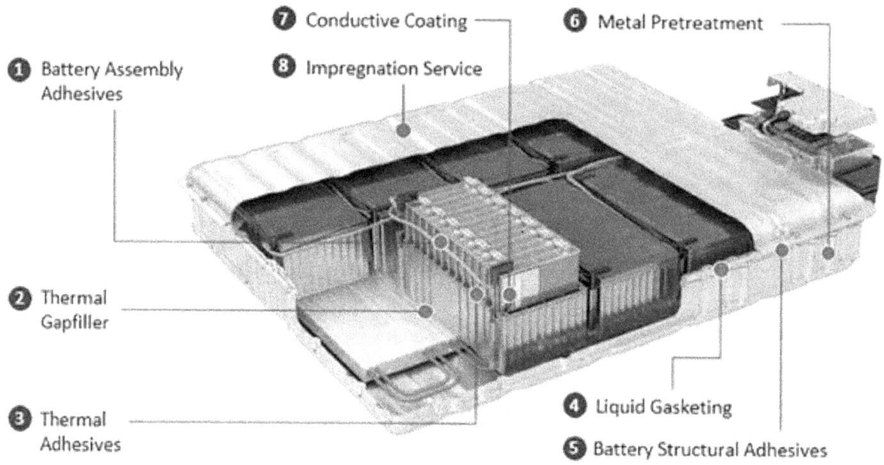

4.1 Phase Change Materials (PCM)

Phase change materials (PCM) are used in EVs for passive thermal management by absorbing and releasing heat during phase transitions, such as from solid to liquid or vice versa. PCMs are particularly effective in maintaining stable temperatures in EV battery packs, where overheating can lead to safety risks and reduced battery life.

PCMs absorb excess heat when the battery operates at high temperatures and release stored heat as the battery cools down, helping to regulate the temperature without requiring active cooling systems. This makes PCMs an energy-efficient and cost-effective solution for temperature management. Common PCMs used in EV applications include paraffin waxes and salt hydrates, which offer a range of phase change temperatures suitable for different thermal management needs in batteries (Mittal et al., 2023). PCMs are also being explored for other EV components that require temperature stabilization, such as in power electronics.

The use of PCMs in battery packs not only enhances thermal management but also contributes to improving battery safety by reducing the likelihood of thermal runaway, a dangerous condition where excessive heat leads to uncontrollable battery failures.

4.2 Thermal Interface Materials (TIM)

Thermal interface materials (TIM) play a crucial role in transferring heat between different components in an EV, particularly between heat-generating components and heat sinks. Efficient heat transfer is essential for maintaining optimal performance, especially in power electronics and battery systems, where excessive heat can reduce efficiency and lead to component failure.

Graphite sheets, metal foils, and thermally conductive polymers are commonly used as TIMs in electric vehicles. Graphite-based TIMs are particularly effective due to their high thermal conductivity, flexibility, and ability to conform to irregular surfaces, providing superior heat transfer capabilities (Mittal et al., 2023). Thermally conductive polymers are also gaining attention for their lightweight properties and ease of manufacturing.

TIMs are typically used in high-power electronic components such as inverters, converters, and battery cells. These materials enhance the thermal connection between the components and their respective heat sinks, ensuring that excess heat is efficiently dissipated and the components remain within safe operating temperatures. As electric vehicles continue to evolve, effective thermal management systems will remain a key focus, especially in high-performance applications. The development of advanced thermal management materials, such as PCMs and TIMs, will be essential to ensuring the safety, reliability, and efficiency of EV systems, particularly as battery technologies and power electronics become more advanced.

5. POWERTRAIN AND MOTOR MATERIALS

The electric motor and powertrain are the core components of an electric vehicle (EV), converting electrical energy from the battery into mechanical energy to drive the vehicle. The materials used in these systems must deliver high efficiency, durability, and heat resistance to ensure optimal performance, particularly under demanding conditions such as high speeds or heavy loads. The choice of materials significantly impacts the overall efficiency, performance, and longevity of the powertrain and motor systems. The following materials are crucial in EV motor and powertrain technologies:

5.1 Rare Earth Magnets

Rare earth magnets are commonly used in the electric motors of EVs due to their ability to generate strong magnetic fields, which is essential for achieving high motor efficiency and performance. Permanent magnets based on rare earth elements like neodymium (Nd), dysprosium (Dy), and praseodymium (Pr) are particularly valuable in electric vehicle motors because they can maintain their magnetic properties even at high temperatures, making them ideal for high-performance applications.

Rare earth magnets, especially neodymium-based ones, offer excellent magnetic strength, which enhances motor efficiency and allows for the production of smaller, lighter motors without compromising performance. This is particularly important in EVs, where weight reduction is critical to increasing vehicle range and efficiency. Despite their advantages, rare earth magnets are expensive and have limited availability, with much of the global supply controlled by a few countries. This has raised concerns about the supply chain's stability, leading to research into alternative motor designs that minimize or eliminate the need for rare earth materials (Mittal et al., 2023). These alternatives include motors that use ferrite magnets or designs that rely on electromagnetic induction instead of permanent magnets. The exploration of alternative designs is crucial for reducing dependency on rare earth materials while maintaining the high efficiency that EV motors demand.

5.2 Silicon Carbide (SiC)

Silicon carbide (SiC) is a semiconductor material that has garnered significant attention in the electric vehicle industry for its superior thermal and electrical properties compared to traditional silicon. SiC-based power electronics, such as inverters and converters, are becoming increasingly common in EVs due to their ability to handle higher voltages, withstand extreme temperatures, and operate with greater efficiency.

SiC offers much lower electrical losses than traditional silicon-based components, which leads to higher efficiency in power conversion systems. This is particularly beneficial for EVs, where energy efficiency directly impacts vehicle range and performance. SiC components can switch at higher frequencies, allowing for smaller and lighter power electronics, which further contributes to weight reduction and efficiency improvements in electric vehicles (Cheng et al., 2015).

SiC can operate at much higher temperatures than silicon, which makes it ideal for use in high-power applications such as fast charging and high-performance motors. The increased thermal conductivity of SiC enables better heat dissipation, reducing the need for complex cooling systems and improving the overall durability and reliability of the powertrain (Mittal et al., 2023). SiC's advantages in terms of efficiency and heat resistance make it a key material for the future of EV powertrain systems, particularly as manufacturers strive to increase the range, performance, and reliability of their vehicles.

The ongoing development of materials like rare earth magnets and silicon carbide is essential for the continued advancement of electric vehicle motors and powertrains. These materials not only enhance the performance of EVs but also help to address some of the industry's most pressing challenges, such as energy efficiency, cost reduction, and supply chain resilience.

6. FUTURE TRENDS AND CHALLENGES

The development of advanced materials for electric vehicles (EVs) is critical for the industry's continued growth and innovation. As EV adoption accelerates, new challenges and opportunities emerge, particularly regarding sustainability, cost, and supply chain management. Several trends are shaping the future of materials science for EVs, with a focus on improving performance, reducing costs, and addressing environmental concerns.

6.1 Recycling and Sustainability

As the demand for EVs grows, so does the need for sustainable materials and recycling processes. The transition to electric mobility is driven by a desire to reduce carbon emissions and environmental impacts, making the sustainability of materials used in EV production a key concern. The use of materials that can be recycled or reused at the end of a vehicle's life cycle is becoming increasingly important. This is particularly relevant for batteries, which contain valuable and finite materials such as lithium, cobalt, and nickel. Developing efficient recycling processes for these materials can reduce the need for new mining operations and lower the overall environmental footprint of EV production (Mittal et al., 2023).

In addition to recycling, the adoption of sustainable manufacturing processes is a growing trend in the EV industry. This includes the use of bio-based or recycled composite materials in vehicle construction, as well as more energy-efficient production methods. For instance, advancements in the recycling of carbon fibre composites could make this lightweight material more widely available for use in mass-market EVs, reducing both environmental impacts and costs (Cheng et al., 2015). As sustainability becomes a higher priority, the development of eco-friendly materials and recycling technologies will play a central role in shaping the future of the EV industry.

6.2 Cost Reduction

One of the main challenges in the EV industry is reducing the cost of advanced materials without compromising performance or safety. Many of the materials that offer significant advantages in terms of weight reduction, energy efficiency, and thermal management are expensive, which limits their widespread adoption, particularly in mass-market vehicles.

Rare earth magnets are commonly used in EV motors, but the high cost and limited supply of elements such as neodymium and dysprosium have prompted research into alternatives. Finding cost-effective substitutes or designing motors that minimize the use of rare earth elements is crucial for reducing production costs and ensuring the economic viability of EVs (Wenting et al., 2023).

Carbon fibre reinforced polymers (CFRP) offer exceptional strength-to-weight ratios but are prohibitively expensive for use in most EV models. Ongoing research aims to develop more cost-effective carbon fibre manufacturing processes and recycling techniques that could make these materials more accessible for mainstream EV production (Manzetti & Mariasiu, 2015). The ability to lower the costs of advanced materials will be key to making EVs more affordable for consumers and increasing their adoption on a global scale.

6.3 Supply Chain Resilience

The EV industry's reliance on critical materials, such as lithium, cobalt, and rare earth elements, exposes it to supply chain vulnerabilities. Ensuring stable and secure supply chains for these materials is essential for the industry's long-term growth and success. As the demand for EV batteries grows, securing a stable supply of lithium and other battery materials is becoming increasingly challenging. Geopolitical factors, mining regulations, and environmental concerns all contribute to supply chain risks. To mitigate these risks, manufacturers are investing in the development of alternative battery chemistries, such as sodium-ion batteries, which rely on more abundant and less costly materials (Mittal et al., 2023).

Diversifying the supply chain by sourcing materials from multiple regions and developing alternatives to critical materials can enhance supply chain resilience. For example, research into motors that reduce or eliminate the need for rare earth magnets can decrease dependency on these materials and reduce exposure to supply chain disruptions. Developing robust, flexible, and diversified supply chains will be vital to ensuring the availability of key materials for electric vehicle production, particularly as global demand continues to rise.

The future of electric vehicle materials lies at the intersection of sustainability, cost efficiency, and supply chain management. Advancements in recycling processes, cost reduction strategies, and supply chain resilience will play a pivotal role in shaping the next generation of EVs. As the industry continues to evolve, the development of innovative materials and processes will be crucial in addressing these emerging challenges and meeting the growing demand for electric vehicles.

7. CONCLUSION

Advanced materials are central to the ongoing development and success of electric vehicles (EVs), contributing to improvements in energy efficiency, safety, and the overall longevity of the vehicles. The selection and innovation of materials directly impact the performance of critical EV components, including batteries, powertrains, and thermal management systems. These innovations enable electric vehicles to meet the increasingly stringent demands for sustainability, energy efficiency, and safety, which are essential for the widespread adoption of EVs as a viable alternative to traditional internal combustion engine vehicles.

Advancements in battery technology, particularly the use of materials like lithium-ion, solid-state electrolytes, and silicon-based anodes, have improved the energy density, safety, and lifespan of EV batteries. These improvements have been key to extending the driving range of EVs and ensuring safe, reliable performance. Similarly, the use of lightweight materials, such as aluminium alloys and carbon fibre composites, has contributed significantly to weight reduction, further enhancing the efficiency and range of electric vehicles. Thermal management systems, utilizing advanced materials like phase change materials (PCM) and thermal interface materials (TIM), ensure that EV components, especially batteries and power electronics, operate within optimal temperature ranges. This is crucial for maintaining performance and preventing overheating, which can lead to failure or safety hazards.

The ongoing development of advanced materials will continue to drive the evolution of electric vehicles, particularly in the areas of sustainability, cost reduction, and supply chain resilience. As the EV market grows, the use of recyclable materials, sustainable manufacturing processes, and diversified supply chains will be critical to overcoming the challenges associated with mass production and global adoption. The innovation and application of advanced materials will remain at the forefront of electric vehicle design, shaping the future of transportation by providing safer, more efficient, and more sustainable solutions for the evolving needs of the industry.

REFERENCES

Ahmad, A., & Kumar, P. (2024). Optimization of electric vehicle lithium-ion battery parameter by using Taguchi methodology. In *Smart Electric and Hybrid Vehicles* (pp. 76–94). CRC Press., DOI: 10.1201/9781003495574-5

Andwari, A. M., Pesiridis, A., Rajoo, S., Martinez-Botas, R., & Esfahanian, V. (2017). A review of battery electric vehicle technology and readiness levels. *Renewable & Sustainable Energy Reviews*, 78, 414–430. DOI: 10.1016/j.rser.2017.03.138

Armand, M., & Tarascon, J. M. (2008). Building better batteries. *Nature*, 451(7179), 652–657. DOI: 10.1038/451652a PMID: 18256660

Bandhauer, T. M., Garimella, S., & Fuller, T. F. (2011). A critical review of thermal issues in lithium-ion batteries. *Journal of the Electrochemical Society*, 158(3), R1–R25. DOI: 10.1149/1.3515880

Bhattacharya, R., Banerjee, R., & Mukherjee, A. (2016). Aerogels: Promising thermal insulators for electric vehicles. *Energy Reports*, 2, 56–63. DOI: 10.1016/j.egyr.2016.04.002

Chen, S., Li, Q., & Pan, F. (2020). High-strength and high-conductivity copper alloys for electric vehicle applications. *Materials Science and Engineering R Reports*, 142, 100579. DOI: 10.1016/j.mser.2020.100579

Chen, Z., & Qin, Y. (2020). Next-generation anode materials for high-performance electric vehicle batteries. *Advanced Energy Materials*, 10(2), 1902343. DOI: 10.1002/aenm.201902343

Cheng, M., Sun, L., Buja, G., & Song, L. (2015). Advanced electrical machines and machine-based system for electric and hybrid vehicles. *Energies*, 8(9), 9541–9564. DOI: 10.3390/en8099541

Chong, W. T., Poh, S. C., Fazlizan, A., & Bakar, M. H. A. (2015). The assessment of electric vehicle performance on battery sizing and charging infrastructure. *Energy Procedia*, 75, 2223–2228. DOI: 10.1016/j.egypro.2015.07.420

Chung, D. D. L. (2012). Review: Carbon materials for structural self-sensing, electromagnetic shielding and thermal interfacing. *Carbon*, 50(10), 3342–3353. DOI: 10.1016/j.carbon.2012.01.031

Dabra, V., Dubey, A. K., Kumar, A., & Kumar, P. (2024). Cylindrical lithium-ion battery. Smart Electric and Hybrid Vehicles: Design, Modeling, and Assessment by Industry 4.0 Approaches, 32. https://doi.org/DOI: 10.1201/9781003502470-3

Gao, H., Zhou, T., Zheng, Y., & Tang, Z. (2017). Performance and life cycle analysis of solid-state batteries for electric vehicles. *Journal of Power Sources*, 359, 126–136. DOI: 10.1016/j.jpowsour.2017.05.073

Gibaja, C., Marquez, F., & Ruiz, M. C. (2020). A review of thermal management technologies for electric vehicle battery systems: Mathematical models. *Renewable & Sustainable Energy Reviews*, 117, 109504. DOI: 10.1016/j.rser.2019.109504

González-Cortés, S. L., Pérez-Díaz, J. A., & Lozano-Miralles, J. J. (2022). Thermal management in electric vehicles using phase change materials: A review. *Applied Thermal Engineering*, 203, 117924. DOI: 10.1016/j.applthermaleng.2021.117924

Goodenough, J. B., & Kim, Y. (2010). Challenges for rechargeable Li batteries. *Chemistry of Materials*, 22(3), 587–603. DOI: 10.1021/cm901452z

Guo, Y., Zhao, X., & Xu, M. (2019). Magnesium-based batteries for electric vehicles: A review of current challenges and future perspectives. *Journal of Energy Chemistry*, 39, 38–55. DOI: 10.1016/j.jechem.2018.12.003

Harris, S. J., Lu, P., & Manthiram, A. (2017). The role of solid electrolytes in lithium batteries. *Journal of Power Sources*, 352, 66–75. DOI: 10.1016/j.jpowsour.2017.03.028

Hirsch, A., Dietrich, F., Berger, S., & Zaeh, M. F. (2019). Manufacturing technologies for lightweight electric vehicle structures. *Journal of Manufacturing Processes*, 38, 377–389. DOI: 10.1016/j.jmapro.2019.01.039

Kawamura, K., & Yasuda, K. (2016). Advances in silicon carbide (SiC) technology for electric vehicles. *Semiconductor Science and Technology*, 31(7), 073002. DOI: 10.1088/0268-1242/31/7/073002

Kim, S., Lee, Y., & Jang, Y. (2020). Recycling of electric vehicle batteries: A review of current trends and future prospects. *Renewable & Sustainable Energy Reviews*, 117, 109477. DOI: 10.1016/j.rser.2019.109477

Kumar, A., Kumar, P., & Liu, Y. (2024). *Industry 4.0 Driven Manufacturing Technologies*. Springer Nature., DOI: 10.1007/978-3-031-68271-1

Kumar, A., Liu, Y., & Kumar, R. (Eds.). (2024). *Handbook of Intelligent and Sustainable Manufacturing: Tools, Principles, and Strategies*. CRC Press. DOI: 10.1201/9781003405870

Kumar, A., Rajak, D. K., Kumar, P., & Kumar, A. (Eds.). (2024). *Advances in Sustainable Biomaterials: Bioprocessing 4.0, Characterizations, and Applications* (1st ed.). CRC Press., DOI: 10.1201/9781003434313

Kumar, A., Rajak, D. K., Kumar, P., & Rathee, S. (Eds.). (2024). *Smart Electric and Hybrid Vehicles: Design, Modeling, and Assessment by Industry 4.0 Approaches* (1st ed.). CRC Press., DOI: 10.1201/9781003502470

Kumar, A., Rajak, D. K., Kumar, P., & Rathee, S. (Eds.). (2024). *Smart Electric and Hybrid Vehicles: Fundamentals, Strategies and Applications* (1st ed.). CRC Press., DOI: 10.1201/9781003495574

Liu, T., Zhang, Y., Lin, L., & Wang, J. (2022). Trends and challenges in recycling of automotive lightweight materials. *Journal of Cleaner Production*, 346, 131161. DOI: 10.1016/j.jclepro.2022.131161

Lu, L., Han, X., Li, J., Hua, J., & Ouyang, M. (2013). A review on the key issues for lithium-ion battery management in electric vehicles. *Journal of Power Sources*, 226, 272–288. DOI: 10.1016/j.jpowsour.2012.10.060

Luo, W., Li, H., Shi, W., & Zhang, Y. (2021). Lightweight magnesium alloys for automotive applications: A review. *Materials Science and Engineering A*, 808, 140928. DOI: 10.1016/j.msea.2021.140928

Manzetti, S., & Mariasiu, F. (2015). Electric vehicle battery technologies: From present state to future systems. *Renewable & Sustainable Energy Reviews*, 51, 1004–1012. DOI: 10.1016/j.rser.2015.07.010

Mittal, S., Korde, P., Shenbagalakshmi, S., & Sharma, K. (2023). *A systematic review on recent advancement in electric vehicle materials*. IRJEAS., DOI: 10.55083/irjeas.2023.v11i04006

Nigam, A., Kumar, P., & Kumar, A. (2024). Smart electric and hybrid vehicle architecture. In *Smart Electric and Hybrid Vehicles* (pp. 1–11). CRC Press., DOI: 10.1201/9781003502470-1

Olivetti, E. A., Ceder, G., Gaustad, G. G., & Fu, X. (2017). Lithium-ion battery supply chain considerations: Analysis of potential bottlenecks in critical metals. *Joule*, 1(2), 229–243. DOI: 10.1016/j.joule.2017.08.019

Park, M., Zhang, X., Chung, M., Less, G. B., & Sastry, A. M. (2010). A review of conduction phenomena in Li-ion batteries. *Journal of Power Sources*, 195(24), 7904–7929. DOI: 10.1016/j.jpowsour.2010.06.060

Pillai, V. K., & Sundaram, V. (2021). Recent advances in magnesium alloys for structural applications in electric vehicles. *Journal of Magnesium and Alloys*, 9(2), 432–452. DOI: 10.1016/j.jma.2020.08.011

Pusateri, A., & Risser, S. M. (2021). Silicon carbide inverters: A key technology for electric vehicle performance improvements. *IEEE Transactions on Power Electronics*, 36(5), 5845–5854. DOI: 10.1109/TPEL.2020.3048558

Qiu, Y., Li, J., & Li, B. (2019). Thermal management in lithium-ion batteries for electric vehicles: Current and future trends. *Energy Conversion and Management*, 182, 262–282. DOI: 10.1016/j.enconman.2018.12.051

Rawal, K., Patel, R., Gupta, E., Patel, G., Sharma, A., Sahu, N., & Kumar, A. (2024). Fabrication of test rig to analyze the effects of variable belt tension in a CVT system. In *Smart Electric and Hybrid Vehicles* (pp. 41–63). CRC Press., DOI: 10.1201/9781003495574-3

Ritu, K. R., Wadhwani, A. K., Kumar, A., & Kumar, P. Energizing the future: Electric vehicle charging infrastructure, integration, and policy perspectives. In *Smart Electric and Hybrid Vehicles* (pp. 95–127). CRC Press., DOI: 10.1201/9781003495574-6

Scrosati, B., Hassoun, J., & Sun, Y. K. (2011). Lithium-ion batteries. A look into the future. *Energy & Environmental Science*, 4(9), 3287–3295. DOI: 10.1039/c1ee01388b

Song, Z., & Li, X. (2015). Carbon fiber composite materials: New advancements for electric vehicle structural applications. *Composite Structures*, 131, 111–121. DOI: 10.1016/j.compstruct.2015.04.025

Tarascon, J. M., & Armand, M. (2001). Issues and challenges facing rechargeable lithium batteries. *Nature*, 414(6861), 359–367. DOI: 10.1038/35104644 PMID: 11713543

Tege, S., & Kumar, P. Intelligent and Sustainable Manufacturing Applications in the Automotive Industry. In Handbook of Intelligent and Sustainable Manufacturing (pp. 305-335). CRC Press. https://doi.org/DOI: 10.1201/9781003405870-17

Wenting, S., Sun, Z., Ding, R., & Wang, X. (2023). Graphene-based materials for the separator functionality in lithium-ion batteries. *Materials (Basel)*, 16(12), 4449. DOI: 10.3390/ma16124449 PMID: 37374632

Wood, E., Alexander, M., & Bradley, T. (2011). Investigation of battery end-of-life conditions for plug-in hybrid electric vehicles. *Journal of Power Sources*, 196(11), 5147–5154. DOI: 10.1016/j.jpowsour.2011.02.025

Wu, W., Zhu, H., & Xie, L. (2021). Recent developments in magnesium-based materials for electric vehicle applications. *Journal of Alloys and Compounds*, 867, 159074. DOI: 10.1016/j.jallcom.2021.159074

Yang, Z., Zhang, J., Kintner-Meyer, M. C., Lu, X., Choi, D., Lemmon, J. P., & Liu, J. (2011). Electrochemical energy storage for green grid. *Chemical Reviews*, 111(5), 3577–3613. DOI: 10.1021/cr100290v PMID: 21375330

Zhang, H., Huang, Y., & Liu, Z. (2020). Current progress in advanced materials for high-performance sodium-ion batteries. *Energy Storage Materials*, 24, 1–22. DOI: 10.1016/j.ensm.2019.10.006

Zhang, S. S. (2006). A review on the separators of liquid electrolyte Li-ion batteries. *Journal of Power Sources*, 164(1), 351–364. DOI: 10.1016/j.jpowsour.2006.10.065

Zhao, H., Gao, X., & Hu, J. (2020). A review on next-generation lithium batteries for electric vehicles: Advances in cathode materials. *Energy Storage Materials*, 24, 329–353. DOI: 10.1016/j.ensm.2019.07.017

Chapter 5
Use of Geogrids for Stabilization of Soil: A Review

Ram Vilas Meena
JECRC University, India

Harsh Kumar Chittora
JECRC University, India

Ankit Singh Beniwal
https://orcid.org/0000-0001-8655-903X
JECRC University, India

ABSTRACT

The process of increasing the soil's engineering properties and making it more stable is known as soil stabilization. When the construction soil is not suitable for the intended use, soil stabilization is required. Stabilization, in its broadest sense, encompasses many similar processes, including compaction, pre-consolidation, drainage, and many others. However, the term "stabilization" is typically limited to actions that change the soil's physical makeup in order to enhance its properties. For the purpose of stabilizing a natural soil, a cementing substance or chemical is added or geosynthetic material is used. Permeability and compressibility of the soil mass in earth structures can be reduced and shear strength can also be increased by soil stabilization. It can also be used for enhancement of bearing capacity of foundation soil. Although, soil stabilization is mainly used for improvement of natural soil properties in highway and embankment construction.

DOI: 10.4018/979-8-3693-5513-8.ch005

1. INTRODUCTION

Geosynthetics have been extensively used to fulfill a variety of functions that give significantly good performance of roadways. They consist of the features like separation, stiffening, filtration, reinforcement, barrier, drainage, and protection. Many different types of geosynthetic material emerged and have been developed for used in construction, it includes: Geotextile, Geostrap, Geogrid, Metal Strip etc. (Kumar et al., 2024) Soil reinforcing technique has been broadly used in construction of highways, railways, earth dams and other civil construction works. Geosynthetic material are extensively used in engineering practice to enhance the strength properties of pavements, slopes, foundations etc. One type of hard surface (Kumar et al., 2024) used for carrying vehicular or pedestrian traffic is pavement, which is a durable surface material that has been installed. Its main objective is to distribute the applied vehicle loads among the several subgrade layers. The road's pavement should provide sufficient grip, a comfortable ride, good light reflection, and low noise emissions. Lowering the vehicle transmitted load is intended to prevent it from exceeding the sub-bearing grade's capacity. The development of any construction depends greatly on the road paving. The two most common types of road pavement are rigid and flexible pavements. Soil stabilization is deals in pavement construction like-highways, dam, bridges, railway etc. Thus, soil stabilization can be used in different mean of construction technology. Table 1 show the application and type of geosynthetics grid construction. Geosynthetics are used for following application in different type of construction.

Table 1. Application and Type of Construction

S. No	Application	Type of Construction
1	Separation/Confinement/Distribution of Loads	Roads, Lane, hillsides, banks, stream access points
2	Reinforcement of soil	Soil walls, box culverts/bridges, bridge abutments, and soil arches
3	Soil movement (Piping)	Back fill around water intakes and drainage systems
4	Controlling of water pressure	Foundation walls
5	Erosion	Slopes and embankments
6	Filtration	Retaining walls

(Venkatappa Rao, G. et al, Geosynthetics Testing-A Laboratory Manual)
Note - Some individuals may carry out multiples of these tasks at once.

There are different types of geosynthetic materials available with different trade names. According to their uses in construction applications, they can be classified into the following groups:

- **Geotextiles** - A permeable geosynthetic comprised solely of textiles. Generally of two type -Woven (threads that are weaved into cloth-like materials at perpendicularly to one another) and non-woven (materials like felt with irregularly arranged fibres). They are used for separation, drainage and reinforcement.
- **Geogrids** - A grid-like polymeric structure formed from intersecting ribs (tensile elements) joined at their junctions with resulting apertures and used for stabilization and reinforcement.
- **Geocells** - A 3-dimensional, permeable, polymeric honeycomb or web structure made of strips of geotextiles, geogrids or geomembranes linked alternatingly and used for stabilization.
- **Geomembranes** - Is utilised in contact with soil or rock as a moderately impermeable polymeric sheet that is either synthetic or natural. It could be of bituminous, elastomeric or plastomeric materials. And used for creating an impermeable layer.

A summary of the wide variety of the geosynthetic applications is presented in Table 2

Table 2. Detailed Summary of Application, Primary Function and Products

S.No	Application	Primary Function	Products
1	Rail road Track bed Stabilization	Drainage Separation Filtration	Geotextile/Geogrid
2	Subgrade/Pavement Stabilization	Separation Reinforcement Filtration	Geotextile/Geogrid
3	Soil Reinforcement: Embankment Steep Slopes Vertical Walls	Reinforcement	Geotextile/Geogrid
4	Asphalt Overlay	Stress relieving layer Water proofing	Geotextile/Geogrid / Glass Grids
5	Subsurface Drainage (French drains)	Filtration	Nonwoven Geotextile

continued on following page

Table 2. Continued

S.No	Application	Primary Function	Products
6	Subsurface Drainage	Filtration Fluid Transmission	Prefabricated drainage Composites
7	Erosion Control Filter (under rip-rap)	Filtration Separation	Nonwoven Geotextile
8	Surficial Erosion Control	Turf Reinforcement	Erosion control mats /Biomats
9	Canal/Pond Lining	Moisture Barrier	Geomembrane
10	Landfills	Filtration Separation Drainage Reinforcement Barrier	Geogrids/ Geotextiles/ Geomembranes/ Geosynthetic Clay Liners

(Venkatappa Rao, G. et al, Geosynthetics Testing-A Laboratory Manual)

1.1 Material Properties

1.1.1 Black Cotton Soil

A significant soil type in India is black cotton soil. The black cotton soil appears to be a dark hue. It exhibits the typical phenomenon of expanding when wet and contracting when dry. As a result, these soils grow and become sticky during the rainy season and shrink during the dry season, resulting in deep soil cracks. Black cotton soils contain lots of clay. Black cotton soils are chemically made up of lime, magnesium, iron, alumina, and potash, but it lack organic matter, nitrogen, and phosphorus. They are good for cotton production because they can hold water. The soil is known as black cotton soil. Table 3 presented the properties of black cotton soil.

Table 3. Properties of Black Cotton Soil (Landmark Testing Laboratory, Jaipur)

Properties	Value
Clay Content, %	36
Silt Content, %	64
Specific Gravity	2.70
Liquid Limit, %	62
Plastic Limit, %	35
Shrinkage Limit, %	8

continued on following page

Table 3. Continued

Properties	Value
Maximum Dry Denisty, gm/cm^3	1.90
Optimum Moisture Content, %	18
California Bearing Ratio (CBR) Value, %	1
Cohesion Intercept (UU), kg/cm^2	1.8
Angle of Internal Friction (UU), deg	2

1.1.2 Polypropylene Biaxial Geogrid

Superior polypropylene grades are used in the precision punching and drawing operations to create these products. Table 4 show the physical properties of geogrid. The placement of TechGrid PP Biaxial Geogrids (manufactured by Techfab India) lead to enhanced structural capacity, strong contact, mostly through the interlocking of particles within the geogrid apertures, in a layer of granular material. The mechanical improvement of the soil-geogrid system results from the development of four distinct mechanisms, as follows:

1. Separation
2. Lateral Restraint
3. Tensioned Membrane
4. Enhanced Bearing Capacity

Table 4. Physical Properties of Geogrid (www.techfabindia.com)

Product Code	Mechanical Properties							Dimensions	
	Unit Tension (kN/m)						Junction Efficiency	Aperture Size (mm)	
	Ultimate		Load@2.0%		Load@5.0%				
	MD	TD	MD	TD	MD	TD		MD	
TechGrid PP 2020	**20**	**20**	**7.5**	**7.5**	**15**	**15**	**≥95%**	**38**	**38**
TechGrid PP 2020L	20	20	7.5	7.5	15	15	≥95%	65	65
TechGrid PP 3030	30	30	11.0	11.0	21.6	21.6	≥95%	38	38
TechGrid PP 3030L	30	30	12.5	12.5	21.6	21.6	≥95%	65	65
TechGrid PP 4040	40	40	14.5	14.5	28	28	≥95%	38	38
TechGrid PP 4040L	40	40	14.5	14.5	28	28	≥95%	65	65

2. LITERATURE REVIEW

Various studies had been done on the effects of geosynthetic and other ingredient for soil stabilization. The compendium of such studies is tabulated in Table 5.

Table 5. Research from the past on using geosynthetic in road construction.

S.No	Type of Soil	Geosynthetic material used	Properties Evaluated	Outcome	Reference
1	Soft Subgrade Soil & Fine Grained Soil	Different type of geosynthetic depending upon function	Mitigation of crack, separation, stabilization of road base and soft subgrade, lateral drainage	Decrease of cracks, effective separation of layer and better distribution of load hence less pavement thickness and more serviceability and life of roadway	(Zornberg, 2017)
2	Well graded gravelly high plastic soil	Geogrid	CBR value of subgrade and pavement thickness	CBR value of origin soil increased from1.67% to 1.8%, 2.5% & 3.91% by placing the geogrid at H/4, H/2 & 3H/4 from the bottom respectively	(Madhu et al., 2018)
3	Lateritic And Clay	Non-woven Geotextile	CBR	Improved the CBR value from 4% to 15% in Sample A and from 7% to 21% in sample B	(Ogundare et al., 2019)
4	Soil (ASTHO A-4)	Geogrid & chemical stabilization	CBR	The CBR value increased from 10% to 56% in soaked condition and from 25% to 74% in unsoaked condition.	Ogunkunbi & Jimoh, 2019)
5	Black cotton soil	Geogrid	CBR	The soil sub-grade strength is substantially improved by addition of geo-grids as reinforcement in the soil. It was evident that CBR values increased when geo grids were placed at 3/5 the distance from the base than at 2/5 and 4/5 distances from the base.	(Thangaraj et al., 2019)
6	Sand	Geogrid	Settlement and Bearing Capacity	In comparison to the unreinforced embankment, the reinforced embankments with a single geogrid layer had a 7.14% increase in bearing capacity and an 11.24% decrease in settlement. The results achieved for the third to fifth series of embankments were respectively in the order of (19, 36.14), (26.3, 52.8) and (28.9, 53.42) %.	(Esmaeili et al., 2018)
7	Clayey Soil	Geogrid	Field CBR	The CBR (soaked) of Soil sample-1 increased by 2.15 &1.90 times and Soil sample-2 increased by 2.62 & 2.24 times respectively using Strata Grid-1 & Strata Grid-2 respectively.	(Rajesh et al., 2016)

continued on following page

Table 5. Continued

S.No	Type of Soil	Geosynthetic material used	Properties Evaluated	Outcome	Reference
8	Loose Granular Soil	Geogrid	CBR	CBR value increased from 9 to 14.5, 30.6, 22.6 & 17.3 with providing geogrid at 0.1H, 0.15H, 0.2H & 0.25H respectively from the base.	(Fakhraldin, 2016)
9	Very soft / Soft soil	Geogrid	Bearing capacity (plate load test)	Bearing capacity increased from 0.6 kg/cm^2 to 4.2 kg/cm^2 using geogrid combination with five layers.	(Al-Sinaidi and Ali, 2006)
10	CI-MI (inorganic clay & silt of intermediate Plasticity)	Geotextiles, geonets and geogrids, Non woven polypropylene fibers	Unconfined Compression Test & CBR	UCC strength (Unconfined Compression Strength) increase from 149.3 kPa to 211.5 kPa and CBR value increased from 6.82% to 10.2% with 1% PF (Polypropylene fiber). Soil reinforced with geosynthetics showed improved strength from 149.3 kPa to 214 kPa and CBR value increased from 6.82% to 11.7% with providing geosynthetics at 3D/4 from top of the soil	(Lakshmi et al., 2018)
11	Laterite Soil	Coir and Geogrid	CBR	CBR value increased from 7.3% to 20.49% using geotextile and from 7.3% to 25.62% using geogrid.	(Unni et al., 2017)
12	Clay of intermediate Compressibili-ty (CI)	Biaxial Geogrid	CBR & UCS (unconfined Compression Strength)	Geogrid is applied in a single layer (i.e., 0.2H) and in double layers (i.e., 0.2H and 0.4H) from the top of the specimen, maximum improvements of 36% and 41% in CBR and 62% and 70% in UCS values are seen.	(Mittal and Shukla, 2019)
13	Black Cotton Soil	Geogrid	CBR	CBR value increased from 2.8% to 8.7% when the geogrid applied at 0.2H height from the top.	(Sai et al., 2020)
14	Silty clay soil and sand	Biaxial geogrid	Bearing Capacity of Soil	The bearing capacity of the two-layer soil increases by an average of 16.67% when geogrid is placed at the soil's interface and 33.33% increased when geogrid is placed in the middle of the sand layer.	(Munawir, 2020)
15	Granular Soil	Geogrid	Interaction coefficient	Soils with smaller particle sizes improve less, soils with bigger particle sizes improve more, and soils with coarser particles interact more with geogrids.	(Minažek et al., 2012)

continued on following page

Table 5. Continued

S.No	Type of Soil	Geosynthetic material used	Properties Evaluated	Outcome	Reference
16	Sandy Clay	Geogrid	CBR	CBR Value is increased from 3.4 to 4.3% (26%) when the geogrid placed at 1st layer from the top and increased to 5.5% i.e. 62% increment when the geogrids are placed at several layers (four layers).	(Zumrawi and Mansour, 2016)
17	Granular Soil & Silty Clay	Geogrid	CBR	Substantial increase in the CBR values observed for soaked condition. Geogrids introduced at 2/5(12.8%, 11%, 8.8%), 3/5(15.1%, 14%, 12.2%) and 4/5(11%,10%, 8.3%) respectively as when compared with the initial soaked CBR (3,9,6%) without geogrids.	(Akolade and Olaniyan, 2014)
18	Soil of different group (Silty Sand, Inorganic clays of high Plasticity, & Silty clay of low plasticity)	Geogrid	Durability and Serviceability of pavement	When geo-grid reinforcement is added to soil, surface penetration and deformation are reduced, longitudinal cracks does not occur, increase in bearing capacity and the stress distribution on the soil sample is improved.	(Ramteke, 2014)
19	Lateritic Soil	Geogrid	CBR	CBR value increased from 26% to 45%, from 12% to 18% and from 11 to 16% of original soil, 10% mixing with clay powder & 20% mixing of clay powder respectively using geogrid.	(Adams et al., 2016)
20	Black Cotton Soil	Cement (2,4,6 & 8%) & Geogrid	CBR	The CBR strength improves from 7.15% to 17.59% with the use of woven biaxial geogrid in soft soil.	(Neetu et al., 2014)
21	Black Cotton Soil	Geogrid & Sandy Gravel Soil	CBR	The CBR improves from 6.8% to 10.12% when geogrids are used in gravely sand. This is the optimum limit in improving the CBR strength.	(Keerthi and Kori, 2018)
22	CL	Jute Geogrid (Untreated & treated with 0.1N NaOH)	CBR	CBR value increased by 50.07% with treated geogrid as compared to untreated geogrid when placed at 0.2 H from the top of the samples in soaked conditions.	(Nair and Deepthy, 2017)
23	Lateritic Soil	Polyethylene Geogrid	CBR	CBR value is increased from 4.91% to 6.99% on reinforcing with geogrid.	(Patil and Bhutekar, 2017)

Zornberg, (2017)explore the different capabilities of geosynthetic materials in various highway construction applications. According to published research, geosynthetics can be used for various highway applications such as: - Stabilizing road bases, stabilizing soft subgrades, lateral drainage, preventing cracking in asphalt toppings and separation etc. Madhu et al., (2018) study showed that the aggregate thickness is reduced by 40% if geogrid reinforcement is used in soil subgrades and it benefits economically and ecologically. In this study CBR value of origin soil was 1.67% which were increased to 1.8%, 2.5 & 3.91 by placing the geogrid at H/4, H/2 & 3H/4 from the bottom respectively. The study showed that soil properties can be improved using geogrid in the subgrade material at different levels. According to this study, we can increase the load carrying capacity of the subgrade soil, increase the service life of road, reduce the pavement thickness and reduce the cost of road construction.

Ogundare et al., (2019) study the soil stabilization using geotextile and concluded that geotextile reinforced soil performs better than ordinary soil under dynamic stress. These are non- biodegradable, long lasting and increases serviceability duration of the pavement. Regardless of the depth at which the non-woven geotextile is placed inside the thickness of the subgrade, the use of non-woven geotextile generally increases the strength of the subgrade soil as measured by the CBR. The non-woven geotextile, however, performs better at a depth of H/4 from the bottom surface since it offers the improve in strength from 4% to 15% and from 7% to 21% of different soil samples, which can help reduce the value of the pavement thickness.

Ogunkunbi and Jimoh, (2019) according to their research inferred that the CBR value increased under unsoaked and soaked conditions when the sample was reinforced with the geogrids for the unsoaked and soaked condition at different depths within the CBR mould. The CBR value increased from 10% to 56% in soaked condition and from 25% to 74% in unsoaked condition.

Thangaraj et al., (2019) showed that when soil is reinforced with geosynthetic, it becomes stronger and stiffer than unreinforced soil. By incorporating geotextiles into the soil, the SBC of soil is increased, which further will increase the pavement's life. In the study showing that an improvement in subgrade strength that is consistent with the theoretical research done. When utilised in pavements, geosynthetics serve a variety of purposes, including strengthening, separation, drainage, and filtration. In the study Geo-grids are placed at 3/5 the distance from the base, showed greater CBR value when compared to those when placed at 2/5 and 4/5 distances from the base.

Esmaeili et al., (2018) in their study stated that when geogrid is provided in the railway embankment, bearing capacity of soil will increase and bearing capacity of soil will increase up to maximum three layers of geogrid provided in the embankment at different level after that bearing capacity of embankment soil will not be increase due to extra layer of geogrid provided. In comparison to the unreinforced embankment, the reinforced embankments with a single geogrid layer had a 7.14% increase in bearing capacity and an 11.24% decrease in settlement.

Rajesh et al., (2016) studied that when two different type of geogrid is used with two different type of soil then soil properties of less finer and less plasticity soil will increase with higher grade of geogrid than the lower grade of geogrid. In this study it is shown that other than CBR value soil properties like- OMC (optimum moisture content), MDD (maximum dry density) also improve with the use of geogrid. The soaked CBR value of sample-1 increased by 2.15, 1.90 times and sample-2 increased by 2.62, 2.24 times respectively by using different geo-grids respectively for grade SG-1 (Strata grid-1) and SG-2(Strata Grid-2).

Fakhraldin, (2016) stated that CBR value will increase after use of geogrid than the without use of geogrid in the soil. In this study it is shown that CBR value is conducted with and without use of geogrid and geogrid is provided at different height from the base of CBR mould. In this study geogrid is provided at 0.1H, 0.15H, 0.2H and 0.25H from the base of CBR mould and it is concluded that CBR value increase up to 0.15H and after that it will continuously decrease with the increase of height.

Al-Sinaidi and Ali, (2006) suggested that when geogrid is provided in five layer from the base and each layer of soil is 400 mm than soil bearing capacity will increase 7 times than the natural soil and 1.5 times than the stone column provided. Cost of construction can also be reduced up to 1.5% of total cost and construction time will also reduce using geogrid than the stone column technique.

Lakshmi et al., (2018) stated that maximum unconfined compression strength 211.5 kPa is observed with 1% of polypropylene fibers and maximum CBR value is observed 10.2%. UCS (unconfined compression strength) and CBR value increased by 41.66% and 1.5 times respectively that the unreinforced soil. In this study it is found that maximum strength, maximum CBR is observed using geogrid, geotextile and geonet at three forth the depth from the top of soil.

Unni et al., (2017) as per this research article geogrid and geotextile was used in laterite soil for stabilization of subgrade soil. Geotextile was used at H/2, H/3 and H/4 from the top of the soil and it is found that maximum CBR value is achieved at H/4 from the top of soil. Geogrid was also used at 3H/5, 2H/5 & 4H/5 from the top of the surface and it is found that maximum CBR value was achieved at 2H/5 from the top of the surface.

Mittal and Shukla, (2019) suggested that using biaxial polyester geogrid at different level with the combination of one or two layer of geogrid we can improve the strength of poor subgrade soil. In this research MDD, OMC, UCS and CBR value of poor subgrade soil is analyzed with and without geogrid of geogrid at different level. From this research it is concluded that maximum CBR value with single layer of geogrid was achieved at 0.2H height from the top of the mould and with double layer of geogrid it was achieved at 0.2H & 0.4H combination of height from the top of the mould.

(Sai et al., 2020) study that CBR value of black cotton soil can be increased with using of geogrid. In this research geogrid was used at different level (0.2H, 0.4H, 0.6H & 0.8H) of height from the top of the CBR mould and it is found that at 0.2H height, maximum CBR value was achieved.

(Munawir, 2020) study that soil bearing capacity of silty clay can be improved by replacing the top silty clay with sand. Bearing capacity of silty clay also can be improved by using the geogrid in single layer or double layer of geogrid. The bearing capacity of the two-layer soil increases 16.67% using one geogrid layer at interface of soil (i.e., sand and silty clay soil) with the u/B equal to 0. 667. The increase in bearing capacity for soil with two layers and u/B and h/B ratios of 0.33 was 44.44%, 61.11%, and 72.22% for geogrid layers with two, three, and four numbers, respectively.

(Minažek et al., 2012) study about the soil-geogrid interaction coefficient with different type of geogrid and different type of granular soil. As per their research in the reinforced soil many factors influence the soil reinforcement interaction. These factors are shape & size of the soil particle, moisture content and type of geogrid. Soil having different size & shape of particle develop the different interaction coefficient with the different type of geosynthetics.

(Zumrawi and Mansour, 2016) carried out the experimental work to determine the soil improvement of soft soil using geogrid at different level of height. As per this research it is concluded that we can improve the CBR of soft soil by using the geogrid and position of geogrid is also influences the bearing capacity of soil. In this research at 1st layer, they achieved the maximum improvement in the bearing capacity of soft soil.

(Akolade and Olaniyan, 2014) conducted the experimental study with three type of soil and geogrid. They provided the geogrid at 2/5H, 3/5H & 4/5H height from the top of the surface of CBR mould and it is concluded that strength of poor subgrade soil can be improved by using the geogrid material in the subgrade. As per published research maximum strength was achieved at 3/5H height.

(Ramteke, 2014) in their study they prepared the soil sample with geogrid and geonet at different height with and without the combination and multilayer in the CBR mould and test were conducted. As per their research it is concluded that CBR value is increased in both the cases comparatively non reinforced soil. CBR value is increased with geonet is more than the geogrid.

(Adams et al., 2016) studied that CBR value can be increase in marginal soil with geogrid placement in soaked and unsoaked condition. They provided geogrid in single layer at different depth and near placement from the top of the surface of the soil showing more strength. They also provided geogrid in combination of two layer and its shows more strength or improvement in CBR value than the single layer provided. Now as per published research we can conclude that we can increase the soil strength in terms of CBR by providing the geogrids in the soft marginal soil subgrade.

(Neetu et al., 2014) in their study they conducted various test of soil like- specific gravity, atterberg's limits, proctor test with mixing of cement in various proportion (2%, 4%, 6% & 8%) and found the improvement in black cotton soil properties from each proportion separately. Further they conducted CBR test with providing the cement and geogrid in the black cotton soil. Geogrid is provided at 2/5H from the top of the surface of the soil. As per their research maximum CBR value is achieved with 6% cement content and geogrid.

(Keerthi and Kori, 2018) provides geogrid at different height in the CBR mould (0.2H, 0.4H, 0.6H & 0.8H) with black cotton soil and sandy gravel soil and they found that CBR value is maximum at 0.4H height from the top of the surface with both the soil and improvement in strength is more in the sandy gravel soil than the black cotton soil. So it is concluded that CBR value can be increased by providing the geogrid as soil reinforcement at 0.4H height from the top of the surface.

(Zumrawi et al., 2016) this study shows that the CBR values of soil samples with polyethylene geogrids are higher than soil samples without geogrids. Comparing the two different samples, we find that the soil with polyethylene geogrid increases his CBR value significantly more than the soil without geogrid. CBR value is increased from 4.91% to 6.99% on reinforcing with geogrid. The geogrid used in this study is a waste plastic material that can contaminate the soil if not disposed of properly. Soil contamination can be reduced by using this waste for soil stabilization. A soil sample with a polyethylene geogrid was experimented to have more potential for an applied load than a soil sample without a geogrid. These CBR results correlate with flexible pavement thickness. Therefore, in this study, we concluded that using polyethylene geogrid as a soil stabilizer could enhance his CBR in lateritic soils.

3. SUMMARY AND CONCLUSION

The research chapter abbreviates the use of geosynthetic material to stabilize or improve the engineering properties of subgrade and poor soil on the basis of earlier published research and test results. From this research we learn about the stabilization of soil using geosynthetic materials in different application. In view of earlier research, we found the different techniques for stabilize the subgrade or soil using different type of geosynthetic materials based on the application.

REFERENCES

Adams, C. A., Tuffour, Y. A., & Kwofie, S. (2016). Effects of Soil Properties and Geogrid Placement on CBR Enhancement of Lateritic Soil for Road Pavement Layers "Effects of Soil Properties and Geogrid Placement on CBR Enhancement of Lateritic Soil for Road Pavement Layers. *American Journal of Civil Engineering and Architecture*, 4, 62–66. DOI: 10.12691/ajcea-4-2-4

AKOLADE, A.S and OLANIYAN, O.S., (2014). Application of Geogrids on the Geotechnical Properties of Subgrade Materials under Soaked Condition 6, 12–19.

Al-Sinaidi, A. R., & Ali, A. H. (2006). Improvement in bearing capacity of soil by geogrid -an experimental approach. *Iaeg*, 2006, 1–5.

B.M.Patil1, S.B.B., (2017). Improving Soil Stability By Using Polyethylene Geogrid. Int. J. Res. Eng. Technol. 06, 70–72. DOI: 10.15623/ijret.2017.0606012

Esmaeili, M., Naderi, B., Neyestanaki, H. K., & Khodaverdian, A. (2018). Investigating the effect of geogrid on stabilization of high railway embankments. *Soil and Foundation*, 58(2), 319–332. DOI: 10.1016/j.sandf.2018.02.005

Fakhraldin, M. K. (2016). Improvement of Loose Granular Soil By Using Geogrid Reinforcement. *Kufa J. Eng.*, 7(3), 66–79. DOI: 10.30572/2018/KJE/731201

Keerthi, N., Kori, S., (2018). Study on Improvement of Sub Grade Soil using Soil - Reinforcement Technique 13, 126–134.

Kumar, A., Liu, Y., & Kumar, R. (Eds.). (2024). *Handbook of Intelligent and Sustainable Manufacturing: Tools, Principles, and Strategies*. CRC Press., DOI: 10.1201/9781003405870

Kumar, A., Rajak, D. K., Kumar, P., & Kumar, A. (Eds.). (2024). *Advances in Sustainable Biomaterials: Bioprocessing 4.0, Characterizations, and Applications* (1st ed.). CRC Press., DOI: 10.1201/9781003434313

Lakshmi, S. M., Gayathri, S., Susanna, K. D., Ammaiappan, M., & Manoharan, M. (2018). An Experimental Study On Soil Strength Enhancement Using Geosynthetics. *Int. J. Innov. Res. Explor.*, 5, 433–442.

Madhu, P., Yadav, G., Bharath, S., Kumar, M. M., Reddy, M. N., & Reddy, G. C. (2018).. . *International Journal of Engineering Sciences & Research Technology Usage of Geogrids in Flexible Pavement Design*, 7, 144–153.

Minažek, K., Kopić, Ž., & Mulabdić, M. (2012). Simple Procedure of Geogrid-Soil Interaction Efficiency Assesment. *Tech. Gaz.*, 19, 399–405.

Mittal, A., & Shukla, S. (2019). Strength improvement of poor subgrade soil reinforced with polyester biaxial geogrid. *Jordan J. Civ. Eng.*, 13, 214–225.

Nair, K.S., Deepthy, B.L., (2017). Improvement of Soft Subgrade Using Geogrid Reinforcement 181–183.

Ogundare, D.A., Familusi, A.O., Osunkunle, A.B., Olusami, J.O., (2019). American Journal of Engineering Research (AJER) Utilization Of Geotextile For Soil Stabilization 224–231.

Ogunkunbi, G., & Jimoh, Y. (2019). Design and Economic Analysis of a Flexible Pavement on a Geosynthetic Reinforced Subgrade. *Journal of Applied Science & Environmental Management*, 23(1), 121. DOI: 10.4314/jasem.v23i1.18

Rajesh, U., Sajja, S., & Chakravarthi, V. K. (2016). Studies on Engineering Performance of Geogrid Reinforced Soft Subgrade. *Transportation Research Procedia*, 17, 164–173. DOI: 10.1016/j.trpro.2016.11.072

Ramteke, N.B., (2014). A Review : Effect of Geo-grid reinforcement on soil ISSN : 2348 9510 Geo-grid : 1, 35–47.

Sai, P.N.V., Surendra, S., Arjun, N.V.K., Sai, M.S., Rao, C.H.H., (2020). Reinforcement of Black Cotton Soil by Using Geogrid Reinforcement of Black Cotton Soil by Using Geogrid 0193–4120.

Thangaraj, P., Janakiraman, G.G., Paramaguru, P., Pratheep, T., Scholar, P.G., (2019). Stabilization of Subgrade Using Geosynthetics. Artic. J. Adv. Res. Dyn. Control Syst. 12.

Unni et al., (2017), 2017. Efficacy of Geosynthetics in Subgrade Stabilization-A Comparative Study in Laterite Soil 6709–6716. https://doi.org/DOI: 10.15680/IJIRSET.2017.0604087

Zornberg, J. G. (2017). Functions and Applications of Geosynthetics in Roadways. *Procedia Engineering*, 189, 298–306. DOI: 10.1016/j.proeng.2017.05.048

Zumrawi, M. M. E., & Mansour, N. (2016). Laboratory Evaluation of Geogrid Used forStabilising Soft Subgrade. *Int. J. Civ. Environ. Eng.*, 10, 418–421.

Chapter 6

Computational Insights Into Municipal Solid Waste Pyrolysis in Bhopal, India: A Comprehensive Review

Nishant Vibhav Saxena
https://orcid.org/0000-0001-7829-3307
University Institute of Technology RGPV, Bhopal, India

Sharad Kumar Pradhan
NITTTR, Bhopal, India

R. S. Rajput
RGPV, Bhopal, India

ABSTRACT

Rapid urbanization, exemplified by Bhopal City in Madhya Pradesh, India, poses significant challenges in municipal solid waste (MSW) disposal. Pyrolysis, a thermal decomposition process, offers potential for transforming MSW into biochar, syngas, and bio-oil while mitigating environmental impacts. This review provides an overview of computational studies on MSW pyrolysis in Bhopal City, highlighting recent advances, limitations, and future research directions. The review begins by highlighting the strain on conventional waste disposal systems in Bhopal City due to escalating waste generation. It emphasizes the essential role of computational modeling in optimizing pyrolysis processes. Key aspects of MSW pyrolysis, including reactor design, reaction kinetics, and feedstock composition, are explored. Computational studies improve our understanding of waste behavior during pyrolysis,

DOI: 10.4018/979-8-3693-5513-8.ch006

enabling process optimization for resource recovery and reduced environmental impact.This review paper provides a concise summary of computational research on MSW pyrolysis in Bhopal, India.

1. INTRODUCTION

In the context of global development shifting towards increased urbanization, cities have become pivotal in humanity's remarkable transformation. (Kumar, A et al., 2023) covered a comprehensive coverage of environmental pollution and health hazards caused by differing types of waste, its recycling and other e-waste management strategies, and required political and legal interventions. It also provides carbon-recycling routes available and investigates how within the long-term they might be applied to reinforce waste management in industrial countries as well as developing and emerging economies.The rapid urbanization phenomenon, marked by the migration of individuals from rural to urban areas, characterizes our current era. Bhopal, situated in Madhya Pradesh, India, serves as a prominent example of this notable shift. Like several other urban areas, Bhopal grapples with complex challenges stemming from its rapid growth. This chapter examines significant concerns related to urbanization in Bhopal, with a specific focus on MSW disposal. It explores the potential of pyrolysis as an innovative waste-to-resource approach to address these challenges and offers substantial environmental benefits (Varun Singh Bundela, et. Al, 2022).

The total quantity of SWG in the country is 160038.9 TPD of which 152749.5 TPD of waste is collected at a collection efficiency of 95.4%. 79956.3 TPD (50%) of waste is treated and 29427.2 (18.4%) TPD is land filled. 50655.4 TPD which is 31.7% of the TWG remains un-accounted (Annual Report on Solid Waste Management, 2020-21).

1.1 Trends in Solid Waste Management

- SWM information for the last six years (2015-21) has been examined and following are the observations:
    - **Per capita Solid waste generation:** Per capita solid waste generation has been calculated for the last six years. Marginal decreasing trend is observed in per capita solid waste generation over the last six years (Arun Patel et al.,2022).

The table below is for the generation of solid wastes per capita in grams per day for six years from 2015-16 to 2020-21. Per capita waste generation was high at 118.68 gm/day during 2015-16. Generation was highest at 132.78 gm/day during 2016-17 and was highest for the whole record period. It had taken a significant dip to the lowest in this period by reaching 98.79 gm/day in 2017-18. The subsequent years saw stabilization of waste generation around 121.54 gm/day in 2018-19, then the subsequent years showed slight reduction at 119.26 gm/day in 2019-20 and finally 119.07 gm/day in 2020-21. This overall trend suggests an initial rise and fluctuations thereafter to around 119 gm/day, presumably reflecting changes in the waste management practices or varying external influences over these years.

Table 1. SWG Per Capita

Year	**2015-16**	**2016-17**	**2017-18**	**2018-19**	**2019-20**	**2020-21**
Solid Waste Generation Per Capita(gm/day)	118.68	132.78	98.79	121.54	119.26	119.07

Figure 1. SWG Per Capita (gm/day)

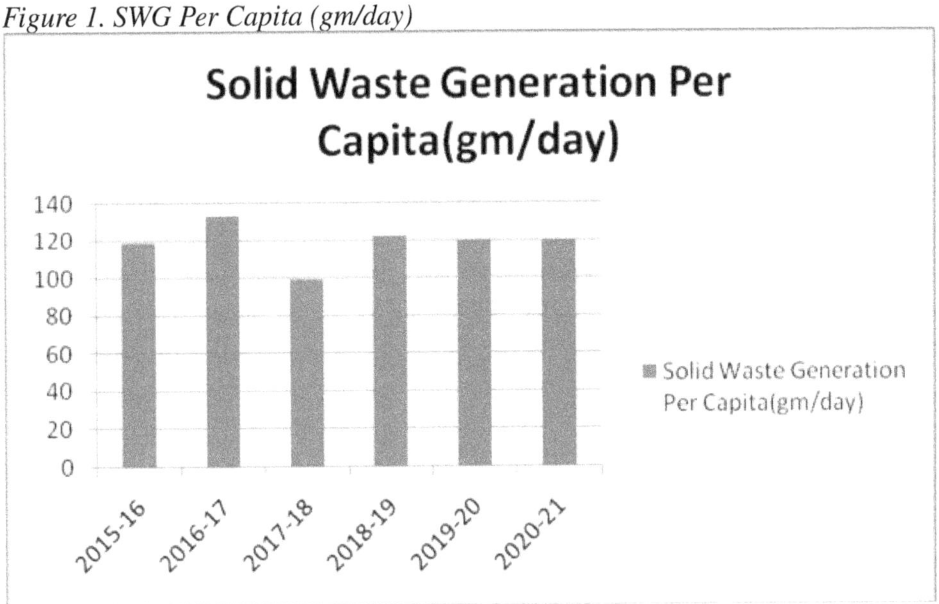

The bar graph represents the generation of solid waste per capita in grams per day for six years from 2015-16 to 2020-21. The bars show a fluctuating trend, starting at around 118.68 gm/day in 2015-16, which shoots up to 132.78 gm/day in 2016-17, which is the peak in this period. In 2017-18, it shows a significant drop to around 98.79 gm/day, which is the lowest value in the dataset. The following years present a steady rise and stabilization, at 121.54 gm/day in the year 2018-19, followed by 119.26 gm/day in the year 2019-20 and 119.07 gm/day in the year 2020-21. Altogether, the graph reflects an initial rising curve, a sharp falling curve, and then steady curves around the 119 gm/day mark in the later years. This pattern might reflect changes in practices in waste management or some other socio-economic factors within the observed period.

- **Trend in gap in Solid waste management:** Percentage gap in solid waste management has been calculated for the period 2015-21. Decreasing trend in gap in solid waste management has been observed up to 2019-20 during the last six years wherein percentage gap in Solid Waste Management has decreased from 40.99% in 2015-16 to 25.82% in 2019-20 and slightly increased to 31.70% in 2020-21 (Rajeev Singh Parihar et al., 2018).

The given table presents the gap in solid waste management as a percentage across six years between 2015-16 and 2020-21 to depict variations over this period. For the year 2015-16, the gap stands at 40.99% that could be viewed as relatively moderate on the scale. This gap sharply rocketed up to 79.78% for the year 2016-17-the highest in the dataset; it signified that quite a struggle was going to be undertaken to manage that year. However, by 2017-18, there was a significant improvement as the gap dropped sharply to 9.38%, reflecting a period of enhanced waste management effectiveness. In 2018-19, the gap rose again to 30.35%, followed by a slight reduction to 25.82% in 2019-20. The final year, 2020-21, saw a minor increase in the gap to 31.7%. These fluctuations indicate inconsistent performances in solid waste management characterized by periods of improvement coupled with increased inefficiencies at different times, possibly occasioned by changes in generation of waste, policy intervention or resource allocation for managing waste during these years.

Table 2. Gap in Solid Waste Management

year	2015-16	2016-17	2017-18	2018-19	2019-20	2020-21
Gap in Solid Waste Management (%)	40.99	79.78	9.38	30.35	25.82	31.7

Figure 2. Gap in SWM in %

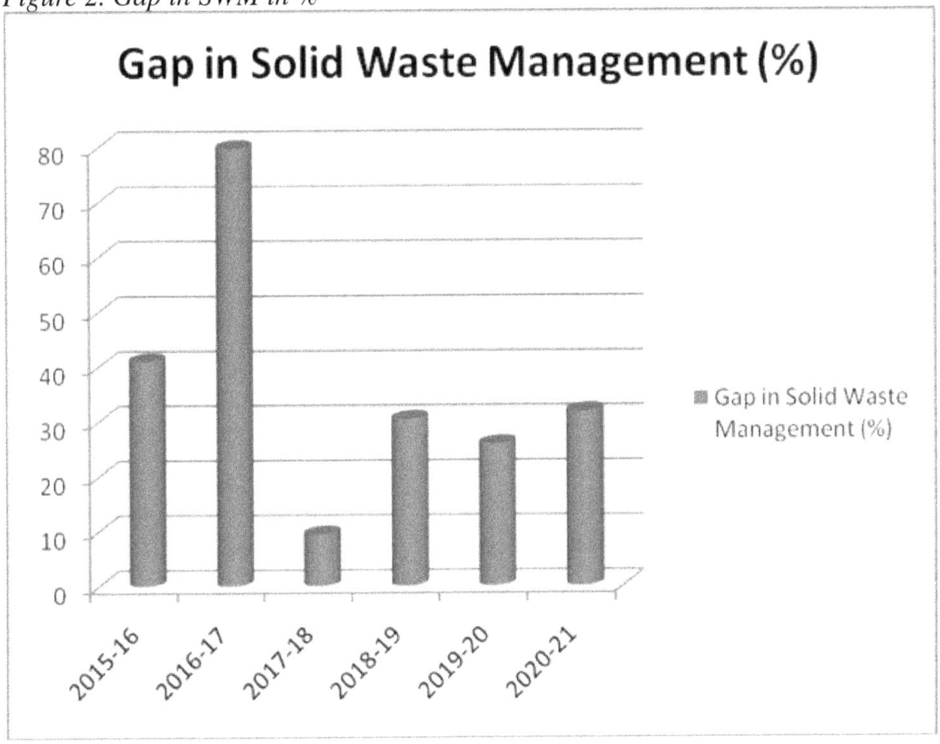

The bar graph depicts the comparison of solid waste management percentage for six years, between 2015-16 and 2020-21. From the above data, it is clear that there have been tremendous highs and lows in waste management efficiency during this period. In 2015-16, the gap was at 40.99% - a gross inefficiency in managing waste. Inefficiency in managing waste increased to a high 79.78% - the highest in the database - which clearly depicts it was a very trying period during that year. Dignified improvement is found in the year 2017-18, since it has narrowed down the gap to 9.38%, which proved to be a good period of effective waste management; however, it again rises to 30.35% in the next year, 2018-19, it decreases to 25.82% in 2019-20, and further increases marginally up to 31.7% in the last year, 2020-21. The same pattern shows how the rate of inefficiency in dealing with solid wastes is also varied, with occasional high peak inefficiency interspersed with marked improvements, which suggests that the trends may well be influenced by changing practices or external events over such periods.

1.2 Challenges Arising from Rapid Urbanization in Bhopal City

Bhopal, located in central India, has undergone significant transformations in recent decades. The population has experienced substantial growth, resulting in infrastructure expansion and increased economic activity. Rapid urbanization has brought numerous opportunities, including expanded employment opportunities and improved access to essential services. However, it has also given rise to a multitude of challenges that demand immediate attention(S. D. Diwevdi et al., 2014).

One of the foremost challenges arising from rapid urbanization is the strain on urban infrastructure and services. Bhopal's infrastructure, initially designed for a smaller population, struggles to meet the needs of its growing residents. The city's inhabitants face overcrowded public transit, inadequate sanitation facilities, and a scarcity of affordable housing. These difficulties not only affect individual well-being but also place significant pressure on local governance (Cheng, Y., He et al., 2019).

Furthermore, growing urbanization has led to an unprecedented increase in trash generation. The rise in population and economic activities has resulted in a substantial surge in MSW production. Managing this waste has become a formidable challenge for local authorities, who struggle to meet the rising demand for waste disposal services (Fawzy, M et al., 2018). Inadequate MSW management poses significant risks to the environment, public health, and the aesthetic appeal of urban areas.

Municipal solid waste management is an important as it impacts health, Environment and aesthetic society if it is not managed properly. Hence to improve quality and standard of living, the Bhopal Municipal Corporation (BMC) has proposed to strengthen the covering collection, segregation, recycling, transportation, processing and disposal with option for composting, waste to energy, disposal in 8 Urban Local Bodies (ULBs) in Bhopal(ZairaZaman Chowdhury et al., 2017).The BMC intends to institutionalize a holistic Integrated, sustainable environment and eco-friendly MSWM System in the urban local bodies (ULBs)(Al-Salem et al., 2009). Rapid industrialization and population explosion in India has led to the migration of people from villages to cities, which generate thousands MT of municipal solid waste daily, which is one of the important contributors for environmental degradation at national level. Improper management of municipal solid waste (MSW) causes hazards to inhabitants. Therefore, BMC has initiated the project that involves setting up an Integrated Solid Waste Management System (ISWM) for a cluster comprising of 8 ULBs within a radius of 80 km of Bhopal (Bridgwater, A. V., 2012).

In Bhopal District, the total quantity of Solid Waste Generation is around 856 tons / day. The bulk of this waste originates within the BMC, according for roughly 850 MT/Day. Bhopal Municipal Corporation (BMC) has taken various measures for collection, transportation, treatment & disposal of Municipal Solid waste & Construction & Demolition wastes as stated here in above (Chen, D et al., 2018).

Total population of Bhopal is about 2390000 and the total area under BHOPAL MUNICIPAL CORPORATION (BMC) is 648.24 km^2 the city is divided into 19 zones and the entire 19 zone is sub divided into 85wards. There are of garbage is too much high due to huge population so BMC set plan for managing waste. BMC garbage cart collecting garbage door to door every day in morning to evening from all the 85 ward of Bhopal. There are more than 420 garbage cart in Bhopal for collecting door to door garbage(Chen, Y et al., 2020). Firstly, BMC garbage cart goes door to door and collect wet, dry, electronic and all types of mixed waste and then dump it into sub dumping station (Bhopal city there are 6 sub dumping station 1. Kolar 2. Danapani 3. Bhadbhada 4. Transport nager 5. Arifnager 6. Shahjahani). On these dumping station segregate all the garbage into wetand dry waste. Again from here dry waste is loaded into truck and wet waste is loaded into capsule type of garbage cart and then its dumped at near AdampurChhawani which is situated at outside of Bhopal (Demirbas A., 2004).

Figure 3. Process of MSW

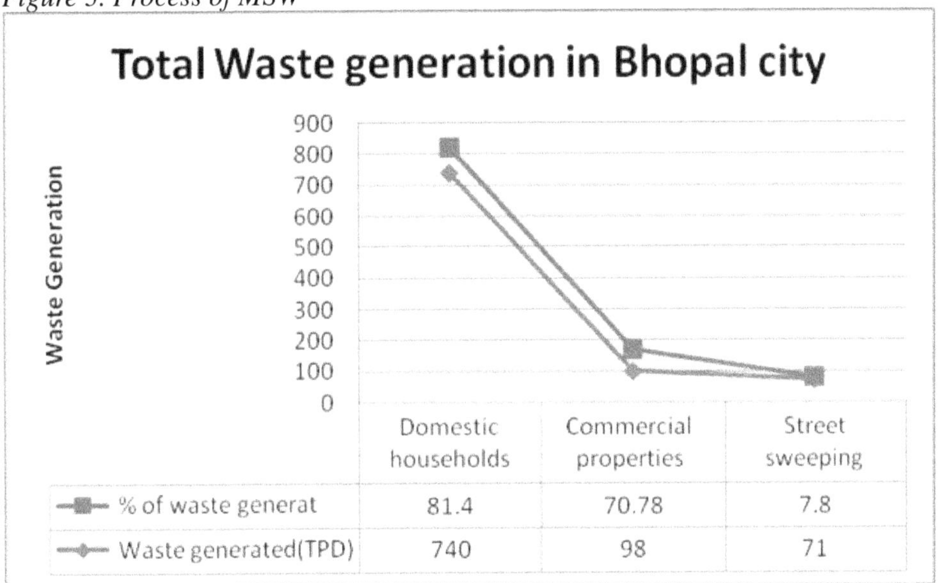

The figure above is a sequential process for municipal solid waste management, starting with collection door to door, where waste is collected directly from households or businesses. Then, the collected waste is transported for disposal to a municipal sub-station, which acts as an intermediary point for temporary storage or preliminary processing. At this stage, the wastes are classified into dry and wet, which is significant since it facilitates even more efficient recycling, composting, or disposal means depending on the type of waste. Finally, it is taken to the primary municipal station where it might be further processed, treated, or disposed of depending on the type of the waste. Schemes of structured waste management; efficient collection and staging will have to be backed with supporting waste segregation, and ultimately, sustainable practices of disposing waste.

The given table provides the details the quantum of waste generated across zones and wards. There are 19 zones and 85 wards, which have 850 TPD in aggregate. This brings the average to about 10 TPD per zone and 10 TPD per ward. This gives an idea of the quantum of waste being handled within the municipal structure and can be used to guide resource allocation and waste management planning across different areas.

Table 3. Waste generation from Bhopal city

Total Zones	Total Wards	Waste generation (TPD)
19	85	850

Figure 4. Total Waste generation in Bhopal

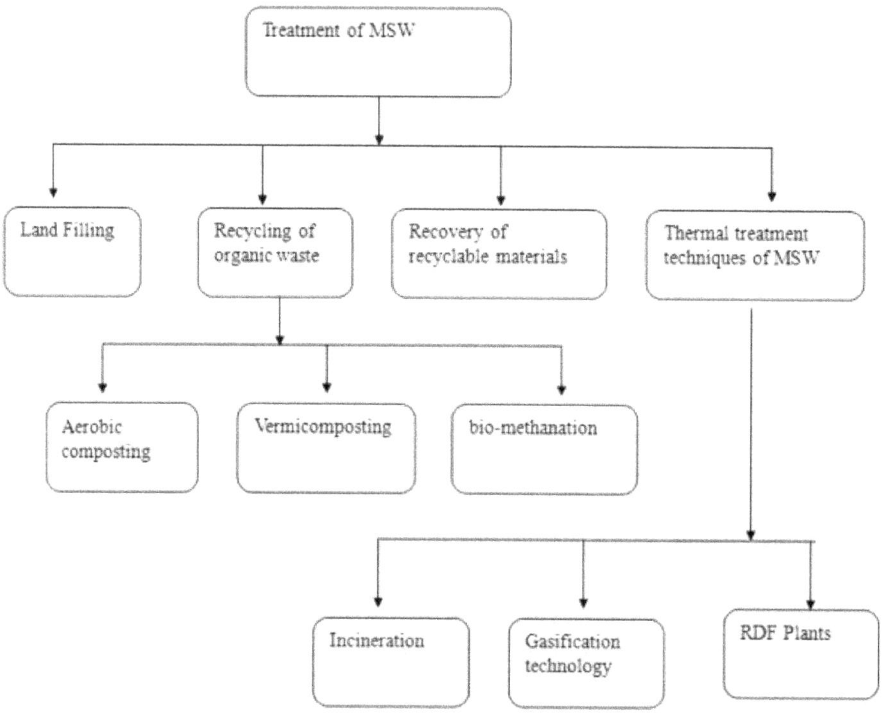

The graph shows the production of total wastes in the city of Bhopal that have been divided into domestic houses, commercial properties, and street sweeping. The domestic houses produce the largest wastes at 81.4%, equivalent to 740 TPD. The commercial properties account for 70.78% of the waste stream, which amounts to 98 TPD. Street sweeping has the smallest share, with 7.8% of the total waste generated at 71 TPD. This categorization clearly indicates that the maximum percentage of waste in Bhopal is generated from household households, followed by commercial properties, and street sweeping is the lowest. These results suggest that the focus should be on household waste management because it can have the most impact on the overall reduction of waste in the city.

1.3Types of disposal treatment system (MSW)

The two leading innovative mechanisms of waste disposal being adopted in India include

- composting (aerobic composting and vermi-composting) and
- waste-to-energy (WTE) (incineration, pelletisation, biomethanation).

The flowchart below discusses various methods of disposing of and treating Municipal Solid Waste (MSW) which are categorized under the general main category of MSW treatment. The first method under this category is land filling, where waste is put directly into designated areas for disposal as a first-generation approach. Other types of organic waste recycling process include two, namely: aerobic composting, where organic wastes are broken down in the presence of oxygen to form compost and vermicomposting, wherein earthworms break up the waste to produce nutrient-rich compost. The last one is the recovery of recyclable materials process in which reusable materials are derived from waste to reduce reliance on new resources. Thermal treatment techniques for MSW include heat application in the processing of waste. Under this category, bio-methanation is a biological process of converting organic waste into methane gas for fuel. Incineration is a process that involves burning waste at high temperatures, thereby reducing waste volume to a significant level and producing energy. Gasification technology is another thermal process where waste is converted into synthetic gas for energy production. Finally, RDF plants treat combustible fractions of waste into fuel. In total, all these technologies form part of an integrated system for the management and treatment of MSW, covering biological recycling, recovery, and thermal processing to better treat waste.

Different methods for the disposal and treatment of MSW are-

Figure 5. Methods for the disposal and treatment of MSW

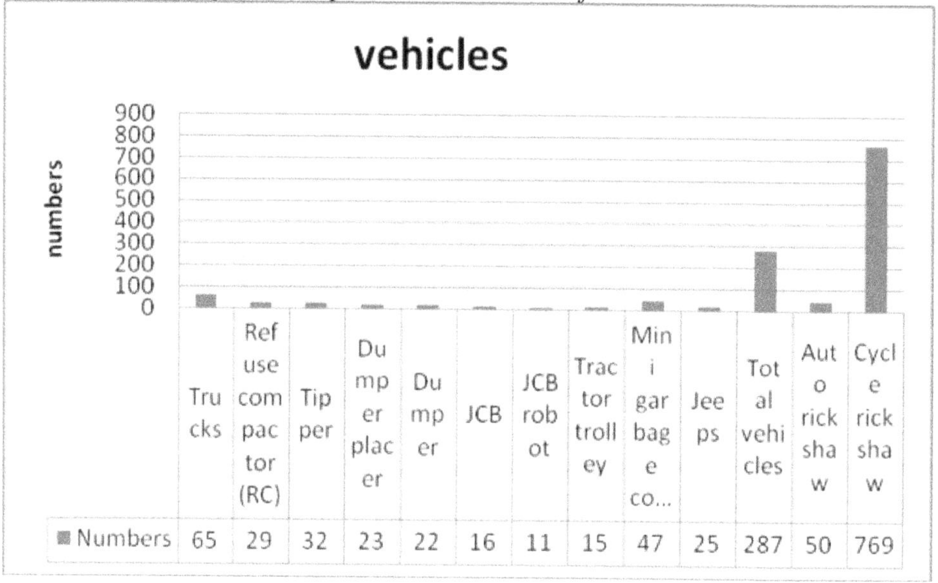

During the collection of municipal solid waste samples the major collection sites are identified which are covering a larger size of population. Based on the type of area such as residential, commercial, industrial, market, slum etc. sampling points are distributed uniformly all over the study area. There are 7 sampling points are considered which are listed as follows:

1. AIIMS Zone
2. Kolar road
3. DIG Bungalow
4. BhanpuraKunti
5. Sehore road Beraghat
6. Badbada
7. Danapani

1.4 Existing waste management practices in Bhopal

(Devi L et al., 2003)

Figure 6. Existing vehicles for waste collection and transportation in Bhopal

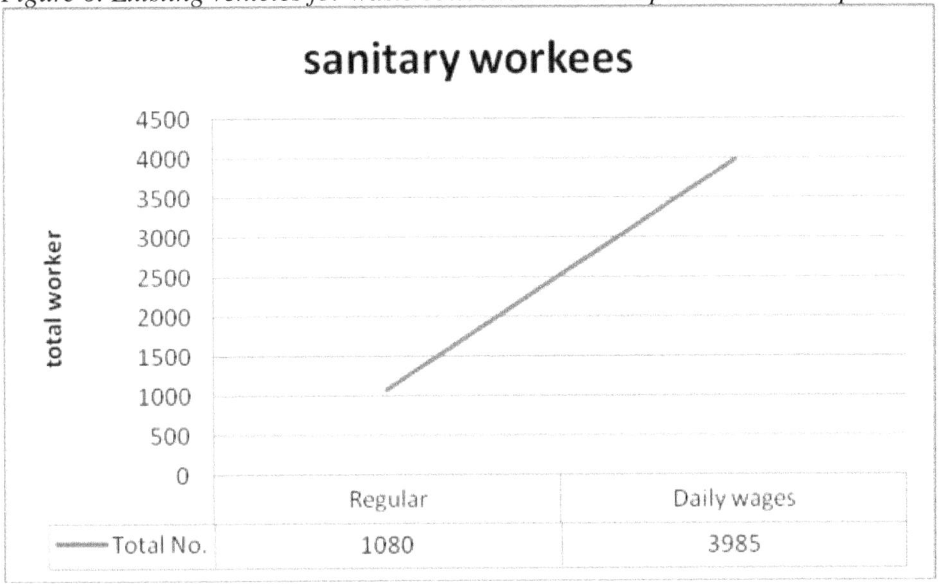

The bar graph is gives an overview of the kind and quantity of vehicles in use with specific counts in each category. Cycle rickshaws are the most widely used vehicle at 769, a figure far exceeding any other and, by extension, suggesting heavy use in operations. The rest has 287 total vehicles under the general category. If it is the sum of other categories, or added extra vehicles is not shown. There are also so many mini garbage collectors as many as 47 operational units. Other types of vehicles like trucks 65, refuse compactors 29, tippers 32, dumper placers 23, dumpers 22, JCBs 16, JCB robots 11, tractor trolleys 15, jeeps 25, and auto rickshaws 50 have fewer counts; they are more specialized and therefore less often used. That therefore indicates a system in which cycle rickshaws and mini garbage collectors probably do the routine, or door-to-door or even localized waste collection while all the other vehicles take any support roles, or high and large-scale transportation and eventually disposal.

Figure 7. Existing workers for waste collection in Bhopal

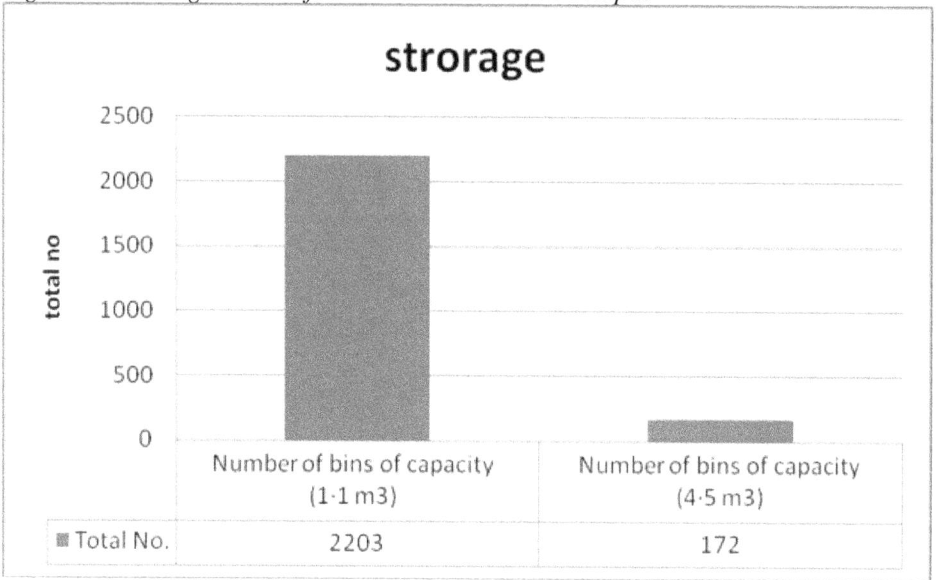

The above graph gives information regarding the number of sanitary workers employed. These sanitary workers are categorized into two different groups: regular and daily wage. There are 1,080 regular workers. They probably work permanently on a regular basis with stable pay and probably some benefits in addition. There are 3,985 daily wage workers in contrast, which shows the workforce is much more voluminous and employed on a temporary or as-and-when basis, and this would be earning most likely based on the number of days worked. Such strong dependence on daily wage workers-about four times the number of regular employees-strongly indicates flexibility in the labor structure which allows scaling labor resources on demand, which is generally common in sectors like sanitation where labor needs fluctuate.

Figure 8. Existing bins for waste collection in Bhopal

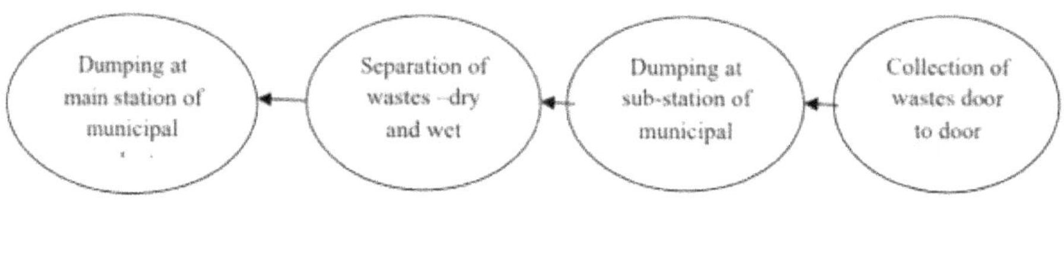

The given graph shows the number of waste storage bins that were distributed based on their capacity in cubic meters, m^3. Most bins have a capacity of 1-1 m^3. In total, there are 2,203 available bins. The smaller bins are probably distributed throughout different areas to allow for easy disposal of waste by residents and businesses. There are 172 bins, but with a capacity of 4-5 m^3, which means that these should be strategically located in busy areas or collection centers in order to deal with bigger volumes of waste. In this arrangement, it shows emphasis on accessibility by having a higher number of smaller bins and providing larger storage as needed to deal with larger waste loads.

1.5 Challenges Linked to MSW Disposal in the Context of Urbanization

Managing MSW is a complex and multifaceted issue affecting urban areas worldwide. The challenge is exacerbated by the rapid development in Bhopal City. The challenges related to MSW disposal in the urban context are numerous and require prompt attention (González J. F et al., 2009).

Firstly, the sheer volume of trash produced in Bhopal strains the current waste management system. Current methods of collecting, transporting, and disposing of waste materials struggle to handle the daily influx effectively (Hassan M. A et al., 2006). As a result, uncollected garbage accumulates on streets and in open landfills, leading to unsightly and unhygienic conditions, as well as potential health hazards for residents.

Additionally, it's crucial to note that conventional approaches to MSW management, namely landfilling and open dumping, are not sustainable in the long term. These disposal methods have severe consequences, including land shortages, environmental degradation, and groundwater contamination (KarthikeyanO. P et al., 2013). Sustainable biomaterials are used as substitutions for traditional materials in aerospace, automotive, civil, mechanical, environmental engineering, medical, and other industries (Kumar A el al., 2024). There is an urgent need for innovative and ecologically sustainable waste management solutions, and this chapter explores the potential of pyrolysis as a viable alternative.

1.6 Pyrolysis as a Waste-to-Resource Solution and Its Environmental Benefits

In the pursuit of sustainable and efficient waste management strategies, pyrolysis emerges as an attractive option. Pyrolysis is a thermal degradation process that converts organic materials, including various forms of waste, into economically valuable products such as biochar, bio-oil, and syngas. This technology has gained significant attention for its ability to address multiple challenges simultaneously (Li Y et al., 2015).

One of the key benefits of pyrolysis is its capacity to significantly reduce waste volume. It efficiently converts organic waste materials, a substantial component of MSW, into high-value biochar. Biochar, rich in carbon, has the potential to enhance soil fertility, sequester carbon, and mitigate greenhouse gas emissions when applied in agricultural settings (Lua A. C., 2008). Pyrolysis not only reduces the burden on landfills but also contributes to soil restoration and climate change mitigation.

Moreover, pyrolysis offers an environmentally-friendly method for producing bio-oil and syngas, providing potential alternatives to conventional energy sources. These products can reduce reliance on fossil fuels, leading to lower greenhouse gas emissions and decreased air pollution (Mohan, D et al., 2006). In the context of Bhopal, a city experiencing rapid urbanization and growing energy demands, adopting bio-oil and syngas from pyrolysis holds significant potential for enhancing energy security and environmental conditions (Ng, Y. L et al., 2011).

In addition to waste reduction and energy generation, pyrolysis offers a solution to hazardous waste management challenges (Ren X et al., 2019). Certain categories of MSW, such as plastics and electronic waste, contain hazardous components that pose significant risks to public health and the environment if mishandled. Pyrolysis can transform these hazardous substances into safer forms, mitigating the environmental risks associated with their disposal (ShafieeSet al., 2019).

This chapter investigates the viability of pyrolysis as an environmentally-friendly method for converting waste into valuable resources, particularly in the context of rapid urbanization in Bhopal City, located in Madhya Pradesh, India. Through a comprehensive examination of urbanization complexities, MSW disposal challenges, and the environmental benefits of pyrolysis, our goal is to provide insights into a potentially sustainable and environmentally-friendly urban future. By conducting an extensive analysis of existing scholarly literature and empirical research, the chapter also aims to clarify the practical feasibility and socio-economic implications of pyrolysis technology in Bhopal and similar metropolitan contexts.

1.7 Types of Combustion Process and Pyrolysis Process

Combustion process - Combustion is a chemical process in which a substance reacts rapidly with oxygen and gives off heat. The original substance is called the fuel, and the source of oxygen is called the oxidizer. During combustion, new chemical substances are created from the fuel and the oxidizer (Singh N et al., 2017).

Types of the combustion process-

Figure 9. combustion process

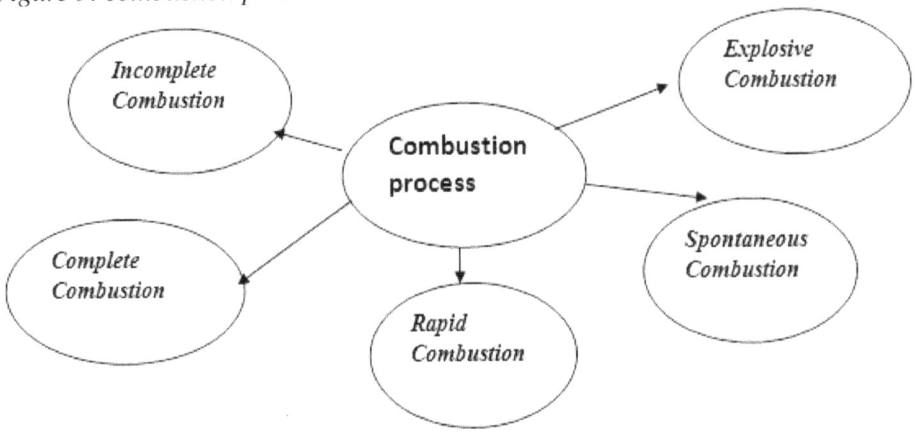

The above diagram classifies the types of combustion processes under the central concept of combustion process. There are five different types of combustion. Incomplete combustion results from insufficient oxygen, thereby leading to unburned fuel and possibly harmful emissions. Complete combustion occurs with the presence of abundant oxygen, allowing all fuel to be fully burned and maximum energy produced. Rapid combustion: It is defined as a combustion process that takes very high rates for a burning reaction in generating both high heat and visible light; it generally can be found in control conditions of laboratory-flame and gas burner. Explosive combustion: This term is used to define its onset without external aid, usually in form of ignitiation, typically because of an increase in temperatures within some given material that becomes susceptible to self-heat buildup. Each of these combustion types represents different conditions and reactions in which combustion can take place.

Itis defined as the form of processing bio- fuels by burning under very high temperatures without oxygen, which could cause complete combustion. This causes irreversible physical and chemical changes. The word is resulting from two Greek words: 'pyro', which means fire, and 'lysis', which means disintegration into integral parts. It is a simple, inexpensive technology for processing a wide variety of feed stocks. It reduces waste going to landfill and greenhouse gas emissions. It reduces the risk of water pollution (Singh, R. K et al., 2012).

Pyrolysis, which is also the first step in gasification and combustion, occurs in the absence or near absence of oxygen, and it is thus distinct from combustion (burning), which can take place only if sufficient oxygen is present. The rate of Pyrolysis increases with temperature (Sonne, C et al., 2015).

Types of Pyrolysis-

1. Depending on the heating rate Pyrolysis process can be classified as-
   i) Slow Pyrolysis ii) Fast Pyrolysis

Slow Pyrolysis- In slow Pyrolysis process, the time of heating the biomass substrate to Pyrolysis temperature is longer than the time of retention of the substrate at characteristic Pyrolysis reaction temperature.

Fast Pyrolysis- in fast Pyrolysis, the initial heating time of the precursors is smaller than the final retention time at Pyrolysis peak temperature

2. Based on mediumPyrolysis process can be classified as
   i) Hydrous Pyrolysis
   ii) Hydro-Pyrolysis

1.8 The Role of Computational Modeling in Waste Management

Efficiently addressing the challenges of MSW disposal, particularly in the context of a rapidly growing urban area like Bhopal, requires a multidisciplinary approach that incorporates advanced technology and computational methods. Computational modeling, a powerful tool in engineering and environmental sciences, plays a pivotal role in optimizing waste-to-resource processes like pyrolysis (Stark N. M. et al., 2017).

Computational modeling allows researchers and waste management professionals to simulate and analyze complex processes involved in pyrolysis. By developing mathematical models that represent the physical and chemical interactions within a pyrolysis reactor, it becomes possible to predict and optimize key parameters such as temperature, residence time, feedstock composition, and reaction kinetics (StrezovV et al., 2011). These insights are invaluable in designing efficient pyrolysis systems that can maximize resource recovery and minimize environmental impact.

With industry 4.0 technologies big data, digital twin, AI, machine learning and blockchain, waste management could see a revolution. In collection processes, the tool assists in real-time monitoring of waste and predictions about effective resource recovery. Hence the whole process of collecting wastes, sorting, recycling etc., becomes efficient with help of cyber-physical system and ICT. ICT coordinates accountability, and Blockchain-based transparency is enabled while following waste tracking. Together, these innovations form a more efficient, data-driven, and sustainable waste management system that responds to changes in demand and regulatory requirements, benefiting industries, municipalities, and the environment (Kumar A et al., 2023).

Furthermore, computational modeling enables a comprehensive assessment of various waste feedstocks. Different types of waste materials exhibit unique properties that affect their behavior during pyrolysis. Through simulations, researchers can evaluate the suitability of different feedstocks and identify the most efficient combinations for achieving desired outcomes, whether it's biochar production, energy generation, or pollutant minimization (SudhakarC et al., 2016).

1.9 Overview of the Review

This comprehensive chapter aims to provide a thorough examination of computational studies related to MSW pyrolysis in Bhopal City, India. We delve into recent advances, identify limitations, and suggest future research directions.

2. KEY ASPECTS OF MSW PYROLYSIS

Key aspects of municipal solid waste (MSW) pyrolysis encompass fundamental components such as reactor design, reaction kinetics, and feedstock composition:

Reactor Design: Computational modeling significantly contributes to optimizing the pyrolysis reactor. Computational models help determine the ideal reactor geometry, size, and configuration for efficient heat transfer, residence time control, and temperature profiles. Various reactor designs, such as fixed-bed, fluidized bed, and rotary kilns, are explored to find suitable options (Tagliaferro A et al., 2019).

Reaction Kinetics: Understanding pyrolysis reaction kinetics is crucial for process optimization. Computational models clarify the rate at which different feedstock components break down into valuable products like biochar, bio-oil, and syngas. This knowledge aids in fine-tuning operating conditions to maximize product yields and quality (Tang Y et al., 2019).

Feedstock Composition: MSW is a heterogeneous mix of organic and inorganic materials. Computational studies assist in characterizing the composition of MSW and its influence on the pyrolysis process. Different waste components necessitate varying conditions for efficient decomposition, and computational models predict how various feedstock compositions impact product distribution (Thakur P et al., 2016).

3. CHEMISTRY OF MSW PYROLYSIS

The chemistry of MSW pyrolysis is intricate, encompassing numerous reactions that transform organic materials into valuable products and, in some cases, pollutants:

Pollutant Formation Mechanisms: Computational studies elucidate the mechanisms behind pollutant formation during pyrolysis. These pollutants can include volatile organic compounds (VOCs), particulate matter, and toxic gases (TianN et al., 2019). Understanding these pathways enables researchers to develop strategies for pollutant minimization.

Valuable Product Production: Computational models optimize the production of valuable products like biochar, bio-oil, and syngas (Vasile Cet l., 2016). They predict how various process parameters influence product yields and properties, aiding in waste reduction and minimizing environmental impact.

4. ENVIRONMENTAL SUSTAINABILITY ASSESSMENT

Assessing the environmental sustainability of MSW pyrolysis is essential for evaluating its suitability as a waste management solution:

Environmental Impacts: Computational models quantify the environmental impacts associated with MSW pyrolysis. This includes assessing energy and resource requirements, as well as estimating emissions of greenhouse gases and other pollutants, providing a basis for comparing environmental performance (Wang J et al., 2016).

Carbon Footprints: Calculating the carbon footprint of MSW pyrolysis is crucial for addressing climate change. Computational models track carbon emissions and sequestration, especially when biochar is used in agriculture, reducing the overall carbon footprint.

Life Cycle Assessments: Computational tools facilitate comprehensive life cycle assessments (LCAs) to evaluate the entire environmental impact of MSW pyrolysis, from feedstock collection to product utilization and waste disposal. LCAs offer a holistic view of sustainability.

5. BARRIERS AND CHALLENGES IN MSW PYROLYSIS

MSW pyrolysis faces several barriers to widespread adoption, particularly in urban areas like Bhopal City (Xu J et al., 2015):

Regulatory Hurdles: Computational models assist in navigating regulatory complexities, predicting environmental compliance, and aiding in emissions control.

Scalability: Scaling up pyrolysis to industrial operations is complex. Computational modeling supports the design and optimization of large-scale facilities.

Economic Viability: Computational models assess the economic feasibility of MSW pyrolysis projects, including capital and operational costs, and revenue generation from product sales.

6. FUTURE DIRECTIONS

- **Model Refinement:** Future research should focus on refining computational models by incorporating more detailed and accurate data, enhancing predictive capabilities.
- **Validation with Real-world Data:** Validating computational models with real-world data from pyrolysis facilities will improve their reliability.
- **Local Context Consideration:** Adapting computational models to local contexts, accounting for waste composition, climate, and regulations, is essential [37].
- **Interdisciplinary Collaboration:** Collaboration among experts from various fields will drive innovation and address complex urban waste management challenges.

7. CONCLUSION

In conclusion, this comprehensive review has delved into the multifaceted realm of municipal solid waste (MSW) pyrolysis, emphasizing the pivotal role of computational modeling in advancing our understanding of this waste-to-resource conversion process. By analyzing various facets of MSW pyrolysis and its implications, we have underscored its potential as a sustainable solution for urban waste management, particularly in rapidly growing regions like Bhopal, India.

The exploration of key aspects, such as reactor design, reaction kinetics, and feedstock composition, has revealed how computational modeling assists in optimizing the pyrolysis process. These insights enable us to design efficient systems that reduce waste volume and transform organic materials into valuable biochar, bio-oil, and syngas.

Examining the intricate chemistry of MSW pyrolysis, we have elucidated the formation mechanisms of pollutants and the pathways to producing valuable products. Computational models have emerged as indispensable tools for achieving optimized yields and minimizing emissions, ensuring that waste management aligns with environmental sustainability goals.

Environmental sustainability assessment has emphasized the importance of evaluating the holistic impact of MSW pyrolysis. Computational methods enable the quantification of environmental footprints, carbon emissions, and life cycle assessments, ensuring that this approach is not only resource-efficient but also environmentally responsible.

Despite the promising prospects of MSW pyrolysis, this review acknowledges the existence of barriers and challenges, including regulatory hurdles, scalability issues, and economic considerations. Computational modeling is indispensable in navigating these challenges, offering insights for process optimization and techno-economic analysis.

Looking forward, the future of MSW pyrolysis research lies in refining computational models, validating them with real-world data, adapting them to local contexts, and fostering interdisciplinary collaboration. These efforts will not only enhance the efficiency of waste management but also promote a sustainable and environmentally friendly urban future.

In summary, this chapter has provided a comprehensive overview of MSW pyrolysis in the context of Bhopal, India, highlighting the transformative potential of this technology. By harnessing the power of computational research and addressing the complexities of waste management in rapidly urbanizing regions, we can move closer to realizing a circular economy that benefits both the environment and society.

REFERENCES

Al-Salem, S. M., Lettieri, P., & Baeyens, J. (2009). Recycling and recovery routes of plastic solid waste (PSW): A review. *Waste Management (New York, N.Y.)*, 29(10), 2625–2643. DOI: 10.1016/j.wasman.2009.06.004 PMID: 19577459

Annual Report on Solid Waste Management (2020-21), CPCB, Delhi

Bridgwater, A. V. (2012). Review of fast pyrolysis of biomass and product upgrading. *Biomass and Bioenergy*, 38, 68–94. DOI: 10.1016/j.biombioe.2011.01.048

Chen, D., & Yin, L. (2018). Pyrolysis technologies for municipal solid waste: A review. *Waste Management (New York, N.Y.)*, 79, 277–299.

Chen, Y., & Sharma, B. K. (2020). Recent advances in biochar applications in agricultural soils: Benefits and environmental implications. *Clean Technologies and Environmental Policy*, 22(5), 989–1005.

Cheng, Y., He, P., Yang, H., Wang, H., & Zhang, H. (2019). A review of biomass pyrolysis and hydro-pyrolysis for high-value chemicals. *Fuel Processing Technology*, 196, 106194.

Demirbas, A. (2004). Pyrolysis of municipal plastic wastes for recovery of gasoline-range hydrocarbons. *Journal of Analytical and Applied Pyrolysis*, 72(1), 97–102. DOI: 10.1016/j.jaap.2004.03.001

Devi, L., Ptasinski, K. J., & Janssen, F. J. (2003). A review of the primary measures for tar elimination in biomass gasification processes. *Biomass and Bioenergy*, 24(2), 125–140. DOI: 10.1016/S0961-9534(02)00102-2

Diwevdi, S. D. (2014). *Sheeraz Ahmad Wagay, Solid Waste Management InBhanpur Bhopal: Status And Need- A Case Study. Chemical And Process Engineering Research Issn 2224-7467 (Paper) Issn 2225-0913* (Vol. 18). Online.

Fawzy, M., & Sanaa, E. (2018). Review on fast pyrolysis of biomass: A potential technology for renewable energy and soil amendment. *Sustainable Environment Research*, 28(6), 283–288.

González, J. F., Encinar, J. M., & Canito, J. F. (2009). Pyrolysis of different organic waste from agro-industrial activities: Characterization and energy evaluation. *Journal of Chemical Technology and Biotechnology*, 84(8), 1242–1249.

Hassan, M. A., & Al-Malack, M. H. (2006). A review on oil mist lubrication systems. *International Journal of Machine Tools & Manufacture*, 46(12-13), 1279–1288.

Karthikeyan, O. P., Visvanathan, C., & Thayanithy, K. (2013). Management of urban biodegradable solid waste through biogas production: A review. *Renewable & Sustainable Energy Reviews*, 21, 978–988.

Kumar, A., Mittal, R. K., & Goel, R. (Eds.). (2023). *Waste Recovery and Management: An Approach Toward Sustainable Development Goals*. Press., DOI: 10.1201/9781003359784

Kumar, A., Rajak, D. K., Kumar, P., & Kumar, A. (Eds.). (2024). *Advances in Sustainable Biomaterials: Bioprocessing 4.0, Characterizations, and Applications* (1st ed.). CRC Press., DOI: 10.1201/9781003434313

Li, Y., Wang, B., Wang, Y., Yang, M., & Sun, R. (2015). Influence of pyrolysis temperature on properties of biochar obtained from pyrolysis of paper mill sludge. *Journal of Analytical and Applied Pyrolysis*, 111, 173–180.

Lua, A. C. (2008). Aromaticity and the degree of aromatic condensation of char. *Carbon*, 46(12), 1606–1612.

Mohan, D., Pittman, C. U.Jr, & Steele, P. H. (2006). Pyrolysis of wood/biomass for bio-oil: A critical review. *Energy & Fuels*, 20(3), 848–889. DOI: 10.1021/ef0502397

Ng, Y. L., & Ng, D. K. S. (2011). A review on the utilization of fly ash. *Progress in Energy and Combustion Science*, 37(4), 512–537.

Rajeev Singh Parihar. Anil Sharma, PrashantBaredar(2018) MSWMinBhopal city: a critical analysis and a roadmap for its sustainable management. Article in ICE Proceedings Municipal Engineer · February 2018 DOI: DOI: 10.1680/jmuen.17.00011

Ren, X., Li, X., Wang, S., Ma, X., & Zhang, Y. (2019). Recent advances in biomass pyrolysis for biochar, bio-oil, and bioenergy production. *Critical Reviews in Environmental Science and Technology*, 49(1), 40–71.

Shafiee, S., & Topal, E. (2009). When will fossil fuel reserves be diminished? *Energy Policy*, 37(1), 181–189. DOI: 10.1016/j.enpol.2008.08.016

Singh, N., & Balasubramanian, R. (2017). A comprehensive review on the pyrolysis of lignocellulosic biomass. *Renewable & Sustainable Energy Reviews*, 73, 68–86.

Singh, R. K., & Pant, D. (2012). Kinetics of thermal decomposition of municipal solid waste components. *Waste Management (New York, N.Y.)*, 32(6), 1120–1132.

Sonne, C., González, M. R., Solevic-Knudsen, T., Wenzel, H., Mosbæk, H., & Astrup, T. F. (2015). Life cycle assessment of thermal waste-to-energy technologies: Review and recommendations. *Waste Management (New York, N.Y.)*, 37, 104–115. DOI: 10.1016/j.wasman.2014.06.011 PMID: 25052337

Stark, N. M., & Rowlands, R. E. (2017). Pyrolysis as a recycling technique for plastics. *Current Opinion in Green and Sustainable Chemistry*, 6, 12–16.

Strezov, V., Evans, T. J., & Hayman, C. (2011). Thermal processing of sewage sludge to recover valuable metals: A critical review. *Renewable & Sustainable Energy Reviews*, 15(1), 471–481.

Sudhakar, C., & Linga, P. (2016). A review on thermal and catalytic pyrolysis of plastic solid waste (PSW). *Environmental Progress & Sustainable Energy*, 35(3), 779–799.

Tagliaferro, A., Freda, C., & Sánchez-Silva, L. (2019). A critical review on the recycling of liquid crystal displays. *Journal of Cleaner Production*, 236, 117596.

Tang, Y., & Niu, M. (2019). The characteristics and environmental effects of plastics additives. *Environmental Chemistry Letters*, 17(2), 541–558.

Thakur, P., Anand, P., Kumar, A., & Vijay, V. K. (2016). Impact of feedstock characteristics on bio-oil yield and product distribution during fast pyrolysis. *Energy Conversion and Management*, 108, 373–382.

Tian, N., Deng, S., Zhang, S., Wang, H., Li, Z., & Zhang, Y. (2019). Biochar made from hydrothermal liquefaction of biomass: A review. *Renewable & Sustainable Energy Reviews*, 108, 268–279.

Bundela, V. S., & Patel, D. A. (2022). MSW Management and Best Suitable Option Virtually Waste to Energy Plant for Resource and Energy Recovery for Bhopal (MP) India. *International Journal of Management and Humanities*, 8(11), 1–15.

Vasile, C., Stan, E. L., & Popa, M. V. (2016). Biochar—A sustainable approach for improving plant growth and soil properties. *Journal of Soils and Sediments*, 16(3), 883–897.

Wang, J. Y., Zhang, J., Bai, X. M., & Luo, Y. (2016). Catalytic pyrolysis of lignin for the production of phenolic-rich bio-oil. *Bioresource Technology*, 216, 71–76.

Xu, J., Zhang, X., Zhou, Y., & Jiang, G. (2015). Fast pyrolysis of biomass for bio-oil production: A review. *Journal of Engineering for Gas Turbines and Power*, 138(10), 101801.

ZairaZaman Chowdhury. Sagadevan Suresh (2017) Pyrolysis: A Sustainable Way to Generate Energy from Waste. , pp 1-36DOI: 10.5772/intechopen.69036

Chapter 7
A Literature Review on Recycle Asphalt Pavement Material Used in Infrastructure

Ram Vilas Meena
JECRC University, India

Mohd Waseem
https://orcid.org/0009-0002-9651-2918
JECRC University, India

Ankit Singh Beniwal
https://orcid.org/0000-0001-8655-903X
JECRC University, India

Kuldeep Singh Kulhar
https://orcid.org/0000-0002-2993-8858
JECRC University, India

ABSTRACT

In the highway or infrastructure, the requirement of raw material is increase day by day as per traffic load and design requirement with sustainable development. In the flexible pavement the life cycles an approximately 4 to 5 years and also required maintenance, that all activity increases the cost of project. To reduce the cost of project use "Reclaimed Asphalt Pavement Material" (RAP). The RAP material can be used in bearing coarse and sub-base coarse. By using the RAP material minimize the cost of project also use of useless material or eco-friendly part of construction. Use of RAP material in BC (Bituminous Concrete) & DMB (Dense Bituminous

DOI: 10.4018/979-8-3693-5513-8.ch007

Macadam) by replacing the virgin raw material (Aggregate & bitumen) and design the BC layer & DBM layer for road pavement. The review research "A Literature Review on RAP material use in surface, base and sub-base of flexible pavement" is presented in this document.

1. INTRODUCTION

The waste from bituminous roads that makes up the reclaimed asphalt pavement material is significant. Old materials were removed during the renovation and resurfacing of the road surface in order to restore the existing asphalt pavement. Bitumen and aggregate make up the top surface material that is removed, resulting in recycled asphalt pavement (RAP). Construction and demolition waste are generated from various construction and demolition activities like renovation, buildings, roads, repair, bridges, flyover etc (Chouhan et al., 2019). Construction and demolition garbage is typically improperly disposed of in a disorganized way in developing nations like India, which contributes to environmental problems. This material can be used for sub-base of pavement with soil to improve the bearing capacity and permeability properties of soil as well as for resurfacing, repaving, and reconstruction of pavement in BC and DBM layer (Vilas, Kumar, Singh, et al., 2022). The most common building projects, requiring thousands of tons of aggregates annually, are highway constructions. Natural resources are severely strained by it (Vilas, Kumar, & Singh, 2022). In which sand and aggregate are obtained from disintegration of rocks. The major difference between sand and aggregate was their particle size. If the particle size is greater than 4.75 mm called coarse aggregate and less than 4.75 mm called fine aggregate (Meena et al., 2023). Despite the scarcity of natural resources, research have shown that using recycled asphalt in pavement building is highly beneficial. If a pavement fails or becomes unusable, the aggregates that are currently in it are natural resources taken from the environment, and they shouldn't just be thrown away at dump sites because doing so will harm the environment and make it worse by forcing the extraction of additional new aggregates from the environment to replace the wasted materials(Vilas Meena, Singh Chouhan, et al., 2021) [2]. Utilizing the bitumen that is already present in the pavement provides economic justification because it reduces the need for additional bitumen (Kumawat et al., 2023). Crushed stone aggregates are blended in the proper proportion to provide the required strength since RAP by itself does not meet IRC requirements for the strength of base and sub-base material because of the gap gradation of aggrade sand. This model of recycling has a positive effect on the environment as it decreases the volume of deposited waste and control excess mining of aggregates (Vilas Meena, Kumar Jain, et al., 2021). It is noted that RAP materials can replace the fresh aggregate for road pavement construction up to 40%-60% with satisfied performance (Sunarjono & Hidayati, 2019).

2. LITERATURE REVIEW

(Xu et al., 2021) According to author the RAP material use in concrete and check the mechanical properties for cold resign, the RAP material 30% used and 5.5% cement dosage use and prepared a reclaimed asphalt aggregate concrete and testing the mechanical properties of concrete like axial compressive strength, shrinkage limit in cold resign (-10 to 0°C) and spilt tensile strength of the concrete and observed at the 30% RAP material and 5.5% cement dosage strength and shrinkage limit good as compare to other proportion. The cement dosage if increase then the heat of hydration increases and shrinkage limit affected. (Rahman et al., 2020)According the author, potholes or patch work should be done by the RAP material, However, the maintenance coast will be reducing and natural resources also save, the RAP material mixture has more stability, flow value and sufficient air voids to maintain the stability of pavement. (Raouf Dar & Singh, 2020) By study this paper the useless material or waste material use in semi dense bituminous concrete (SDBC) replaced with bitumen content made the bituminous concrete and other building purpose. The range of waste material 5% to 9% used for bituminous concrete at the 5% stability and flow maximum but by increasing the dosage flow increase and stability decrease at 9%. By use of plastic waste increase the property of bituminous concrete and reduce the cost of project and also it is eco-friendly activities. (Sunarjono & Hidayati, 2019)According to the author the physical properties are differ from fresh material and also described the aged bitumen change the property of RAP material as well as water absorption low, density low, specific gravity low and also shape of aggregate is rounded due to aged bitumen on the aggregate. Mostly the grain size distribution in the RAP material the coarser affected but finer is not affected which is same as fresh material for blending in job mix. The author suggested the RAP material used in dense bitumen macadam. (Mazzoni et al., 2018) In which paper described rejuvenators used in bitumen and analysis the rheological properties of virgin bitumen and RAP bitumen content. Also the viscosity and phase angle of the bitumen in virgin and RAP bitumen content analysis and observed the phase angle is high in RAP bitumen content. (Bida et al., 2016) In which study and work, the RAP material use by replacing fine aggregate in concrete 0 to 100%. The workability, density and compressive strength determine and compare with specified standard values. The water cement ratio in concrete decrease with increase of RAP material Percentage in concrete. (Zaumanis et al., 2016) According to the author the RAP material is use approximately 40% replacement in flexible pavement and some other industry innovative worked on 100% replacement for constructed a flexible pavement and also fixed rejuvenator percentage used in hot mix reclaimed asphalt pavement material to produce workable mixture at the site. The author mixes the rejuvenator in performance grade (PG) bitumen and rheological analysis and observed if rejuvenator is used un 100% RAP material we can produce satisfactory performance as well as virgin bitumen mix. (El-Maaty & Elmohr, 2015)In this paper the author described the RAP material uses and benefit of the uses in the construction like to minimize the cost of project by use of RAP material replacement by other virgin material. And also tested the Marshall Stability and flow value of bituminous mix. (Mathias et al., 2004) According to the author the RAP material can be used in concrete road construction with vibrator compaction. The properties of RAP material is affected by layer thickness of pavement. And also says the RAP material uses is rehabilitation not recycle. The RAP material use in concrete has more advantage in road construction and structure. The characteristic of the concrete decrease with increase of RAB martial which is replaced by total mass of aggregate, however, the compressive strength, tensile strength, modulus of elasticity, crack test and fatigue test in lean concrete at different temperature shows by graphs. (Mary chyne et al., 2019) According to the author the physical properties of the RAP material is weak as compared to the natural material but some properties is satisfactory as compare to the virgin material. For example, the specific gravity is high natural aggregate and less in RAP material aggregate. And the strength properties decrease with increase of RAP material percentage. But in the slump test or workability is more as compare to the natural aggregate and the water permeability of RAP used concrete is high which cause corrosion in RCC if used in RCC. (Sunarjono & Hidayati, 2019) According to the author the physical properties are differ from fresh material and also described the aged bitumen change the property of RAP material as well as water absorption low, density low, specific gravity low and also shape of aggregate is rounded due to aged bitumen on the aggregate. Mostly the grain size distribution in the RAP material the coarser affected but finer is not affected which is same as fresh material for blending in job mix. The author suggested the RAP material used in dense bitumen macadam. Another study by (Zeeshan et al., 2017) found that the compressive strength decreased with time when they used crushed brick and marble as a partial aggregate replacement for concrete. Asphalt binder and blends were examined by (Bajaj et al., 2020) using rheological and chemical techniques, and different recycling agents were categorized into three groups. To forecast the efficacy of different recycling agents, a unique metric was created. In order to investigate the behavior of recycled asphalt pavement (RAP) aggregate concrete, (Sigurdsson Houghton B. & Wedge, 2000) used a variety of combinations, including natural aggregate, reclaimed coarse aggregate, reclaimed coarse and fine aggregate, and reclaimed coarse aggregate with 30% fly ash. chosen to determine the best possible performance. According to the findings, the RAP aggregate somewhat degraded the concrete's mechanical properties, including its tensile and compression strengths. Nevertheless, fly ash and RAP aggregate had a positive impact on the microstructure and ductility characteristics.

3. FUTURE SCOPE

The main purpose of this study is the utilization of wastes RAP 100 percentage substitutions as a substitute material to the natural aggregate in pavement construction. Even these properties with RAP tile are still missing. RAP material use with another waste material like marble slurry, waste Kota and other waste materials give batter result.

4 CONCLUSION

Based on earlier published research we can conclude that using the RAP material we can minimize the cost of project and sustainable development. The RAP material also can be used in concrete. Surface of flexible pavement and base of highway pavement by replacing of aggregate and bitumen binder. We can also use in bituminous concrete and dense bituminous macadam in 10% to 30% by replacing of virgin material. For further research work, we will use RAP material above 30% in bituminous concrete & bituminous macadam and we will test this mix material and try to achieve specification criteria as per rural roads guidelines. If we will succeed in our research then we can reduce more cost of construction of rural pavement.

REFERENCES

Bajaj, A., Epps Martin, A., King, G., Glover, C., Kaseer, F., & Arámbula-Mercado, E. (2020). Evaluation and classification of recycling agents for asphalt binders. *Construction & Building Materials*, 260, 119864. DOI: 10.1016/j.conbuildmat.2020.119864

Bida, S. M., Danraka, M., & Mohammed, J. (2016). Performance of Reclaimed Asphalt Pavement (RAP) as a Replacement of Fine Aggregate in Concrete. [IJSR]. *International Journal of Science and Research (Raipur, India)*, 5(4), 65–67. DOI: 10.21275/v5i4.NOV162395

Chouhan, H. S., Kalla, P., Nagar, R., & Gautam, P. K. (2019). Gainful utilization of dimensional limestone waste as fine aggregate in cement mortar mixes. *Construction & Building Materials*, 221, 363–374. DOI: 10.1016/j.conbuildmat.2019.06.097

El-Maaty, A. E. A., & Elmohr, A. I. (2015). Characterization of recycled asphalt pavement (RAP) for use in flexible pavement. *American Journal of Engineering and Applied Sciences*, 8(2), 233–248. DOI: 10.3844/ajeassp.2015.233.248

Kumawat, V., Meena, R. V., Beniwal, A. S., & Gupta, G. (2023). *Materials Today : Proceedings " Utilization of Kota stone slurry along with demolition and construction waste in concrete ". xxxx*. DOI: 10.1016/j.matpr.2023.06.103

Mary chyne, J., Kiran Sepuri, H., & K, T. H. (2019). A Review on Recycled Asphalt Pavement in cement concrete. *International Journal of Latest Engineering Research and Applications (IJLERA), 04*(February), 9–18. www.ijlera.com

Mathias, V., Sedran, T., & Larrard, F. de. (2004). *Recycling reclaimed asphalt pavement in concrete roads. November 2004*, 66–75.

Mazzoni, G., Bocci, E., & Canestrari, F. (2018). Influence of rejuvenators on bitumen ageing in hot recycled asphalt mixtures. [English Edition]. *Journal of Traffic and Transportation Engineering*, 5(3), 157–168. DOI: 10.1016/j.jtte.2018.01.001

Meena, R. V., Beniwal, A. S., Jain, A., Choudhary, R., & Mandolia, R. (2023). Evaluating resistance of ceramic waste tile self-compacting concrete to sulphuric acid attack. *Construction & Building Materials*, 393(March), 132042. DOI: 10.1016/j.conbuildmat.2023.132042

Rahman, M. T., Mohajerani, A., & Giustozzi, F. (2020). Recycling of waste materials for asphalt concrete and bitumen: A review. *Materials (Basel)*, 13(7), 1495. Advance online publication. DOI: 10.3390/ma13071495 PMID: 32218261

Raouf Dar, A., & Singh, E. S. (2020). Use of Recycled Materials in Pavement Construction. *IJSDR2008038 International Journal of Scientific Development and Research, 5*(8), 295–304. www.ijsdr.org

Sigurdsson Houghton, B. M. S. R. H. y S. J. H., & Wedge, F. M. (2000). No Title. *Encyclopedia of Volcanoes., 3*, 662.

Sunarjono, S., & Hidayati, N. (2019). Physical properties of reclaimed asphalt pavement. *IOP Conference Series. Materials Science and Engineering*, 674(1), 012029. Advance online publication. DOI: 10.1088/1757-899X/674/1/012029

Vilas, R., Kumar, J., & Singh, A. (2022). " *Sustainable self-compacting concrete containing waste ceramic tile aggregates : Fresh, mechanical, durability, and microstructural properties ." 57*(March).

Vilas, R., Kumar, J., Singh, H., & Singh, A. (2022). Use of waste ceramics to produce sustainable concrete : A review. *Cleaner Materials*, 100085, 100085. Advance online publication. DOI: 10.1016/j.clema.2022.100085

Vilas Meena, R., Kumar Jain, J., & Singh Chouhan, H., mandolia, R., & Singh Beniwal, A. (. (2021). Impact of waste ceramic tile on resistance to fire and abrasion of self-compacting concrete. *Materials Today: Proceedings*, xxxx. Advance online publication. DOI: 10.1016/j.matpr.2021.12.287

Vilas Meena, R., Singh Chouhan, H., Kumar Jain, J., & Kumar Satankar, R. (2021). Construction and Demolition Waste as an Alternative of Rigid Pavement Material: A Review. *IOP Conference Series. Earth and Environmental Science*, 795(1), 012019. DOI: 10.1088/1755-1315/795/1/012019

Xu, W., Li, W., & Ji, Y. (2021). Mechanical behavior investigation of reclaimed asphalt aggregate concrete in a cold region. *Materials (Basel)*, 14(15), 4101. Advance online publication. DOI: 10.3390/ma14154101 PMID: 34361295

Zaumanis, M., Mallick, R. B., & Frank, R. (2016). 100% Hot Mix Asphalt Recycling: Challenges and Benefits. *Transportation Research Procedia*, 14(June), 3493–3502. DOI: 10.1016/j.trpro.2016.05.315

Zeeshan, K., Mirza, B. A., Student, M., Khan, Z., & Baig, M. A. (2017).. . *Utilization of Demolish Brick and Marble Waste in Concrete For Partial Replacement of Coarse Aggregate.*, 3, 1702–1707.

Chapter 8
Investigation of Pozzolanicity in Waste Excavation Soils From Cappadocia and Prospective Industrial Applications of Waste Earth:
Pozzolanicity in Cappadocia Waste Soils and Industrial Uses

Serife Ozata
https://orcid.org/0000-0003-1993-7592
Samsun University, Turkey

Serhan Ulukaya
Yildiz Technical University, Turkey

Nabi Yuzer
Yildiz Technical University, Turkey

ABSTRACT

This study assesses the pozzolanic activity of Cappadocia volcanic waste soils (CWs) using established methodologies and explores their potential for diverse applications. Targeted research aims to reduce visual pollution, and repurpose waste into

DOI: 10.4018/979-8-3693-5513-8.ch008

economic resources and engineering materials, thereby safeguarding the cultural and natural heritage of the region. While Frattini test results confirm pozzolanic traits in most CWs, discrepancies arise in the strength activity index that necessitate further investigation. Compressive strength and electrical conductivity tests emerge as suitable alternatives, albeit with varying durations. Despite methodological variations, all CW samples exhibit pozzolanic properties, with CW4 showing the highest pozzolanicity. This study features the versatility of CWs, emphasizing their role and economic feasibility in construction, environmental engineering, agriculture, and industrial applications. Harnessing these materials promotes sustainable waste management and heritage conservation, offering a promising solution to the region's waste challenges.

INTRODUCTION

The comprehensive understanding of sustainable engineering materials is paramount in addressing the contemporary global challenges of environmental degradation and resource depletion. Thus, there is a pressing need for advancements that facilitate the transition towards pollution-free environments and reduce the environmental impact of current engineering materials. In this context, the utilization of waste excavation earth from Cappadocia as a pozzolanic material presents an innovative solution that resonates with the principles of environmental stewardship and resource efficiency. Pozzolan can be defined as a siliceous or siliceous-aluminous material that, on its own, has minimal or no cementitious value. However, when finely divided and in the presence of water, it can react with calcium hydroxide (lime) at ambient temperatures to produce compounds with cementitious properties (Becerra-Duitama & Rojas-Avellaneda, 2022). The mass percentage of $SiO_2 + Al_2O_3 + Fe_2O_3$ in this raw excavated earth from Cappadocia meets the Turkish Standards Institute [TS] requirement of exceeding 70% for natural pozzolans (TS 25, 2008). Thus, this study aims to explore the potential of utilizing waste excavation soil as a pozzolanic material in industrial applications, with a primary focus on investigating the material's pozzolanic properties. Additionally, this research seeks to contribute to the discourse on sustainable material solutions by emphasizing environmental responsibility, particularly through on-site waste reduction and cost-effectiveness.

The buildings and construction sector, responsible for an astounding 37% of global greenhouse gas emissions, significantly contributes to the carbon footprint through the manufacturing and utilization of various building materials (United Nations Environment Programme & Yale Center for Ecosystems + Architecture, 2023). Among these, cement production alone accounts for 5-8% of global CO_2 emissions and is a major contributor to the sector's greenhouse gas output (Mikulčić et al.,

2016). This analytical research and scientific inquiry not only strive to integrate sustainable waste materials into engineering and architectural practices, thereby promoting environmental responsibility, but also seek to mitigate climate change by enhancing the utilization of newly identified supplementary cementitious materials (SCMs), specifically intended for use in the construction industry. Supplementary cementitious materials (SCMs), which are generally used by mixing with materials such as cement that have a certain amount of binding properties, are a group that includes materials that show hydraulic and/or pozzolanic behavior. With growing awareness of global warming, numerous strategies have been proposed for using supplementary cementitious materials (SCMs) as alternatives to cement, as well as for producing cement-free concrete, mortar, and various building elements, aimed at promoting cleaner production methods. SCMs, which exhibit hydraulic or pozzolanic behavior, are crucial for reducing anthropogenic greenhouse gas emissions and enhancing the sustainability of construction materials. They comprise a diverse array of materials with varying origins, chemical compositions, mineralogical compositions, and particle properties. Two categories of SCMs can be delineated, including materials of natural origin and those of man-made or artificial origin. The first category comprises materials that can serve as SCMs in their natural state, while the latter category consists of artificial SCMs that have undergone structural changes through production and manufacturing processes. An overview of this classification is illustrated in Figure 1. While natural SCMs typically undergo only sieving and grinding processes for use, artificial SCMs, such as metakaolin, result from the thermal activation of kaolin clays. Materials like blast furnace slag, fly ash, or silica fume are obtained as waste or by-products from production processes involving high temperatures.

The diversity seen in SCMs of volcanic origin in Figure 1 is a reflection of the wide content variability of the outer eruptive rocks and the natural variability in magmatic and metamorphic processes. The widespread occurrence of volcanoclastic layers (Aydan & Ulusay, 2003; Ayhan, 2004), pumice, ashes, and zeolitic tuffs (Özata, 2015, 2018) across the Cappadocia Region indicates its potential as a source of SCMs derived from materials of volcanic origin.

Figure 1. The classification of supplementary cementitious materials (SCMs) (Snellings et al., 2012)

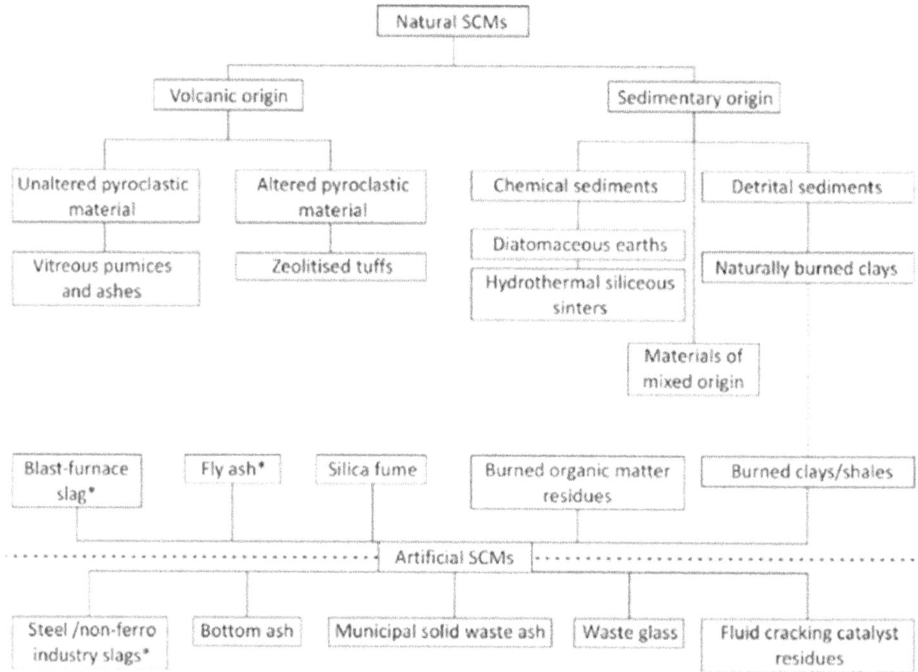

Volcanic rocks, abundant in many countries, are the primary source of natural pozzolans utilized across various sectors. In Figure 2, which marks the location of Cappadocia (in red), the global distribution of volcanic rocks and reported natural pozzolan deposits is depicted. It is evident that these resources are widely distributed worldwide. However, it is important to note that while volcanic rocks are widespread globally, not all of them exhibit pozzolanic activity. Thus, the investigation into the pozzolanicity of waste soils conducted within the scope of this research has the potential to inspire and inform strategies for addressing cement-related challenges and waste management issues in other countries facing similar concerns.

Figure 2. The widespread presence of volcanic rocks, along with deposits of natural SCMs, and the location of Cappadocia, Türkiye

In Türkiye, natural pozzolan reserves span approximately one-fifth of the country's area, totaling approximately 155,000 km^2 (Türkmeno lu & Tankut, 2002). This extensive coverage suggests that volcanic tuff materials, with their abundant reserves in Türkiye, hold significant potential for applications in the construction sector, agriculture, and industry, depending on their composition (Balog et al., 2014; Özata & Arun, 2018; Türkmeno lu & Tankut, 2002).

While the utilization of volcanic materials in construction has a long history, the exact origins of this practice remain uncertain. Kitsopoulos and Dunham (1996) highlight the extensive use of high-silica volcanic ash at the Akrotiri site in Santorini, Greece, which boasts solid archaeological evidence of Minoan culture. However, there are also indications of volcanic material usage in the Mesopotamian Region during the Neolithic period (7000 BC), predating the Ancient Greek and Roman civilizations (Malinowski & Garfinkel, 1991). Further evidence suggests that the deliberate use of siliceous volcanic soil in Santorini by the Ancient Greeks can be traced back to at least 500-400 BC (Efstathiadis, 1978). Approximately five hundred years later, Vitruvius devised the term "*Pulvis Puteolanis*" for this material, describing it as "a type of powder that inherently yields outstanding results" (Vitruvius, 2017[1]). This substance, used to produce concrete with greater strength and durability than lime, as per Vitruvius's definition, transformed Roman concrete into a versatile building material crucial for the empire's development. This innovative mortar-concrete composite found widespread application from Hadrian's Wall in Northern England to the Rhine and Danube Valleys, Turkey, Syria, Israel, Egypt, North African coastal regions, Spain, and Southern France, but most notably in

Rome (Idorn, 1997). Iconic structures such as the Colosseum and the Pantheon stand as a testament to the effectiveness of this material. Despite the restricted access to books in medieval Europe, knowledge found alternative means of dissemination. The combination of lime with specific volcanic materials, pozzolans, imparts hydraulic properties to mortar and concrete, enabling them to harden underwater and significantly increase their mechanical strength. This technique was historically utilized in Italy, particularly for constructing thermal springs and residences dating back to the 2nd century BC (Rispoli et al., 2019). A similar pozzolanic mortar, incorporating natural pozzolan, was identified in the construction of an ancient Roman harbor at Soli-Pompeiopolis (Stanislao et al., 2011). It has also been proven that pozzolan-based hydraulic mortars, previously used in the Hagia Sophia in Turkey and the medieval castle in Rhodes, Greece, can withstand both static and dynamic loads over extended periods (Moropoulou et al., 2004). Natural pozzolanic materials, along with various aggregates and admixtures such as gypsum, stone dust, sand, and organic additives, have been prioritized in modifications to lime-based mortar and plaster preparation. These adaptations, observed in regions such as India, China, and Syria, aim to enhance durability and improve the compressive strength of construction materials (Tripathi et al., 2023). However, with the advent of Portland cement patenting in 1824, natural pozzolan materials which had been utilized since ancient times, began to recede into the background worldwide (Balog et al., 2014). In Türkiye, post the mid-1900s, clinker was gradually replaced with natural and artificial pozzolans in varying proportions, surpassing a 35% utilization rate by 1992 (Kaplan & Binici, 1995). In the 21st century, various types of supplementary cementitious materials (SCMs) are employed in concrete and cement production for their hydraulic or pozzolanic properties, expanding beyond cement to encompass other innovative materials.

The utilization of natural and artificial pozzolanic materials especially in cement and concrete production has been extensively studied to enhance the properties of these construction materials and address environmental concerns. Upon reviewing the literature, it is evident that while numerous studies have extensively explored the properties, performance, and applications of artificial supplementary cementitious materials (SCMs) such as fly ash, ground granulated blast furnace slag, silica fume, and metakaolin (Black, 2016; Manjunatha et al., 2021; Tural et al., 2024), research on natural SCMs remains comparatively limited. The topics explored in the studies, particularly focusing on the natural pozzolan reserves in Türkiye, which have the potential to be used as SCMs, include the following.

Türkmenoğlu and Tankut (2002) investigated the use of tuffs from the Galatean Volcanic Province in central Turkey as admixtures in pozzolanic types of cement. They concluded that the tuffs from this province generally have suitable chemical and mineralogical properties to be used as pozzolanic admixtures in cement, with the degree of alteration being an important factor affecting their performance. Turanli et al. (2004) discussed the impact of different natural pozzolans on the properties of blended cements containing a high volume (55%) of natural pozzolans. Their research found that the early strength of the mortars was significantly influenced by the particle size distribution of the blended types of cement, whereas the long-term strength development was more closely related to the pozzolanic activity of the natural pozzolan. This study underscores the importance of optimizing particle size and pozzolanic activity to achieve desirable cement properties. Pekmezci and Akyüz (2004) focused on the efficiency of using natural pozzolan as a partial replacement for cement in concrete. The study aimed to determine how natural pozzolan affects concrete properties such as compressive strength and durability. The findings indicated that natural pozzolan can improve the long-term strength and durability of concrete, highlighting its potential as a sustainable alternative to traditional cement. Also, the use of blends comprising tincal waste, fly ash, bentonite, and volcanic tuff as cement admixtures. The study assessed various properties, including setting time, expansion, water requirement, specific surface, and compressive strength (Abali et al., 2006). The results demonstrated that these blends could effectively improve the performance of cement, making them viable alternatives for sustainable construction practices. Topçu et al. (2008) utilized artificial neural networks (ANN) and fuzzy logic (FL) systems to predict the strength development of cement mortars containing different pozzolanic materials. The study included the use of natural zeolite (clinoptilolite), Eskişehir trass, Nevşehir trass, fly ash, and ground granulated blast furnace slag at various replacement ratios. The predictive models developed demonstrated high accuracy, showcasing the potential of ANN and FL systems in optimizing the formulation of blended cement. In another research (Kurudirek et al., 2010) the chemical composition and radiation attenuation properties of clinoptilolite-rich natural zeolite (CRNZ) from Turkey were examined. Their research highlighted the suitability of CRNZ not only as a pozzolanic material but also for its potential applications in radiation shielding, thus expanding the scope of its utility in the construction industry. Küçükyıldırım and Uzal (2014) explored the impact of calcination on the properties and performance of natural zeolites when used as cement replacement materials. Their study found that calcination significantly enhances the pozzolanic activity of natural zeolites, making them more effective in improving the mechanical properties and durability of cementitious materials. The feasibility of using zeolite-rich tuffs as pozzolanic materials for manufacturing eco-sustainable blended types of cement was investigated by Liguori et al. (2015). The study identified several advantages of zeolite-rich tuffs, such as their excellent pozzolanic behavior, low cost, and potential for reducing carbon dioxide emissions and improving energy efficiency in cement production. These findings support the use of zeolite-rich tuffs in developing greener cementitious materials.

Globally, numerous studies have explored the use of natural pozzolans across different sectors and their application as components in various materials. Wahedy et al. (2023) investigate the cementitious potential of a vernacular Afghan natural pozzolan, noting that the resulting mortar exhibits superior freezing and thawing resistance as well as impressive pull-off strength compared to conventional cement mortars. Laakri et al. (2019) investigated Algerian natural pozzolan to develop blended cement with improved mechanical and thermal properties for use in construction. A study has also been conducted on the construction of roads utilizing natural pozzolans in Tanzania (Onyango et al., 2007). In a study conducted by Stefanidou (2016), two natural pozzolans from Greece have been demonstrated to be valuable for assessing their reactivity with lime, making them suitable for producing compatible and durable repair mortars intended for conservation purposes. A case study in China explored the use of natural pozzolan-blended materials in the construction of high-speed railway concrete slab tracks (Qu et al., 2024). Another case study conducted by Rangrazian et al. (2023) investigated the engineering properties of pozzolanic concrete reinforced with fibers, utilizing local mineral pozzolans from Iran.

Although numerous studies have been conducted on natural pozzolans in various countries, research on natural supplementary cementitious materials (SCMs) remains limited. The relatively limited research on natural SCMs can be attributed to several factors. Natural SCMs vary in composition and properties based on their geographical source, leading to inconsistent performance and making them harder to standardize compared to artificial SCMs. Economic factors also play a significant role, as the excavation, mining, and processing of natural supplementary cementitious materials (SCMs) can be prohibitively expensive. Additionally, their processing can be complex due to their unique properties. Additionally, artificial SCMs benefit from being industrial by-products, providing a dual benefit of waste management and material utilization, which has attracted more research. Established performance reliability and advanced processing technologies for artificial SCMs further encourage their study. Lastly, the lack of standardized testing methods and potential environmental concerns associated with extracting natural SCMs also contribute to the focus on artificial alternatives.

As per the American Society for Testing and Materials [ASTM] C618 definition [67], pozzolan, typically composed of siliceous or siliceous-alumina materials, lacks inherent binding properties but acquires such properties upon reacting with calcium hydroxide (lime) at room temperature, when finely ground and in the presence of moisture.

As mentioned above, one of the primary reasons natural pozzolans are not commonly used as supplementary cementitious materials (SCMs) is their general absence as waste materials, unlike artificial SCMs. Inspired by Cappadocia's rich geological legacy, famous for its distinct landscapes shaped by volcanic processes, this study examines the pozzolanic potential of vast quantities of waste excavation soil, presumed to act as natural pozzolans because of its chemical content within the region. Thorough investigations were conducted into methodologies for pozzolanicity measurement, employing rigorous experimental techniques to ascertain the pozzolanic activity values of these materials, thereby providing valuable insights into their suitability for engineering applications. Additionally, this inquiry seeks to pinpoint areas within Cappadocia where pozzolans extracted from various locations can be optimally utilized based on the results of pozzolanic activity tests. This endeavor seeks to advance sustainable engineering practices and improve resource efficiency across various projects by clarifying the origins, performance, and characteristics of diverse waste materials, while also conducting a comparative economic analysis to highlight the differences between these materials and their closest competitors.

Cappadocia, the borders of which are seen in Figure 3, listed as a UNESCO World Heritage Site since 1985 for its rich natural and cultural heritage, boasts a diverse array of historical buildings spanning thousands of years. Among these structures, those intricately connected to the region's geological features, notably the rock-cut structures, are indeed unique to the region. Over millennia, the region's landscape has been shaped by volcanic activity emanating from 19 distinct eruption centers (Toprak, 1998), dispersing lava and ash across the area through various volcanic processes. This geological activity has contributed to the formation of magmatic rocks, tuffs, pumices, and deposits with depths exceeding 400 meters (Le Pennec et al., 2005), that characterize the unique terrain of Cappadocia.

Figure 3. Location of Cappadocia, Türkiye, and Cappadocia volcanic waste soil (CW) collection areas in the region

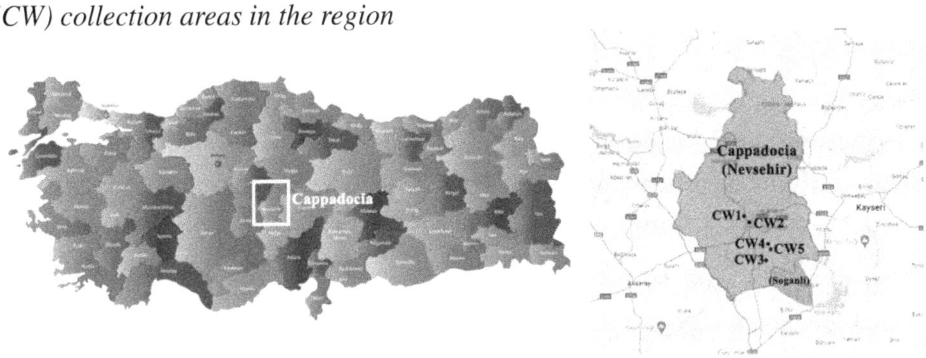

Numerous instances of rock-hewn structures exist (Figure 4a), categorized into three main types based on their intended functions: Religious, military, and civil architectural structures (S. Ozata & Arun, 2018). These structures are found both underground and above ground. These structures are typically adaptable and capable of being expanded and replicated to suit specific needs and the features of the rock formations. Traditionally, rock-hewn structures were carefully constructed by hand, with a skilled stone mason excavating up to 1 m^3 of rock per day, resulting in leftover waste soil. However, over the past five decades, there has been a substantial rise in demand for rock-hewn warehouses, one example of which is shown in Figure 4b. These warehouses are renowned for their ability to preserve fruits and vegetables over extended periods without requiring energy expenditure. Additionally, they enhance the juiciness of stored products by maintaining ideal humidity levels within their environment, effectively trapping moisture within their structures. The surge in production and consumption nationwide has led to an increased demand for additional storage space. Particularly renowned for their unique advantages, rock-hewn warehouses in the Cappadocia Region witnessed a rise in demand post-1970s. These storages are utilized to preserve thousands of tons of harvested potatoes and citrus fruits. Leveraging the natural attributes of the rock formations, storing fruits and vegetables in these warehouses proves to be both cost-effective and beneficial, as it requires no energy consumption and enhances the product's weight. The rising demand prompted the adoption of tunneling machines for warehouse construction in the 1970s with a complete transition to mechanical carving occurring in the 1990s (Özata et al., 2019). Since then, manual carving has declined, except for specialized and detailed workmanship. Over the past three decades, numerous companies in the region have been actively involved in constructing warehouses using excavators one of which is depicted in Figure 4c. As a result, the waste soil from mechanical rock carving, depicted in Figure 4b, has surpassed 40 million tons.

According to the "Regulation on the Design, Calculation, and Construction Principles of Rock-Cut Structures," enacted in 2017 (Ministry of Environment, Urbanisation and Climate Change, 2017), excavation soil waste should be minimized at the source, reused, recycled, and collected separately. Despite these guidelines, rock-hewn construction waste soils, which lack effective evaluation methods, are stored in designated areas, resulting in significant visual pollution. Although not situated within the primary protected areas of the UNESCO World Heritage Listed sites of Cappadocia, these hillocks, as depicted in Figure 4d, pose a threat to buffer zones. Thus, another concern examined in the study is the challenge of assessing waste earth, leading to visual pollution from tuff soil waste generated during the construction of rock-hewn structures in the Cappadocia Region.

Figure 4. (a) Different types of buildings along with natural formations (b) Mechanically carved rock-hewn warehouse (c) Mechanical carving of rock and transportation of waste excavation soil (d) Waste volcanic excavation soils deposited near the protected area's buffer zone

To mitigate these waste soils, exploring alternative solutions becomes crucial. One promising approach involves the use of pozzolanic materials, which can transform excavation soil waste into valuable resources. Understanding the properties and reactivity of pozzolans, particularly in the context of sustainable construction material, is therefore essential. This requires a closer examination of the materials and the development of necessary methods to better understand the waste, as outlined in the following section, paving the way for its potential use in various areas and sectors.

MATERIALS AND METHODS

Pozzolan is defined as any natural or artificial inorganic material that acquires binding properties through a reaction with calcium hydroxide (lime) (McCarthy & Dyer, 2019). According to the ASTM C618 (2019) definition, pozzolan is a siliceous or siliceous-aluminous material that possesses minimal or no binding properties on its own. However, when finely ground and exposed to moisture, it reacts with calcium hydroxide (lime) at room temperature, thereby acquiring binding properties. Pozzolanic activity, encompassing all interactions between the active components of pozzolanic materials, lime, and water, is defined by the material's ability to react with calcium hydroxide (McCarthy & Dyer, 2019). The rate of lime binding due to pozzolanic activity is influenced by various factors, including the structure of active phases in the raw material, the content of pozzolanic material, the lime-to-pozzolan ratio in the mixture, curing time, the specific surface area of the pozzolanic material, and the water-to-binder ratio (Bondar, 2009; McCarthy & Dyer, 2019). Despite the vast diversity and complexity of pozzolanic materials, along with the intricate processes involved during hydration, it is difficult to define a fixed model for determining pozzolanic activity. Nonetheless, numerous methodologies exist in the literature for evaluating the pozzolanic activity of a material.

The strength activity index method is one of them (Donatello et al., 2010; Ni et al., 2022). It measures the impact of the pozzolanic material on the compressive strength of cement mortar, providing an intuitive indication of the material's performance in a composite pozzolan-cement system. However, this method is time-consuming and it is hard to distinguish between the filler effect and the pozzolanic effect. The electrical conductivity test is a rapid method that reflects the consumption of $Ca(OH)_2$ by measuring the conductivity change of the cement-lime slurry system during the pozzolanic reaction. (Böke et al., 2006; Luxan et al., 1989). However, it can be affected by factors like the specific surface area and temperature. The isothermal calorimetry test evaluates activity by measuring the hydration heat of the cement-pozzolanic material system, providing rapid and accurate results with good repeatability, though it has limited versatility and high equipment requirements (Mostafa & Brown, 2005). The thermal analysis method and bound water ratio method quantify activity by analyzing mass changes in the pozzolan-cement or pozzolan-lime system during heating, offering speed and the ability to quantify activity, but thermal analysis relies on assumptions about the hydration reaction that can introduce uncertainty (Alzeer et al., 2022; Yao et al., 2021). The bound water ratio method provides a rapid test process with a short cycle, allowing for quantification of the material's activity. However, it fails to completely separate evaporable water from bound water, and the high temperatures used may alter the physical and chemical properties of the material (Alzeer et al., 2022). Additionally, the method relies on assumed hydration reactions, which may not fully align with the actual maximum reaction of the pozzolanic material. The Frattini test is one of the most applicable approaches for all types of pozzolans (TS EN 196-5, 2012), but it is complicated to operate and time-consuming, which may affect the material's accurate reactivity. The lime absorption method directly measures the consumption of calcium hydroxide, providing a clear indication of the pozzolanic reaction, yet the experimental conditions may not reflect the real-world cement hydration process. The compressive strength test procedure outlined in TS 25 (2008) involves conducting the test on slaked lime-based mortars. Compared to the strength activity index test, this method is particularly more suitable for mortars with a low CaO content. The Chapelle test is easy to quantify and offers a rapid assessment, but it exhibits high discreteness and the complex experimental process may be susceptible to carbonation issues (Šídlová et al., 2023). The acid/alkali dissolution method directly measures the content of active silica and alumina, but the experimental conditions may differ from the actual cement hydration environment (Wang et al., 2023). The microstructure analysis method examines the polymerization degree of the material to infer its pozzolanic activity, nevertheless, the interpretation of the results may be complex and require specialized expertise (Senhadji et al., 2012; Ulukaya et al., 2017). The kinetic model methods provide a comprehensive understanding of the pozzolanic reaction kinetics (Frías et al., 2005), but the development and application of these models can be challenging and may require extensive data collection and analysis.

Some researchers have divided some of the above-mentioned methods for assessing pozzolanic activity into indirect and direct methods (Baki et al., 2020; Donatello et al., 2010; Ulukaya & Yüzer, 2016). In this categorization, direct methods, including chemical titration, X-ray diffraction (XRD), and thermogravimetric analysis/differential thermal analysis are based on the consumption of calcium hydroxide as a result of the pozzolanic reaction. In contrast, indirect methods rely on changes in physical or mechanical properties, such as electrical conductivity, compressive strength, or heat exchange, as indicators of pozzolanic activity.

Consequently, each method for detecting pozzolanic activity has its own set of advantages and disadvantages regarding operation, accuracy, and suitability for various pozzolanic materials. Therefore, based on the advantages and limitations of each method briefly mentioned above, the Frattini test was employed in this study as a direct method. Also, indirect methods such as the Strength activity index test, compressive strength test, and electrical conductivity method were utilized. These methods were chosen to ascertain the presence of pozzolanic activity in the waste tuff soil resulting from the construction of rock-cut structures in Cappadocia.

Before implementing both direct and indirect methods, the ratios of specific oxide contents and the required limit values for certain components, as outlined in both Turkish (TS 25, 2008) and American (ASTM, 2019) standards for natural pozzolans, were verified. Once confirming compliance with these limit values, the experiments were conducted, and the pozzolanic activity of samples collected from various designated waste soil areas in Cappadocia was determined. In the following section, the material properties necessary for some of these experiments will be shared first, followed by a detailed presentation of the methods used.

Materials

Previous research has explored the use of Cappadocia tuff as a natural pozzolan in cement production (Atan, 2015), as well as the repurposing of stone dust, a by-product of stone cutting in the region, for construction purposes (Ekinci et al., 2019). While there have been individual studies on construction waste derived from rock-cut structures in specific areas within the region (Özata et al., 2019; Yuzer et al., 2023), an inclusive study encompassing the entire Cappadocia region is lacking. Thus, to ensure comprehensive results, waste samples were collected from five distinct areas in Nevsehir (a city and the central part of the Cappadocia Region), specifically designated as locations where excavation soil of numerous rock-hewn structures had gathered. These designated sites are marked on the Cappadocia map in Figure 3. The abbreviation CW was used for Cappadocia volcanic waste soil and samples were labeled as CW1, CW2, CW3, CW4, and CW5 to denote each specific area.

The natural pozzolan deposit areas are centered around the following coordinates: CW1 (38.649918°N, 34.703629°E), CW2 (38.647714°N, 34.705775°E), CW3 (38.477521°N, 34.818503°E), CW4 (38.555297°N, 34.811927°E), and CW5 (38.543726°N, 34.827177°E). The natural pozzolan deposit area is approximately 100-300 meters in diameter, centered on these coordinates. The waste tuff soils collected from these coordinates were sieved below 63 μm to facilitate the required analyses and experiments for determining pozzolanic activity.

Chemical Characterization of the Waste Materials

X-Ray Fluorescence (XRF) and X-Ray Diffraction (XRD) analyses were employed to ascertain the chemical composition of the Cappadocia volcanic waste soils (CWs). To verify compliance with defined limits for natural pozzolans according to TS 25 (2008) and ASTM C618 (2019), specific oxides and components were identified. Initially, XRF analysis was utilized to detect the presence and quantity of these oxides and components. The source crystals of the oxides were then determined by comparing the results with XRD patterns, ensuring the accuracy of the findings. For X-Ray Fluorescence (XRF) analysis, the material sieved below 63 μ is compacted and polished before undergoing X-Ray scanning to determine the types and amounts of compounds, oxides, and elements present. The Bruker S8 Tiger device was utilized for this experiment.

This method determines the major oxides in the material along with the loss on ignition (LOI). It enables assessment if the mass percentage of SiO_2+Al_2O_3+Fe_2O_3 meets the requirement of being over 70% for natural pozzolans according to TS 25 (2008). Moreover, it verifies compliance with the specified limits in TS 25 for SO_3 (maximum 3% by mass, or 4% per ASTM C618 (2019)) and chloride (maximum 0.1%

by mass) content. These results aid in identifying the material as a natural pozzolan and determining its type. The reported percentages are determined through semi-quantitative methods, and trace elements below a certain threshold are not included.

In X-Ray Diffraction (XRD) analysis, X-rays are directed towards the test material sieved below 63 μm, characterizing the material through scattering and diffraction data. This experiment detects crystals within the material's internal structure using the 2θ scanning method, enabling the determination of crystal quantities. While this method allows for the determination of crystal quantities, the results are semi-quantitative, making it impractical to provide exact percentages.

The results of the XRF analysis are presented in Table 1. This table, which lists the major oxides present in the materials, indicates that all waste tuffs collected from the region meet the requirement of 70% SiO_2+Al_2O_3+Fe_2O_3 as per the afore-mentioned standards.

Table 1. Major oxide ratios of CWs

Formula	**CW1**	**CW2**	**CW3**	**CW4**	**CW5**
	%				
SiO_2	70.62	68.19	67.55	62.45	61.82
Al_2O_3	12.67	12.78	13.98	14.51	14.83
Fe_2O_3	1.89	2.17	1.73	2.47	2.46
MgO	0.86	1.16	0.99	2.25	2.39
CaO	2.08	1.90	1.75	2.27	2.31
Na_2O	1.83	1.72	1.86	1.02	1.05
K_2O	5.24	4.86	5.81	3.07	3.18
TiO_2	0.22	0.22	0.26	0.29	0.28
P_2O_5	0.06	0.06	0.05	0.06	0.08
MnO	0.08	0.07	0.06	0.08	0.11
Cl	-	0.1	0.15	0.07	0.06
SO_3	0.03	-	-	-	-
Loss of Ignition(LOI)	4.08	6.55	5.64	11.30	11.24
SiO_2+Al_2O_3+Fe_2O_3	**85.19**	**83.14**	**83.26**	**79.43**	**79.11**

It was found that the SO_3 content in all materials does not exceed 3% by mass, while the chloride (Cl) content is slightly above 0.1% in the CW3. Considering the low CaO content along with the specified oxides, it was determined that all waste tuffs qualify as N-class natural pozzolan according to the ASTM C618 standard classification. According to the XRD analysis results in Figure 5, the five samples generally contain minerals such as anorthite, orthoclase, quartz, vaterite, high quartz, high-low cristobalite, calcium manganese oxide, and zeolite.

Vaterite is one of many forms of $CaCO_3$. Feldspars are alumina-silicate group minerals. Orthoclase is a type of K-feldspar (alkali feldspar) while sodium-containing anorthite (anorthite-sodian) is typically classified as Ca-feldspar (plagioclase feldspar type). The presence of anorthite indicates a formation temperature above 850˚C (Cardiano et al., 2004), which is a typical temperature for rocks formed by volcanic activity.

Low-high cristobalite and high quartz (or quartz high) are silicate group minerals and types of quartz. Minerals formed at low temperatures are referred to as "low," while those formed at high temperatures are called "high." The transition from low to high quartz occurs at 573°C, and for cristobalite, it occurs at 200-270°C. Stabilizing cristobalite requires even higher temperatures (Kalemtaş, 2021). The presence of these minerals in the samples indicates that the material formed at various temperatures.

Zeolites belong to the tectosilicate group and are also known as hydrous aluminum silicates. The channels within the structure of zeolites facilitate ion exchange and water removal. When heated, the water in zeolites evaporates as the temperature rises. Zeolites are also known for their pozzolanic properties (Özata, 2015). Carbonated manganese mineral is typically found as a secondary mineral in mineral deposits (Kara, 2019).

Figure 5. XRD analysis graphs of CW (An: Anorthite, Vt: Vaterite, Q: Quartz, Or: Orthoclase, Cr: Cristobalite, Z: Zeolite, Cm: Calcium manganese oxide

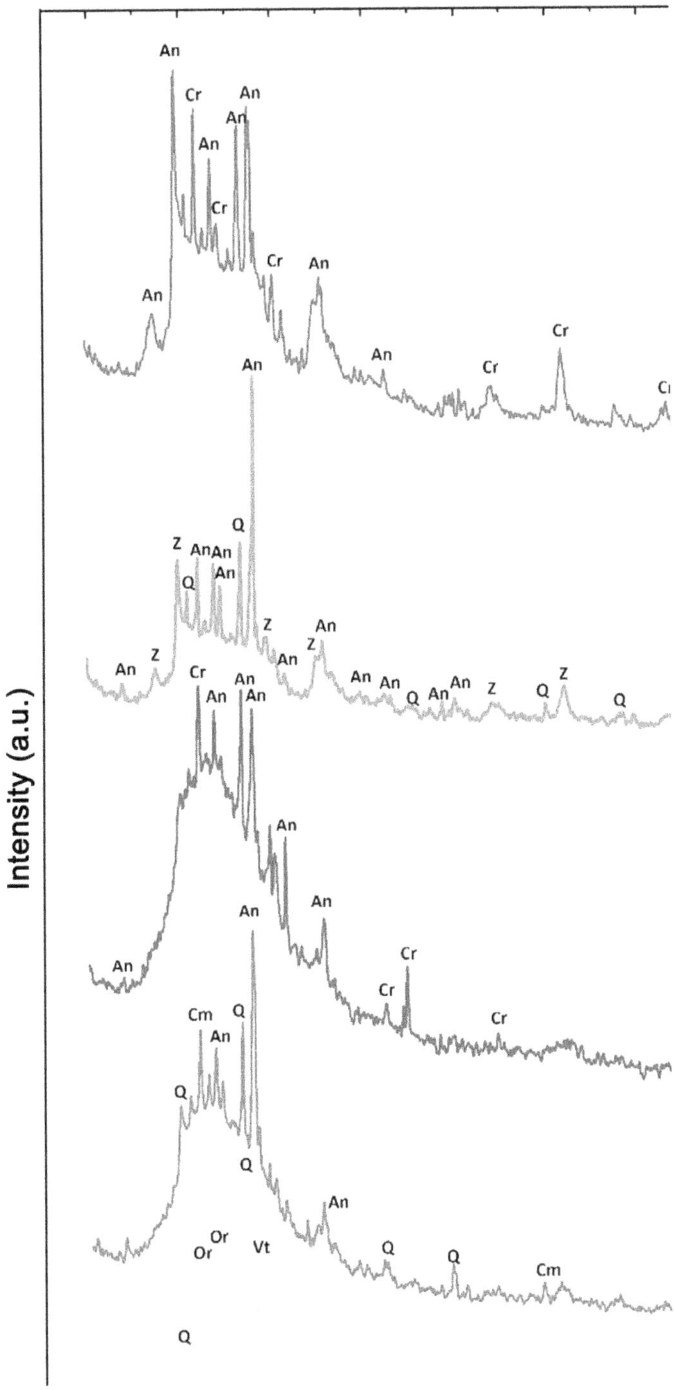

XRD analysis results revealed that all materials are primarily composed of alumina-silicate group minerals, leading to high SiO_2+Al_2O_3 group ratios due to these alumino-silicate-based crystals. It is believed that the SiO_2 identified in the XRF analysis originates from feldspar and quartz group minerals, while the Al_2O_3 comes from anorthite. The XRD analysis results corroborate the oxide ratios obtained from the XRF analysis.

Methods of Pozzolanic Activity Test

The details and procedures of the Frattini test as a direct method, along with the strength activity index test, compressive strength test, and electrical conductivity test as indirect methods, are explained for detecting the pozzolanic activity of five different Cappadocia volcanic waste soils (CWs).

Implementation of the Frattini Test

The Frattini test, as per TS EN 196-5 (2012), involves chemical titration to find out the concentrations of dissolved CaO and OH^- in a solution comprising CEM-I 42.5R type Portland cement and test pozzolan. For the experiment, a mixture of 16 g of the specified cement type (80%) and 4 g (20%) of material sieved below 63 μm was prepared. These 20 g samples were mixed with 100 ml of distilled water, which was kept in an oven at 40±1°C. Then, the solution was sealed in an airtight plastic container and incubated at 40±1°C for 8 days. After incubation, the solution was filtered under vacuum through a Buchner funnel into a vacuum flask using dry double filter paper.

After the filtered material reached room temperature, 50 ml was drawn with a pipette. Five drops of methyl orange indicator were added, causing the solution to turn yellow. Then, 0.1 mol/l dilute hydrochloric acid was added to the solution. The concentration of hydroxyl ions $[OH]^-$ was determined using Equation 1 when the solution turned orange.

$$[OH]^- = \frac{1000x0,1xV1xf1}{50} = 2xV1xf1$$

(mmol/l) (1)

In Equation 1, V_1 represents the volume of hydrochloric acid consumed (ml) and f_1 is the hydrochloric acid solution factor. The pH of the solution, which was titrated with dilute hydrochloric acid, was adjusted to 12.5±0.2 using sodium hydroxide solution and a pH meter. To this basic pH solution, 0.1 g of chalcone indicator was added. The solution was then titrated again using 0.03 mol/l EDTA solution, continuing until the solution turned from pink to blue. The concentration of calcium oxide [CaO] formed as a result of the titration was calculated using Equation 2 given below.

$$\left[CaO\right] = \frac{1000x0{,}03xV2xf2}{50} = 0{,}6xV2xf2$$

(mmol/l) (2)

In Equation 2, V_2 denotes the volume of EDTA solution consumed (ml), and f_2 is the factor for the EDTA solution. The evaluation of the test results relied on a graph provided in TS EN 196-5 (2012). The calcium saturation curve, as outlined in the standard, was derived by expressing the concentration of [CaO] ions from Equation 3 in terms of the hydroxyl ion $[OH]^-$ concentration at 40°C.

$$\max\,[CaO] = \frac{350}{[OH]^- - 15}$$

(mmol/l) (3)

The hydroxyl ion concentration $[OH]^-$ obtained from the titrations using Equation 1 was plotted on the x-axis of this graph, while the calcium oxide [CaO] concentration was plotted on the y-axis. This enabled the examination of whether the intersection of these values fell below the calcium saturation curve. If this point lies under the curve, it is accepted that the calcium ions in the solution are bound by the tested material, indicating sufficient evidence of pozzolanic activity properties and points above the curve indicate the absence of pozzolanic activity.

Strength Activity Index Test Procedure

Strength Activity Indicator (SAI) is a term defined in ASTM C311 (2003), and the test procedures outlined in ASTM C109 (2018) were followed to determine the pozzolanic activity of CWs. The experiment's application and evaluation phase revolve around determining the ratio of the compressive strength of the cement mortar containing the test material to that of the reference cement mortar.

For the reference cement mortar used for comparison, a mixture of cement, water, and sand in the ratio of 1/0.485/2.75 was utilized respectively. This reference mortar was poured into six 5 cm cube molds. Subsequently, the test material, equivalent to 20% of the initial cement weight, was added to produce additional mortars. These mortars, with cement substituted by the test material, were then placed in six cube molds measuring 5x5x5 cm each. The produced cube specimens were kept in a humidity chamber at 95% ± 5% relative humidity and a temperature of 23 ± 2°C for one day (24 hours). Subsequently, they were removed from the molds. The extracted specimens were then saturated in lime-saturated water at 23 ± 2°C until the 7th and 28th days when the experiment was conducted for each CW.

An ELE brand machine, with a precision of 0.1 kN and a capacity of 3000 kN, was utilized for conducting compressive strength tests on samples at both 7 and 28 days, with a minimum of 3 specimens for each. Following the test, the compressive strength of the mortars was determined by averaging the compressive strength values of the specimens, and the SAI of each test material was calculated by proportioning them accordingly. For this calculation, the value of Equation 4 provided in ASTM C311 (2003), was applied. As per the mechanical definition criteria outlined in ASTM C618 (2018), if both the 7-day and 28-day SAI values of the tested material are 75% or higher, the material is classified as pozzolanic.

$$SAI7,28(\%) = \frac{A7,28}{B7,28}x100 \quad (4)$$

Within this Equation 4, $A_{7,28}$ signifies the average compressive strength (MPa) of a minimum of 3 test mortar samples at 7 and 28 days, while $B_{7,28}$ represents the average compressive strength (MPa) of at least 3 reference mortar samples at 7 and 28 days.

Compressive Strength Test Method

The compressive strength test was conducted following TS 25 (2008), using a lime mortar containing pozzolanic material. The quantity of material used in the lime mortars produced for the test is provided in Table 2. The lime used in this experiment is CL-80S hydrated lime. The produced mortars were placed in triple prism molds with dimensions of 4x4x16 cm and covered with a glass plate, then kept at a temperature of 23±2°C for 24 hours. After one day, the mortars were kept in an oven at 55±2°C for 6 days without being removed from the molds. The compressive strengths of the samples removed from the oven were determined on the 7th day of the experiment using an ALSA brand compression-testing machine with a capacity of 60 tons and a precision of 1 N. According to the experiments' results, if the average compressive strength of the air lime-based mortars exceeds 4 MPa, the test material is considered to be pozzolanic.

Table 2. The quantities of materials used in the experiment

Material	Weight (g)
Slaked lime	150
Test sample (TS)	$(\gamma_{TS} / \gamma_{SK})^*$ x 300
CEN standart sand	1350
Su	(150 + TS) x 0.5

Note: γTS is the specific gravity of the tested material and γSK is the specific gravity of lime.

Electrical Conductivity Test Procedure

This indirect method was first proposed by Luxan et al. (1989). Despite the absence of standardization, it is commonly used in pozzolanic activity tests due to its rapid and efficiently obtained results (Böke et al., 2006; Donatello et al., 2010; Tironi et al., 2013). In this experiment, the material quantities recommended by Luxan et al. (1989) were utilized. For monitoring solution conductivity, a recommended duration of 2 minutes was not adhered to; instead, 4 hours was considered, during which the conductivity value was approximately stabilized, and no significant change was observed after this time frame (Ulukaya & Yüzer, 2016).

For this experiment, lime-saturated solutions were prepared in 200 ml volumes, and their electrical conductivity values were measured. Subsequently, 5 g of test material was added to each solution, and the electrical conductivity of the newly formed solution was measured at intervals of 2 minutes, 15 minutes, 30 minutes, 45 minutes, 60 minutes, 2 hours, 3 hours, and 4 hours. Electrical conductivity measurements were led by using a Cond 3210 model device with a precision of 0.01 mS/cm manufactured by WTW. If the difference in electrical conductivity between the reference lime solution and the solution with added test material exceeded 0.4 mS/cm, the material was considered to be pozzolanic (Luxan et al., 1989).

In the subsequent section, the results of the experiments, for which the methods are described in detail, will be presented and discussed.

RESULTS AND DISCUSSION

Frattini Test

The results of the Frattini experiments conducted on at least three samples prepared for five different areas are presented in the graph shared in Figure 6. It has been determined that waste materials collected from all areas exhibit pozzolanic properties as they fall below the calcium saturation curve.

Figure 6. CaO and OH^- quantities obtained from samples according to the Frattini experiment

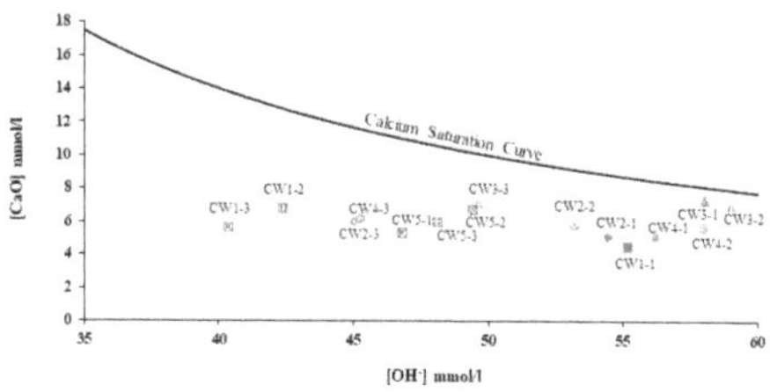

The average percentages of bound calcium oxide in the solution were calculated using Equation 3, and these values were shared in Figure 7. Accordingly, it was established that, except for the CW3 sample, the pozzolanic activities of the other samples are close in comparison.

Figure 7. The amounts of CaO bound in the test materials

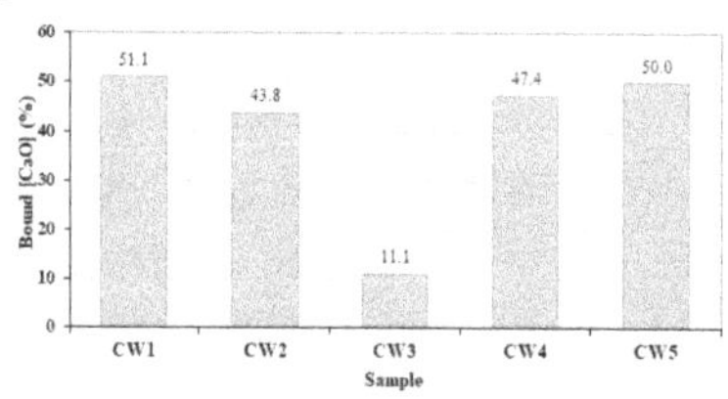

Strength Activity Index Test

The Strength Activity Index (SAI) values for 7 and 28 days are provided in Figure 8. According to the results of the 7-day experiment, CW3 and CW5 samples did not meet the minimum 75% limit value specified in the standard. For the 28-day experiment, only the CW5 sample did not meet the minimum 75% limit value. The

SAI values of CW1, CW2, and CW4 samples exceeded the 75% value both at 7 and 28 days.

In the 7-day activity assessment, the highest value was observed in CW2, while for the 28-day evaluation, CW4 exhibited the highest activity. Considering that compressive strength in cement-based mortars develops over an extended period, the values at 28 days were prioritized. Consequently, CW4 material was recognized as displaying the highest pozzolanic activity.

Figure 8. 7 and 28-day SAI values of tested CWs

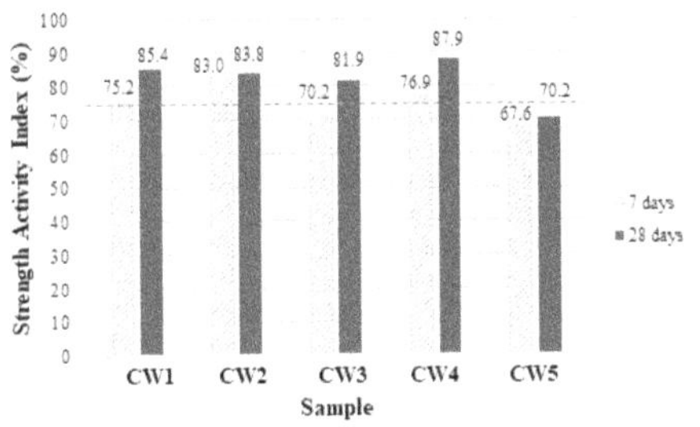

Compressive Strength Test

The experiment results conducted according to TS 25 (2008) are presented in Figure 9. As per these findings, except for the CW3 sample, which did not meet the requirement of at least 4 MPa compressive strength specified in the standard, the other samples were deemed to fulfill the criteria and were identified as pozzolans. The highest compressive strength was achieved in the CW4 sample with 7.7 MPa, followed by CW5 with 7.2 MPa. CW1 and CW2 samples have compressive strengths of 4.2 and 4.1 MPa, respectively, representing the lowest values among the considered pozzolan samples.

Figure 9. The results of the compressive strength test conducted with lime mortar

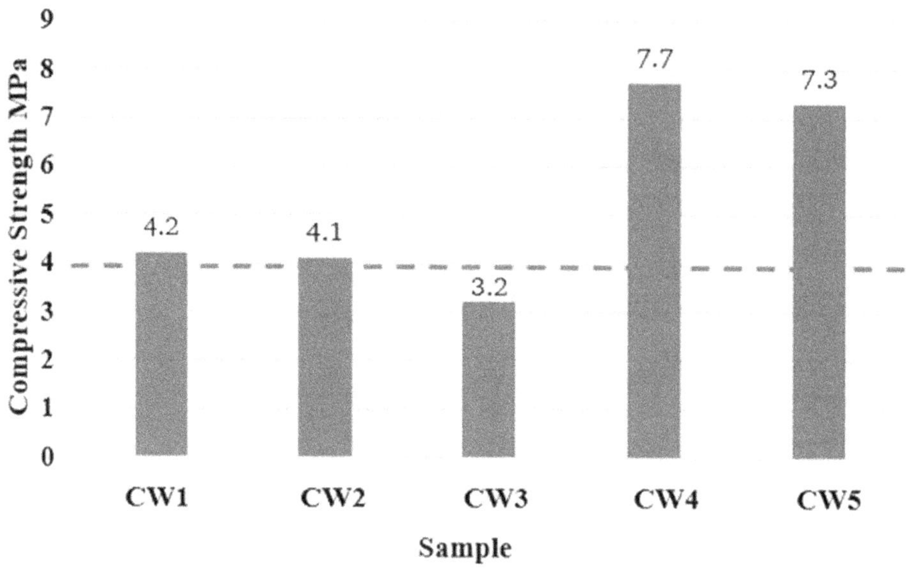

Electrical Conductivity Test

According to the total electrical conductivity differences with the reference lime solution, within the first 2 minutes, only CW4 met the minimum criterion of 0.4 mS/cm among the test materials listed in Table 3. However, after 4 hours, all materials except CW2 (although very close to 0.4 mS/cm) met this criterion.

Table 3. The total electrical conductivity differences between the reference lime solution and the solutions with the test materials

Sample	Electrical Conductivity Difference (mS/cm)	
	After 2 minutes	After 4 hours
CW1	0.22	0.44
CW2	0.33	0.38
CW3	0.20	0.40
CW4	0.40	0.51
CW5	0.33	0.45

Interpretation and Examination of Results

The overall assessment of all direct and indirect pozzolanic activity tests shared in Table 4, along with the supplementary analyses, is as follows:

According to the XRF analysis, the $SiO_2+Al_2O_3+Fe_2O_3$ ratio in all materials is above 70%. The XRD analysis indicates that the materials are generally composed of alumino-silicate-based minerals.

In the Frattini test, the highest CaO binding occurred in the CW1 material at 51.1%, followed by the CW5 material at 50%.

Table 4 . Pozzolanicity classification of CWs

Order	Frattini	28-day SAI	Compressive Strength	Electrical Conductivity
Best	CW1	CW4	CW4	CW4
	CW5	CW1	CW5	CW5
	CW4	CW2	CW1	CW1
	CW2	CW3	CW2	CW3
Worst	CW3	CW5 (x)	CW3 (x)	CW2 (x)

Note: Specimens marked with an (x) did not meet the threshold value for the relevant test.

The 28-day strength activity index showed the best performance from the CW4 material with a rate of 87.9%.

In the lime mortar compressive strength test, the CW4 material reached the highest compressive value of 7.7 MPa among the five materials examined, followed by the CW5 material with a compressive value of 7.3 MPa.

In the electrical conductivity test, the highest difference of 0.51 mS/cm was observed in the solution containing the CW4 material.

Taking the experiments into account, as shown in the order presented in Table 4, the pozzolanic activity ranking from best to worst is as follows: CW4 > CW5 > CW2 > CW1 > CW3. Despite this classification, it is concluded that all the examined Cappadocia volcanic waste soils (CWs) qualify as natural pozzolans and can be utilized at any stage of industrial processes due to their pozzolanic properties.

Natural pozzolans, particularly those of volcanic origin, have multifaceted applications across various sectors, notably in the construction industry. They serve as supplementary cementitious materials (SCMs), enriching concrete and cement properties such as strength, durability, and workability. Noteworthy studies by researchers like Çavdar and Yetgin (2007), Dinç-Şengönül et al. (2024), Gómez Sánchez de Rojas and Frías Rojas (2013), and Ozata et al.(2022) underscore the extensive utilization of natural pozzolans in cement, concrete, mortar, and grout. Furthermore, owing to their resemblance to traditional building materials, natural pozzolans are employed in the restoration of historical buildings and monuments. Additionally, they play a pivotal role in road and infrastructure development, serving as base or subbase materials in road construction projects and contributing to soil stability and fertility enhancement in land reclamation initiatives. Studies conducted by researchers such as Ghahari et al. (2017) and Onyango et al. (2007) corroborate the efficacy of natural pozzolans in pavement and road construction.

In environmental engineering, natural pozzolan is instrumental in the stabilization and solidification of hazardous wastes. Moreover, its porous structure renders it suitable for water treatment applications, facilitating its use in filtration systems for water and wastewater treatment. Notably, natural pozzolans have been explored for their potential in radioactive waste disposal, as evidenced by studies such as those by Kappes (1995).

In agriculture, natural pozzolans serve as valuable soil amendments, enhancing soil properties by improving water retention and nutrient availability. Research, exemplified by studies conducted by al-Swaidani et al.(2016) and Bahadori et al.(2019) focuses on the enhancement of various soil types using natural pozzolans.

Natural pozzolans are applicable across various industrial sectors, notably in ceramics and glass production, where they serve as a fundamental raw material. Research efforts, as seen in some studies (Achiou et al., 2017; Eldera et al., 2023) explore and anticipate the utilization of natural pozzolans in the glass and ceramic industry.

It is evident that Cappadocia volcanic waste soils (CWs) demonstrating verified pozzolanic activities can readily find application across the aforementioned sectors. As detailed, the next section will present an economic analysis of this material, which has evident potential across various sectors, in comparison with its closest counterparts, including SCMs and cement. This analysis will evaluate the material's viability as a pozzolan and assess its economic feasibility.

Comparative Economic Analysis

The literature extensively covers studies on the life cycles of cement-based materials, including natural pozzolans, SCMs, recycled concrete waste, metakaolin, slag, and other alternative (Bajpai et al., 2020; Manjunatha et al., 2021; Miyan et al., 2024; Nguyen et al., 2018) These studies frequently address environmental impacts, greenhouse gas emissions, and cost analyses of these materials. In this study, a basic comparative cost analysis between cement, SCMs, and CW is conducted.

The economic analysis and comparison of CW with other SCMs encompass processes from cradle (raw material extraction) to gate (the stage at which bulk binder material is prepared for packaging or distribution). This analysis considers the feedstock stage, accounting for the inventory of raw materials from their respective sources, and the processing stage, with transportation costs excluded, assuming uniformity of it across all materials. The cradle-to-gate processes for various construction materials, SCMs, and cement are illustrated in Figure 10. All data used in this analysis pertain to the Turkish market. The focus is on activities directly related to the production of cementitious binders, excluding downstream processes such as casting, curing, durability performance, and service life, which also affect the overall product life cycle.

Material costs, including natural pozzolans, vary regionally, with transportation as a significant expense. As a result, SCMs may be more cost-effective than Portland cement in regions close to the source (Cannon et al., 2021). Since the locations of CW, cement, and SCMs within the Turkish market can vary significantly, transportation costs are assumed to be equal when comparing overall costs, as previously mentioned.

Figure 10. The cradle-to-gate processes for various construction materials[2]

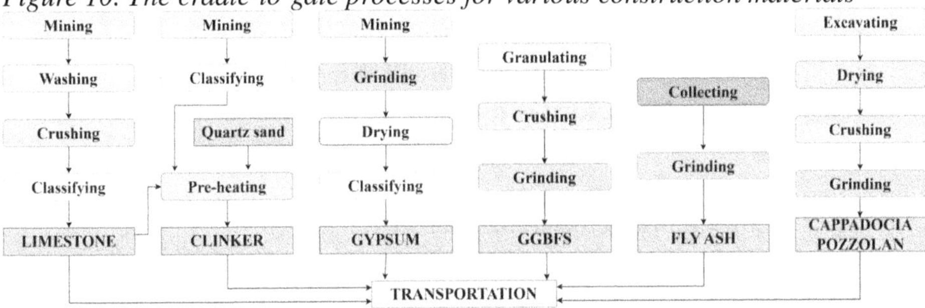

In the comparative cost analysis, limestone has the highest cradle-to-gate cost after cement, at 73.3 USD per ton, with a price very close to that of cement. Slag follows as the next most expensive material, while fly ash is significantly more economical than these three (Figure 11a).

Cappadocia excavation earth, in its raw form, is a free waste material. Since sun drying is sufficient for drying, no additional costs are incurred for this process. However, the cost of 1.2 USD per ton for CW is attributed to the crushing and grinding process, as depicted in Figure 11a, which is part of the cradle-to-gate sequence. This process is foreseen to be done in the excavation area. Slag costs slightly more than half the price of cement, while fly ash is 72.2% cheaper, and CW is 73.9% cheaper (Figure 11b). Based on these results, CW is demonstrated to be more economical than other SCMs and cement.

Figure 11. (a) Prices of some construction materials (USD per ton) (b) Cost of other materials relative to cement

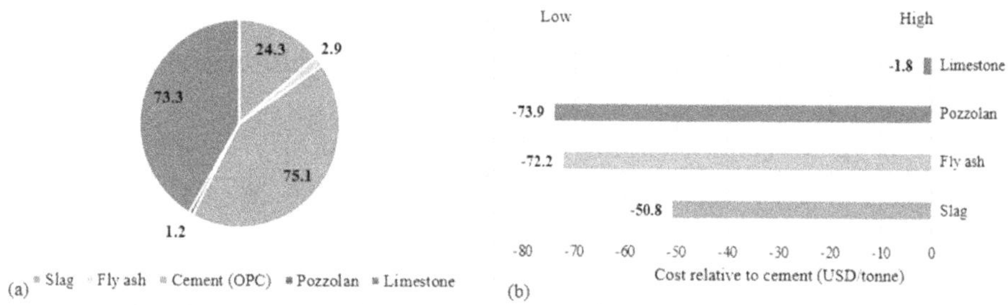

In addition to economic analyses, a comprehensive environmental impact assessment is essential. However, sufficient data for evaluating the environmental impact of the material studied could not be obtained. Previous studies (Bajpai et al., 2020; Manjunatha et al., 2021) indicate that concrete made with natural pozzolans typically outperforms slag- or limestone-based concrete in terms of freshwater ecotoxicity, human carcinogenic toxicity, mineral resource scarcity, and water consumption. It is expected that the material in this study will demonstrate similar benefits, but further detailed research is needed in this area.

CONCLUSIONS AND SUGGESTIONS

This study aimed to evaluate the pozzolanic activity of Cappadocia volcanic waste soils (CWs) through the application of commonly employed test methodologies. Additionally, it endeavored to unveil the latent potential of this waste soil for utilization as a raw material across various sectors, while also comparing its economic feasibility. Through targeted studies, another aim of this research is to preserve the cultural and natural heritage of the Cappadocia Region by addressing the issue of visual pollution caused by the accumulation of millions of tons of waste.

Additionally, it is aimed to transform this waste material into valuable economic resources. The outcomes achieved in pursuit of these objectives are outlined below.

All CW samples, except for CW3, show comparable bounded CaO ratios based on Frattini test results, confirming their pozzolanic characteristics. The Frattini method is a precise testing procedure and is practical due to its minimal sample requirement of 4 grams per test. However, it is relatively time-consuming.

The classification of all tested CW samples, except for CW5, as pozzolans based on the strength activity index (SAI) compliance criteria requires further review. The SAI results show significant discrepancies compared to other tests. Compressive strength and electrical conductivity tests appear more suitable for evaluating natural pozzolans of volcanic origin, due to usage of slaked lime.

The compressive strength test and electrical conductivity test yield almost identical results. However, to obtain clear outcomes, the electrical conductivity test requires a minimum duration of 4 hours, contrary to the duration in the suggested method.

Although the methods used show discrepancies in the ordering of pozzolanic activity, all CW samples exhibit pozzolanic properties to varying degrees. Overall test results indicate that CW4 is the best natural pozzolan, closely followed by CW5 and CW1.

The experiments conducted have demonstrated that the CWs produced during the creation of rock-cut structures exhibit pozzolanic activity, indicating that the waste materials throughout the region can be effectively and economically used as natural pozzolans. This study has confirmed that CWs are highly suitable for use in engineering materials. Cappadocia's natural pozzolans can be extensively utilized in construction, particularly as supplementary cementitious materials, improving the properties of concrete and cement. This study underscores their versatility, showing potential applications in cement, concrete, mortar, grout, historical restoration, and road construction. In environmental engineering, CWs can contribute to hazardous waste stabilization and water treatment, while in agriculture, they can enhance soil properties. Additionally, these natural pozzolans can serve as essential raw materials in ceramics and glass production.

In conclusion, utilizing Cappadocia's volcanic waste soils for various applications, particularly as engineering materials, will promote the use of local natural pozzolans and contribute positively to the economy due to the cost-effectiveness of the material. This approach not only addresses the waste management issue but also aids in protecting a world heritage site from the growing threat of waste accumulation.

REFERENCES

Abali, Y., Bayca, S. U., & Targan, S. (2006). Evaluation of blends tincal waste, volcanic tuff, bentonite and fly ash for use as a cement admixture. *Journal of Hazardous Materials*, 131(1–3), 126–130. DOI: 10.1016/j.jhazmat.2005.09.031 PMID: 16314042

Achiou, B., Elomari, H., Bouazizi, A., Karim, A., Ouammou, M., Albizane, A., Bennazha, J., Alami Younssi, S., & El Amrani, I. E. (2017). Manufacturing of tubular ceramic microfiltration membrane based on natural pozzolan for pretreatment of seawater desalination. *Desalination*, 419(June), 181–187. DOI: 10.1016/j.desal.2017.06.014

Al-Swaidani, A., Hammoud, I., & Meziab, A. (2016). Effect of adding natural pozzolana on geotechnical properties of lime-stabilized clayey soil. *Journal of Rock Mechanics and Geotechnical Engineering*, 8(5), 714–725. DOI: 10.1016/j.jrmge.2016.04.002

Alzeer, M. I. M., Cheeseman, C., & Kinnunen, P. (2022). New synthetic glass-based supplementary cementitious materials derived from basalt composition. *Journal of Building Engineering*, 46, 103699. DOI: 10.1016/j.jobe.2021.103699

American Society for Testing and Materials. (2003). *Standard Test Methods for Sampling and Testing Fly Ash or Natural Pozzolans for Use in Portland Cement Concrete* (ASTM C 311-02). https://www.astm.org/c0311-00.html

American Society for Testing and Materials. (2018). *Standard Test Method for Compressive Strength of Hydraulic Cement Mortars* (ASTM C109/C109M-16a). https://www.astm.org/c0109_c0109m-23.html

American Society for Testing and Materials. (2019). *Standard Specification for Coal Fly Ash and Raw or Calcined Natural Pozzolan for Use in Concrete* (ASTM C618-19). https://www.astm.org/c0618-23e01.html

Atan, M. C. (2015). *Using Cappadocia Tuff As A Natural Pozzolan In the Cement Production* [Unpublished Master's Thesis]. Middle East Technical University.

Aydan, Ö., & Ulusay, R. (2003). Geotechnical and geoenvironmental characteristics of man-made underground structures in Cappadocia, Turkey. *Engineering Geology*, 69(3–4), 245–272. DOI: 10.1016/S0013-7952(02)00285-5

Ayhan, A. (2004). *Geological and Morphological Investigations of the Underground Cities of Cappadocia Using GIS* [Unpublished Master's Thesis]. Middle East Technical University.

Bahadori, H., Hasheminezhad, A., & Taghizadeh, F. (2019). Experimental Study on Marl Soil Stabilization Using Natural Pozzolans. *Journal of Materials in Civil Engineering*, 31(2), 04018363. Advance online publication. DOI: 10.1061/(ASCE)MT.1943-5533.0002577

Bajpai, R., Choudhary, K., Srivastava, A., Sangwan, K. S., & Singh, M. (2020). Environmental impact assessment of fly ash and silica fume based geopolymer concrete. *Journal of Cleaner Production*, 254, 120147. DOI: 10.1016/j.jclepro.2020.120147

Baki, V. A., Nayır, S., Erdoğdu, Ş., & Ustabaş, İ. (2020). Determination of the Pozzolanic Activities of Trachyte and Rhyolite and Comparison of the Test Methods Implemented. *International Journal of Civil Engineering*, 18(9), 1053–1066. DOI: 10.1007/s40999-020-00516-5

Balog, A., Cobirzan, N., Aciu, C., & İlutiu-Varvara, D. A. (2014). Valorification of volcanic tuff in constructions and materials manufacturing industry. *Procedia Technology*, 12, 323–328. DOI: 10.1016/j.protcy.2013.12.493

Becerra-Duitama, J. A., & Rojas-Avellaneda, D. (2022). Pozzolans : A review. *Engineering and Applied Science Research, 49*(4), 495–504. https://doi.org/DOI: 10.14456/easr.2022

Black, L. (2016). Low clinker cement as a sustainable construction material. In Jamal M. Khatib (Ed.), *Sustainability of Construction Materials* (Second Edi). Elsevier. DOI: 10.1016/B978-0-08-100370-1.00017-2

Böke, H., Akkurt, S., İpekoğlu, B., & Uğurlu, E. (2006). Characteristics of brick used as aggregate in historic brick-lime mortars and plasters. *Cement and Concrete Research*, 36(6), 1115–1122. DOI: 10.1016/j.cemconres.2006.03.011

Bondar, D. (2009). Alkali Activation of Iranian Natural Pozzolans for Producing Geopolymer Cement and Concrete. In *Doctoral Thesis,The University of Sheffield* (Issue June).

Cannon, C., Guido, V., & Wright, L. (2021). *Concrete Solution Guise Mix It Up Supplementary Cementitious Materials (SCMs)*. RMI Innovation Center. http://www.rmi.org/concrete- solutions-guide/

Cardiano, P., Ioppolo, S., De Stefano, C., Pettignano, A., Sergi, S., & Piraino, P. (2004). Study and characterization of the ancient bricks of monastery of " San Filippo di Fragalà " in Frazzanò (Sicily). *Analytica Chimica Acta*, 519(1), 103–111. DOI: 10.1016/j.aca.2004.05.042

Çavdar, A., & Yetgin, Ş. (2007). Availability of tuffs from northeast of Turkey as natural pozzolan on cement, some chemical and mechanical relationships. *Construction & Building Materials*, 21(12), 2066–2077. DOI: 10.1016/j.conbuildmat.2006.05.034

Çevre ve Şehircilik Bakanlığı. (2017, October 18). *Kayadan Oyma Yapıların Tasarım, Hesap ve Yapım Esaslarına Dair Yönetmelik*. Resmi Gazete. https://www.resmigazete.gov.tr/eskiler/2017/10/20171018-7.htm

Dinç-Şengönül, B., Bayram, M., Oktay, D., & Yüzer, N. (2024). Valorization of Cappadocia waste earth in the production of sustainable lime-based grouts. *Sustainable Chemistry and Pharmacy*, 38, 101503. DOI: 10.1016/j.scp.2024.101503

Donatello, S., Tyrer, M., & Cheeseman, C. R. (2010). Comparison of test methods to assess pozzolanic activity. *Cement and Concrete Composites*, 32(2), 121–127. DOI: 10.1016/j.cemconcomp.2009.10.008

Efstathiadis, E. (1978). *Greek concrete of three millenniums* (1st ed.). Public Works Research Centre.

Ekinci, E., Türkmen, İ., Kantarci, F., & Karakoç, M. B. (2019). The improvement of mechanical, physical and durability characteristics of volcanic tuff based geopolymer concrete by using nano silica, micro silica and Styrene-Butadiene Latex additives at different ratios. *Construction & Building Materials*, 201, 257–267. DOI: 10.1016/j.conbuildmat.2018.12.204

Eldera, S. S., Aldawsari, S., & Hamzawy, E. M. A. (2023). Utilization of waste glass with natural pozzolan in the production of self-glazed glass-ceramic materials. *Nanotechnology Reviews*, 12(1), 20220565. DOI: 10.1515/ntrev-2022-0565

Frías, M., Villar-Cociña, E., De Rojas, M. I. S., & Valencia-Morales, E. (2005). The effect that different pozzolanic activity methods has on the kinetic constants of the pozzolanic reaction in sugar cane straw-clay ash/lime systems: Application of a kinetic-diffusive model. *Cement and Concrete Research*, 35(11), 2137–2142. DOI: 10.1016/j.cemconres.2005.07.005

Ghahari, S. A., Mohammadi, A., & Ramezanianpour, A. A. (2017). Performance assessment of natural pozzolan roller compacted concrete pavements. *Case Studies in Construction Materials*, 7, 82–90. DOI: 10.1016/j.cscm.2017.03.004

Gómez Sánchez de Rojas, M. I., & Frías Rojas, M. (2013). Natural pozzolans in eco-efficient concrete. In Pacheco-Torgal, F., Jalali, S., Labrincha, J., & John, V. M. (Eds.), *Eco-Efficient Concrete* (pp. 83–104). Woodhead Publishing Limited., https://www.sciencedirect.com/science/article/abs/pii/B9780857094247500049 DOI: 10.1533/9780857098993.2.83

Idorn, G. M. (1997). *Concrete progress from antiquity to third millenium*. Thomas Telford. DOI: 10.1680/cpftatttm.26315

Kalemtaş, A. (n.d.). *Ceramic Materials*. https://depo.btu.edu.tr/dosyalar/metalurji/Dosyalar/ Ceramic_Materials_3_Ayse_Kalemtas.pdf

Kaplan, H., & Binici, H. (1995). Tras ve traslı çimentolar. *Pamukkale Üniversitesi Mühendislik Fakültesi Mühendislik Bilimleri Dergisi*, 1(2–3), 121–127.

Kappes, J. A. (1995). *Characterization of pozzolanic cement for radioactive waste disposal* [Unpublished Master's Thesis]. University of Nevada.

Kara, F. (n.d.). *Manganez (Mangan) Nedir? Nerelerde Kullanılır? Nasıl Üretilir? Özellikleri Nelerdir?*https://malzemebilimi.net/manganez-mangan-nedir-nerelerde-kullanilir-nasil-uretilir-ozellikleri-nelerdir.html

Kitsopoulos, K. P., & Dunham, A. C. (1996). Heulandite and mordenite-rich tuffs from Greece: A potential source for pozzolanic materials. *Mineralium Deposita*, 31(6), 576–583. DOI: 10.1007/BF00196138

Küçükyildirim, E., & Uzal, B. (2014). Characteristics of calcined natural zeolites for use in high-performance pozzolan blended cements. *Construction & Building Materials*, 73, 229–234. DOI: 10.1016/j.conbuildmat.2014.09.081

Kurudirek, M., Özdemir, Y., Türkmen, I., & Levet, A. (2010). A study of chemical composition and radiation attenuation properties in clinoptilolite-rich natural zeolite from Turkey. *Radiation Physics and Chemistry*, 79(11), 1120–1126. DOI: 10.1016/j.radphyschem.2010.06.003

Laakri, M., Oudjit, M. N., & Saidani-Scott, H. (2019). Influence of natural pozzolan and blast furnace slag on hydration and physico-mechanical properties of cement. *International Journal of Sustainable Building Technology and Urban Development*, 10(3), 112–124. DOI: 10.22712/susb.20190012

Le Pennec, J. L., Temel, A., Froger, J. L., Şen, S., & Gourgaud, A.; Boudier, J. L. (2005). Stratigraphy and Age of the Cappadocia Ignimbrites, Turkey: Reconciling Field Constraints with Paleontologic, Radiochronologic, Geochemical and Paleomagnetic Data. *Of Volcanology and Geothermal Research, 141*, 45–64.

Liguori, B., Iucolano, F., De Gennaro, B., Marroccoli, M., & Caputo, D. (2015). Zeolitized tuff in environmental friendly production of cementitious material: Chemical and mechanical characterization. *Construction & Building Materials*, 99, 272–278. DOI: 10.1016/j.conbuildmat.2015.09.035

Luxan, M. P., Madruga, F., & Saavedra, J. (1989). Rapid Evaluation of Pozzolanic Activity of Natural Products by Conductivity Measurement. *Cement and Concrete Research*, 19(1), 63–68. DOI: 10.1016/0008-8846(89)90066-5

Malinowski, R., & Garfinkel, Y. (1991). Prehistory of concrete. *Concrete International*, 13(3), 62–68.

Manjunatha, M., Preethi, S., Malingaraya, Mounika, H. G., Niveditha, K. N., & Ravi. (2021). Life cycle assessment (LCA) of concrete prepared with sustainable cement-based materials. *Materials Today: Proceedings, 47*(xxxx), 3637–3644. DOI: 10.1016/j.matpr.2021.01.248

McCarthy, M. J., & Dyer, T. D. (2019). Pozzolanas and pozzolanic materials. In Hewlett, P. C. (Ed.), *Lea's Chemistry of Cement and Concrete* (5th ed., pp. 363–467). Elsevier., DOI: 10.1016/B978-0-08-100773-0.00009-5

Mikulčić, H., Klemeš, J. J., Vujanović, M., Urbaniec, K., & Duić, N. (2016). Reducing greenhouse gasses emissions by fostering the deployment of alternative raw materials and energy sources in the cleaner cement manufacturing process. *Journal of Cleaner Production*, 136, 119–132. DOI: 10.1016/j.jclepro.2016.04.145

Miyan, N., Omur, T., Amed, B., Özkan, H., Aydın, R., & Kabay, N. (2024). Recycled waste concrete and metakaolin based alkali-activated paste: Characterization, optimization, and life cycle assessment. *Construction & Building Materials*, 416(February), 135233. Advance online publication. DOI: 10.1016/j.conbuildmat.2024.135233

Moropoulou, A., Cakmak, A., Labropoulos, K. C., Van Grieken, R., & Torfs, K. (2004). Accelerated microstructural evolution of a calcium-silicate-hydrate (C-S-H) phase in pozzolanic pastes using fine siliceous sources: Comparison with historic pozzolanic mortars. *Cement and Concrete Research*, 34(1), 1–6. DOI: 10.1016/S0008-8846(03)00187-X

Mostafa, N. Y., & Brown, P. W. (2005). Heat of hydration of high reactive pozzolans in blended cements: Isothermal conduction calorimetry. *Thermochimica Acta*, 435(2), 162–167. DOI: 10.1016/j.tca.2005.05.014

Nguyen, L., Moseson, A. J., Farnam, Y., & Spatari, S. (2018). Effects of composition and transportation logistics on environmental, energy and cost metrics for the production of alternative cementitious binders. *Journal of Cleaner Production*, 185, 628–645. DOI: 10.1016/j.jclepro.2018.02.247

Ni, T., Xu, F., Gao, C., Yang, Y., Ma, W., & Kong, D. (2022). Chemical activation of pozzolanic activity of sludge incineration ash and application as row bonding materials for pervious ecological brick. *Construction & Building Materials*, 329, 127199. DOI: 10.1016/j.conbuildmat.2022.127199

Onyango, M. (1989). Innocent, M., & Busch, C. (2007). se of Naturally Occurring Pozzolans for Road Construction in Tanzania. *Transportation Research Record: Journal of the Transportation Research Board*, (2), 169–177.

Onyango, M., Macha, I., & Busch, C. (2007). Use of naturally occurring pozzolans for road construction in Tanzania. *9th International Conference on Low-Volume Roads Proceedings, 2*, 1–17. DOI: 10.3141/1989-61

Özata, Ş. (2015). *Kapadokya Bölgesi Kaya Oyma Yapı Sorunları ve Çözüm Öneri leri* . [Master's Thesis, Yildiz Technical University]. Institute of Natural and Applied Sciences https://tez.yok.gov.tr/UlusalTezMerkezi/tezSorguSonucYeni.jsp

Özata, Ş. (2018). Sustainability of Rock Carving Techniques in the World Heritage Site, Cappadocia. In A. C. Antunes & Marius Müller (Eds). *New Perspectives in Interdisciplinary Cultural Heritage Studies* (pp. 85-101.). Mazu Press.

Ozata, S., Akturk, B., & Yuzer, N. (2022). Utilization of waste Cappadocia earth as a natural pozzolan in alkali activation: A parametric study. *Construction & Building Materials*, 329, 127192. DOI: 10.1016/j.conbuildmat.2022.127192

Ozata, S., & Arun, E. G. (2018). Damage assessment for rock-cut dwellings and warehouses: A case study in Bahceli village, Cappadocia. *Masonry International*, 30(3), 81–90.

Özata, Ş., & Arun, E. G. (2018). Damage Assessment of Rock-Cut Ortahisar Castle in Cappadocia Region. *Gazi University Journal of Science*, 31(1), 1–11.

Özata, Ş., Yüzer, N., Ulukaya, S., Aktürk, B., & Kızılkanat, A. B. (2019). Alkali Activation and Pozzolanic Activity of Waste Material Excavated from Rock-Cut Buildings of Cappadocia. *10th International Concrete Congress Proceedings*, 42–50.

Pekmezci, B. Y., & Akyüz, S. (2004). Optimum usage of a natural pozzolan for the maximum compressive strength of concrete. *Cement and Concrete Research*, 34(12), 2175–2179. DOI: 10.1016/j.cemconres.2004.02.008

Qu, F., Su, Y., Lu, D., Li, N., Zeng, X., & Li, W. (2024). Anti-cracking and shrinkage performance of sustainable concrete incorporating high-volume natural pozzolans: A case design for high-speed railway concrete slab tracks. *Case Studies in Construction Materials*, 20(February), e03207. DOI: 10.1016/j.cscm.2024.e03207

Rangrazian, M., Madandoust, R., Mahjoub, R., & Raftari, M. (2023). Experimental investigations on the engineering properties of pozzolanic concrete reinforced with fibers and the reduction of CO2 emissions: A case study on local mineral pozzolan. *Materials Today Communications, 34*(October 2022), 105129. DOI: 10.1016/j.mtcomm.2022.105129

Rispoli, C., Bonis, A., & De, . (2019). The ancient pozzolanic mortars of the Thermal complex of Baia (Campi Flegrei, Italy). *Journal of Cultural Heritage*, 40, 143–154. DOI: 10.1016/j.culher.2019.05.010

Senhadji, Y., Escadeillas, G., Khelafi, H., Mouli, M., & Benosman, A. S. (2012). Evaluation of natural pozzolan for use as supplementary cementitious material. *European Journal of Environmental and Civil Engineering*, 16(1), 77–96. DOI: 10.1080/19648189.2012.667692

Šídlová, M., Šulc, R., Škvára, F., Pulcová, K., Formáček, P., & Snop, R. (2023). Pozzolanic activity of stockpile ash: Comparison of test methods. *Case Studies in Construction Materials*, 19(June), 1–11. DOI: 10.1016/j.cscm.2023.e02396

Snellings, R., Elsen, J., & Mertens, G. (2012). Supplementary Cementitious Materials Supplementary Cementitious Materials. *Reviews in Mineralogy & Geochemistry, 74*(December 2013), 211–278. DOI: 10.2138/rmg.2012.74.6

Stanislao, C., Rispoli, C., Vola, G., Cappelletti, P., Morra, V., & De Gennaro, M. (2011). Contribution to the knowledge of ancient Roman seawater concretes: Phlegrean pozzolan adopted in the construction of the harbour at Soli-Pompeiopolis (Mersin, Turkey). *Periodico di Mineralogia*, 80(3), 471–488. DOI: 10.2451/2011PM0031

Stefanidou, M. (2016). Use of natural pozzolans with lime for producing repair mortars. *Environmental Earth Sciences*, 75(9), 1–8. DOI: 10.1007/s12665-016-5444-5

Tironi, A., Trezza, M. A., Scian, A. N., & Irassar, E. F. (2013). Assessment of pozzolanic activity of different calcined clays. *Cement and Concrete Composites*, 37(1), 319–327. DOI: 10.1016/j.cemconcomp.2013.01.002

Topçu, I. B., Karakurt, C., & Saridemir, M. (2008). Predicting the strength development of cements produced with different pozzolans by neural network and fuzzy logic. *Materials & Design*, 29(10), 1986–1991. DOI: 10.1016/j.matdes.2008.04.005

Toprak, V. (1998). Vent distribution and its relation to regional tectonics, Cappadocian Volcanics, Turkey. *Journal of Volcanology and Geothermal Research*, 85(1-4), 55–67. https://pascal-francis.inist.fr/vibad/index.php?action=getRecordDetail&idt=1640484. DOI: 10.1016/S0377-0273(98)00049-3

Tripathi, A., Manohar, S., & Ramasubramanian, K. (2023). Decoding of lime-based binder materials of ancient vernacular Chettinad constructions in Southern India through microanalytical characterization. *Journal of Building Engineering*, 79(September), 107844. DOI: 10.1016/j.jobe.2023.107844

Tural, H. G., Ozarisoy, B., Derogar, S., & Ince, C. (2024). Investigating the governing factors influencing the pozzolanic activity through a database approach for the development of sustainable cementitious materials. *Construction and Building Materials, 411*(November 2023), 134253. DOI: 10.1016/j.conbuildmat.2023.134253

Turanli, L., Uzal, B., & Bektas, F. (2004). Effect of material characteristics on the properties of blended cements containing high volumes of natural pozzolans. *Cement and Concrete Research*, 34(12), 2277–2282. DOI: 10.1016/j.cemconres.2004.04.011

Turkish Standards Institute. (2008). *Doğal Puzolan (Tras) - Çimento Ve Betonda Kullanılan - Tarifler, Gerekler ve Uygunluk Kriterleri* (TS 25).

Turkish Standards Institute. (2012). *Çimento Deney Yöntemleri - Bölüm 5: Puzolanik Çimentolarda Puzolanik Özellik Deneyi*, (TS EN 196-5).

Türkmeno lu, A. G., & Tankut, A. (2002). Use of tuffs from central Turkey as admixture in pozzolanic cements: Assessment of their petrographical properties. *Cement and Concrete Research*, 32(4), 629–637. DOI: 10.1016/S0008-8846(01)00734-7

Ulukaya, S., Yoruç, A. B. H., Yüzer, N., & Oktay, D. (2017). Material characterization of byzantine period brick masonry walls revealed in Istanbul (Turkey). *Periodica Polytechnica. Civil Engineering*, 61(2), 209–215. DOI: 10.3311/PPci.8868

Ulukaya, S., & Yüzer, N. (2016). Assessment of Pozzolanicity of Clay Bricks Fired at Different Temperatures for Use in Repair Mortar. *J. Mater. Civ. Eng., 04016052–1*, 1–9. .DOI: 10.1061/(ASCE)MT.1943-5533.0001560

United Nations Environment Programme & Yale Center for Ecosystems + Architecture. (2023). *Building Materials And The Climate: Constructing A New Future.* https://wedocs.unep.org/20.500.11822/43293

Vitruvius, M. (2017). *Mimarlık Üzerine (De Arquitectura)* (Yayıncılık, A., Trans.). Çiğdem Dürüşken.

Wahedy, M. N., Sharbatdar, M. K., & Rezaifar, O. (2023). Mechanical, environmental, and economic assessment of sustainable cement mortar using Afghan natural pozzolan as a partial replacement for cement. *Construction & Building Materials*, 386(May), 131574. DOI: 10.1016/j.conbuildmat.2023.131574

Wang, H., Liu, X., & Zhang, Z. (2023). Pozzolanic activity evaluation methods of solid waste: A review. *Journal of Cleaner Production*, 402, 136783. DOI: 10.1016/j.jclepro.2023.136783

Yao, G., Wang, Z., Yao, J., Cong, X., Anning, C., & Lyu, X. (2021). Pozzolanic activity and hydration properties of feldspar after mechanical activation. *Powder Technology*, 383, 167–174. DOI: 10.1016/j.powtec.2021.01.042

Yuzer, N., Ozata, S., & Akturk, B. (2023). *Atık Toprak Kapadokya Doğal Puzolanı ile Geopolimer Malzeme Üretimi ve Hazırlama Yöntemi* (Patent No. 2021/019771). Türk Patent. https://www.turkpatent.gov.tr/bultenler

KEY TERMS AND DEFINITIONS

Aluminous: Refers to materials that contain aluminum oxide (Al_2O_3), commonly found in bauxite, clays, and other minerals.

Anorthite: A type of Ca-feldspar found in igneous rocks, indicating high formation temperatures above 850°C.

Blast furnace slag: A by-product of iron production in a blast furnace, commonly used in cement production as an SCM due to its hydraulic properties.

Buffer zone: A neutral area surrounding a protected site, such as a UNESCO World Heritage site, meant to provide an additional layer of protection against external threats.

Calcium hydroxide: Also known as lime ($Ca(OH)_2$), it is a compound used in construction materials, particularly in cement and plaster, due to its reactivity with pozzolanic materials to form a stronger, more durable product.

Calcination: A thermal process used to alter the properties of certain materials, such as natural zeolites, to enhance their pozzolanic activity and make them more effective in cement and concrete applications.

Carbonated manganese mineral: Typically a secondary mineral found in deposits, associated with manganese and carbonates.

Cementitious materials: Materials that possess binding properties similar to those of cement, capable of setting, hardening, and bonding other materials together.

Cementitious properties: The ability of a material to act like cement, binding other materials together when mixed with water, and hardening over time.

Clinker: A solid material produced during the manufacturing of cement, typically used in combination with other materials, such as pozzolans, to create various types of cement.

Cl: Chemical symbol for chloride, a compound that is regulated in pozzolanic materials to prevent corrosion in reinforced concrete structures.

Cristobalite: A high-temperature polymorph of silica (SiO_2). It transforms into a high-temperature form at temperatures between 200-270°C. Stabilizing cristobalite requires even higher temperatures.

Electrical conductivity test: A method to measure pozzolanic activity by tracking changes in the electrical conductivity of a cement-lime slurry during the pozzolanic reaction.

Feldspars: A group of alumina-silicate minerals that are common in the Earth's crust. They are divided into two main groups. K-feldspar (alkali feldspar) includes minerals like orthoclase. Ca-feldspar (plagioclase feldspar) includes minerals like anorthite.

Fly ash: A by-product of coal combustion in power plants, often used as an SCM because of its hydraulic properties.

Hydraulic behavior: The ability of a material to harden or set in the presence of water, similar to the behavior of cement.

Hydraulic activity properties: The ability of a material, such as certain types of mortar or concrete, to harden and gain strength when exposed to water, making it particularly useful for underwater or damp construction conditions.

Isothermal calorimetry test: A technique to measure the heat generated during the hydration of a pozzolan-cement mixture, which indicates the pozzolanic activity.

Magmatic rocks: Rocks formed from the cooling and solidification of magma beneath the Earth's surface. These can include various types of volcanic rocks found in regions like Cappadocia.

Metakaolin: A pozzolanic material produced by the thermal treatment of kaolin clay. It is used in construction to improve the properties of concrete and other cementitious materials.

MPa: Megapascal — A unit of pressure or stress used to measure compressive strength.

Natural pozzolans: Siliceous or siliceous-alumina materials of natural origin that, while not cementitious on their own, can react with calcium hydroxide in the presence of water to form compounds with cementitious properties. They are often used to enhance the strength and durability of concrete.

N-class natural pozzolan: A classification within the ASTM C618 standard that indicates a pozzolan with specific chemical properties, including a high content of reactive silica and alumina.

Orthoclase: A type of K-feldspar, also known as potassium feldspar, commonly found in igneous rocks.

Pumice: A light, porous volcanic rock formed from lava that cools quickly. It is commonly used in construction as a lightweight aggregate and for its insulating properties.

Pozzolanic material: A siliceous or siliceous-aluminous material that has minimal cementitious properties on its own but, when finely divided and mixed with water, reacts with calcium hydroxide to form compounds with cementitious properties. These materials are often used to enhance the properties of concrete.

Pozzolanicity: The ability of a material to exhibit pozzolanic activity, meaning it can react with calcium hydroxide to form cementitious compounds, thereby contributing to the strength and durability of cement-based materials.

Portland cement: A common type of cement used in construction, patented in 1824. It largely replaced the use of natural pozzolans in concrete and cement production due to its ease of use and consistent performance.

Pulvis Puteolanis: A term coined by the Roman architect Vitruvius to describe a type of volcanic ash from the area around Puteoli (modern Pozzuoli, Italy), which was used in Roman concrete to enhance its strength and durability.

Quartz: A common form of silica (SiO_2) with different forms such as low quartz that formed at lower temperatures and high quartz that formed at higher temperatures, transitioning from low to high quartz at 573°C.

Rock-hewn structures: These are architectural structures that have been carved directly into rock formations. Commonly found in areas with soft, workable rock like those in Cappadocia, these structures can serve religious, military, or civil purposes.

Silica fume: A by-product of silicon and ferrosilicon alloy production, used as an SCM to enhance the strength and durability of concrete.

Siliceous: Refers to materials rich in silica (SiO_2), a chemical compound often found in sand, quartz, and certain types of rocks.

Strength Activity Index: A method used to assess the pozzolanic activity of a material by measuring its effect on the compressive strength of cement mortar.

Supplementary cementitious materials (SCMs): They are generally used by mixing with materials such as cement that have a certain amount of binding properties, are a group that includes materials that show hydraulic and/or pozzolanic behavior.

Vaterite: One of the three polymorphs of calcium carbonate ($CaCO_3$), the others being calcite and aragonite. Vaterite forms under specific conditions and is less stable compared to calcite and aragonite.

Volcanoclastic layers: Geological layers composed of fragmented volcanic material, including ash, pumice, and other volcanic debris, which can be used as raw materials in construction.

Waste soil: The by-product of rock excavation, especially during the construction of rock-hewn structures.

X-Ray Diffraction (XRD) analysis: A technique used to identify the crystalline structure of materials by measuring the pattern of X-rays scattered by the internal structure of the material.

X-Ray Fluorescence (XRF) analysis: A method used to determine the chemical composition of materials by measuring the fluorescence (or secondary) X-rays emitted from a material when it is excited by a primary X-ray source.

Zeolite: A group of minerals that are commonly found in volcanic rocks and are known for their ability to absorb water and gases. Zeolites can be used as pozzolanic materials in construction.

ENDNOTES

1 Translated by Çiğdem Dürüşken.

2 Data for materials other than Cappadocia pozzolan in the chart is sourced from (Nguyen et al., 2018).

Chapter 9
Durability of Manufactured Sand (m-Sand) Concrete: A Review

Ankit Singh Beniwal
https://orcid.org/0000-0001-8655-903X
JECRC University, India

Nirbhay Mathur
https://orcid.org/0009-0007-4197-445X
JECRC University, India

Ram Vilas Meena
JECRC University, India

ABSTRACT

Concrete is a homogenous mixture that comprises of binder material like cement, filler material like coarse and fine aggregate and water for preparation of gel like paste to bind the ingredients. Due to growing environmental concerns, the mining of river sand is discouraged lately and is replaced by quarry dust or stone dust or manufactured sand (M-sand). Further the use of pozzolanic material such as fly ash and pond ash is also widely being accepted with the use of Portland Pozzolanic Cement (PPC). Change in behavior is obvious in concrete made by PPC and manufactured sand (m-sand). This research reviews the available literature on effects of concrete's durability due to this replacement. Researchers around the world have come up with a possible solution of mixing steel fibers with concrete to cover durability aspect. This review attempts to identify the potential of m-sand in PPC based concrete to alter the durability and further the significance of steel fibers use

DOI: 10.4018/979-8-3693-5513-8.ch009

to alter behavior of concrete in order to minimizing the undesirable effects of this combination.

1. INTRODUCTION

River sand, a finite and increasingly important resource in the construction industry, is consumed by India more than any other country after China. Sand mining from riverbanks and beds has an influence on the ecosystem, even when done under controlled conditions. To prevent river beds from eroding and to ensure that it serves its natural purpose of protecting the environment, now the governments enforce regulations on sand mining. Government's attempt to control the river sand mining has led to scarcity of fine aggregate for production of concrete. In a country like India, where the construction sector is booming due to rapid urbanization, this demand of fine aggregate has found its way towards other alternatives like quarry sand, stone dust and manufactured sand (M-sand).

Similarly the use of fly ash as pozzolanic material in cement has already in use to reduce the carbon foot prints of cement production industry which hugely produces the CO_2 and contributes to greenhouse gases. Fly ash is therefore a perfect raw material for cement concrete mixtures, which improves the concrete's strength and facilitates smoother pumping. The use of Fly Ash is environmentally sustainable as the waste materials from industry are successfully used to manufacture high-quality concrete. The Portland Pozzolanic Cement (PPC) being cost effective than OPC, is now becoming a preferred choice for commercial concrete production.

Rapid urbanization of cities wherein the concrete in mainly being used in massive construction of infrastructure and CC roads, the commercial aspect of concrete production in view of shortage of natural river sand has compelled the stakeholders to use PPC as cement and manufactured sand as fine aggregate. Further the use of such concrete is being encouraged by every section of society as a sustainable solution as it addresses two major environmental concerns.

The main objective of this research is to review the available literature on influence on characteristics of manufactured sand on PPC based concrete and further the properties enhancement if steel fibers are mixed in different proportions with such cement concrete. The primary characteristic which is accentuated in this paper is the durability aspect of concrete, which is becoming a growing concern now a days. Additionally, it surveys available literature on enhancement of concrete's durability by the appropriate use of different steel fibers.

2. DURABILITY

Durability is the capacity of concrete to endure the conditions for which it is intended for a considerable amount of time without degrading. It typically refers to the lifespan of problem-free operation. Most structures are made with an intended life span of satisfactory service duration and durability of concrete plays a major role in delivery of satisfactory service life under working exposure conditions. Depending on the exposure to environment and the properties of ingredients, different mixes of concrete exhibit a lot of variation in durability levels (Kumar et al., 2024). Apart from cement content, water cement concrete ratio, mix proportions, degree of compaction, service loads pattern and exposure conditions, the factor that majorly affect the durability is the quality of aggregates. The quality of aggregate determines the shrinkage characteristics, thermal stability, moisture absorption & release and permeability of concrete which in turn affect the durability in long term. In concrete, the impacts of particle texture along with shape of fine aggregates are more pronounced than those of coarse aggregates. The strength of concrete may be improved by using angular shape of aggregates as it provides better interlocking. Though it is observed that M-sand particles that have more angularity, when used in cement concrete, causes low workability due to more surface area (Manju & K, 2016).

2.1 Drying Shrinkage

It is the contraction in hardened concrete mix caused by capillary water loss. Before the application of service loads on concrete, this shrinkage increases the tensile stresses, which often leads to cracking, internal warping, and external deflection and finally resulting to loss of durability. Shrinkage while concrete is drying depends on a number of factors. These factors include the characteristics of constituents, their quantities, mixing methodology, presence of moisture throughout the curing process, the environment's dryness, and the size of the member (Zhao et al., 2016). The primary factor influencing drying shrinkage is the overall water content in fresh concrete. Drying shrinkage has a straight linear relationship with water content of fresh concrete (Maggenti et al., 2013).

2.2 Factors Affecting Drying Shrinkage

The decline of water in cement paste along with the capillary water through the slow process of evaporation is the cause of drying shrinkage. The drying shrinkage effects are more pronounced if more water is there in the freshly laid concrete. The main cause of concrete shrinkage is the evaporation of the capillary water that is absorbed by different constituents. The hydrated cement paste shrinks due to moisture loss. This shrinkage of cement paste depends on the composition and fineness of the cement. Relative humidity, temperature, and air circulation are other environmental factors that affect how much material shrinks during drying (ACI, 1997). Concrete in most parts of Indian subcontinents often exposed to hot & dry environment and subjected to greater drying shrinkage than concrete exposed to alternate wetting and drying conditions. Because of higher humidity and slower evaporation at lower temperatures, drying shrinkage is typically reduced. But it is not the case in most parts of the country and concrete remains subjected to hot & dry environmental conditions for long durations. Concrete becomes less compressible and shrink more when the aggregates have high density and high elastic modulus. Hikotsugu Hyodo et al. evaluated the effect of aggregate properties on drying shrinkage of concrete. Multiple regression analysis revealed that aggregate shrinkage, not elastic modulus, was the main cause of concrete shrinkage. A contact-type strain gauge was used to measure the drying shrinkage of a concrete specimen with dimension of 100 x 100 x 400mm in compliance with JIS A 1129. A study conducted for developing the regression model for predicting the drying shrinkage indicated that the moisture absorption is linearly related with shrinkage in aggregates, while depending on the type of rock the relation is different.

3. RIVER SAND AND M- SAND

In concrete when the fine aggregate is replaced by quarry dust, stone dust or manufactured sand the drying shrinkage behavior of concrete is changed substantially as the shape & size properties and other physical properties of alternative sand are different from river sand. The particles of the river sand are fairly angular to spherical due to continuous abrasion in water and mutual collision. Due to attrition, river sand typically have a smoother surface texture and better shape. Whereas M-Sand is crushed stone aggregate that may have rough texture, flaky shaped with sharp edges, produced from hard granite stone. It is washed and graded in equality to be used as a substitute of natural river sand. Due to its filler content and the particles' mutual interlocking mechanism, m-sand-infused concrete has improved mechanical and durability properties.(Shanmuga Priya, 2017). Scanning Electron Microscopy (SEM) study performed by Vijaya B.(Bhoopathy et al., 2018) on manufactured sand containing medium strength concrete revealed that due to attrition from river flow, natural river sand particles are more rounded and have a smoother surface texture than M-sand. On the other hand the M-sand particles have sharp edges and a rough texture, they better interlock with cement particles, which lowers porosity.

The SEM images of M-sand indicate that the particles are elongated and angular in nature with rough surface texture and more micro-roughness. While for river sand fairly spherical shape and smooth surface texture can be seen when observed through SEM images.(Suriya et al., 2023).

The m-sand has lower stiffness compared to river sand and has greater water absorption. Because of their greater capacity to absorb water, m-sand release more water over the time. This makes it possible for the cement to compensate for the absorption of capillary water more effectively, hence minimizing autogenous shrinkage. The amount of water that can evaporate is however increased, which in turn causes drying shrinkage to rise. Additionally, m-sand's reduced stiffness lessens internal resistance to the volumetric contraction of cementous mass, which also encourages drying shrinkage. (Revilla-Cuesta et al., 2022)

3.1 Effect of m-sand on Concrete

Experimental studies suggest that concrete made up of m-sand has 18% to 60% higher compressive strength in contrast to conventional concrete made up by river sand for different concrete grades. Whereas split tensile strength and flexural tensile strength also increased substantially.(Vijaya & Senthil Selvan, 2015) Rapid Chloride Migration, Rapid Chloride Permeability, and Accelerated Carbonation Tests were performed on samples of river sand and manufactured sand in order to compare the evaluation of durability criteria. Test findings showed that the concrete made with m-sand in a chloride-induced environment performed relatively better for durability parameters than those with river sand.

3.2 Effect of m-Sand on Concrete In Context Of Shrinkage

Many experimental studies have been conducted to know the effect on drying shrinkage of concrete by substitution of natural river sand with m-sand or stone dust. Felixkala T and Sethuraman V.S. have tested concrete containing granite powder as fine aggregate with different dosages of admixture and studied its effects on strength and drying shrinkage. The locally available river sand was also used to prepare control mix to compare the results. With different concrete mixes, the variation of shrinkage characteristics, including maximum crack length, maximum crack breadth, and total number of cracks, was investigated. The shrinkage parameters, as well as the plastic and drying shrinkage strains, were experimentally investigated on high performance concrete with fine aggregate replaced by granite powder and partially replaced cement made with 10% fly ash, 10% slag, 7.5% silica fume, and 1% superplasticizer for 0.35 water-cement ratio at 32°C (±2°C) and 38°C (±2°C) for curing of 1, 7, 14, 28, 56, and 90 days. For both 32°C and 38°C curing temperatures, the drying shrinkage strains in the granite powder concrete specimens were greater than those in the conventional concrete specimens (Sethuraman & Kala, 2013).

Ilongowana R. and Mahendrana N. in order to assess the feasibility of using quarry rock dust as a replacement of river sand have evaluated the compressive strength and flexural strength for three different proportions of dust with OPC. Durability studies comparing the conventional concrete with Quarry Rock Dust (QRD) concrete were carried out. For these studies the mix proportions for QRD concrete were adopted as 1:0.99:1.56 and 1:0.91:1.56 after several trials and for the conventional concrete the Indian Standard mix proportions (by weight) were used. The objective of this durability study was to look into deterioration, water absorption, and drying shrinkage for both conventional concrete and QRD concrete in accordance with standards and other requirements. Additionally, the findings of this study demonstrate that the drying shrinkage strains of concrete made with quarry rock dust are significantly higher than those of conventional concrete. However, they have demonstrated equal strain to conventional concrete at a later age. It may be observed here that use of quarry dust resulted in more drying shrinkage strain in first two months period and then got reduced.(Ilangovana et al., 2008)

Mishra R.K. et al. conducted an experimental investigation on early age shrinkage pattern of concrete prepared by replacing fine aggregate with pond ash and granulated blast furnace slag. The binder content was OPC, PPC and PSC types of cement. The objective of this study was to establish relationship between concrete maturity and concrete shrinkage under different materials proportions. The tests were conducted as per IS:4031-1988. Using OPC, PPC, and PSC cement filled in a larger cone in accordance with conventional testing guidelines, shrinkage of concrete with pond ash and granulated blast furnace slag was observed. Observations were taken and plotted for a day depict the shrinkage of concrete using OPC, PPC and PSC with Pond Ash as fine aggregate and slag respectively shows the shrinkage of concrete using OPC, PPC and PSC with Granulated Blast Furnace Slag as fine aggregate. These curves indicate that there is considerable increase in shrinkage pattern when concrete is prepared with PPC cement as well as fine aggregate is replaced with furnace slag (Mishra et al., 2016).

Umamaheswaran V et al. have conducted an experimental programme to study the effect of using m-sand as fine aggregate in high grade cement concrete of M100. The concrete strength and durability properties were compared with river sand also. Adding three mineral admixtures Metakaolin, Silica fume, and Alccofine in both m-sand and river sand, the experimental programme includes testing of materials and designing of mix for M100 grade. River sand concrete has shown to have less drying shrinkage than manufactured sand concrete. However, the decrease was only borderline (0.003%). Among the 3 mineral admixtures Alccofine mixes gave the minimum drying shrinkage value. In this study synthetics polypropylene fibers were also used as fiber reinforcing material. The Alccofine mix with river sand showed the least shrinking and shown superior dimensional stability. The addition of fibers caused the extent of reduced drying shrinkage in m-sand. (Umamaheswaran et al., 2015)

4. USE OF STEEL FIBERS

Here it is worth noting that research work conducted so far on concrete's behavior when mixed with some artificial sand like stone dust or m-sand, indicates occurrence of more drying shrinkage. The drying shrinkage in concrete having m-sand can be considerably reduced if fiber reinforcement is used. As soon as the concrete develops microscopic cracks, steel fibers become active. The fibres can carry stress across the gap, bridge it, and in the process offer some resistance to the fissure's expanding and fracture processes (Kumar et al., 2024). The post-cracking, or residual tensile strength, of a fiber reinforced concrete member is its most crucial feature (Ng & Nyan, 2017).

Harle S.M. performed a comprehensive experimental study and evaluated some key concrete properties such as compressive strength, flexural strength and workability. The study was based on concrete behavior with Hooked steel fiber, Flat crimped steel fiber and rubberized steel fiber. The cement used was OPC type cement with natural stone aggregates. This study revealed considerable improvement in compressive and flexural tensile strength on addition of steel fiber, while workability goes on declined (Harle, 2019).

Some researchers have also worked to assess the behavior of concrete' shrinkage apart from other characteristics when steel fibers are used as reinforcement. One study was done by Zhao S. et al. with steel fiber of milling type, 36 mm length, equivalent diameter of 1.35 mm, and an aspect ratio of 27.1 to assess the drying shrinkage. The fine aggregate used was manufactured sand in continuous gradation with size of 1.6–5 mm and lightweight sand of fine sintering expanded shale along with OPC cement. The test results of drying shrinkage in this study indicate that the increased volume fraction of steel fiber slowed down the development of drying shrinkage. A frame type construction that prevents the shrinkage deformation of concrete is created by the three-dimensional distribution of steel fibers in concrete. Due to the better and better net structure bridged among steel fibbers, this effect becomes stronger as the volume fraction of steel fiber increases. Steel fibers' control of concrete's drying shrinkage was evident due to their high tensile strength and elastic modulus.

Zaid O. et al. performed an experimental study with double hooked end steel fibers 55 mm in length and 62 aspect ratio for concrete production. The cement used in this study was Type I Portland cement as per ASTM C150 which is OPC in fact. While the aggregates were natural stone aggregates and crushed stone aggregates obtained after crushing the stones. Super plasticizers were also used to enhance the workability. Tests conducted for various properties of concrete for different proportions of ingredients including addition of GBFS. The outcomes of test after adding 2% fibers show a decrease in shrinkage strain. As observed in this study, the decrease in drying shrinkage of recycled aggregate concrete compared to natural aggregate concrete was attributed to steel strands (Zaid et al., 2022).

5. SUMMARY AND CONCLUSION

From literature review it is observed that most researchers have conducted experimental studies on various combinations of fine aggregate materials to replace the river sand. The growing restrictions on river sand mining have prompted the demand of alternative solutions. The available research mainly concentrates on performance of concrete made up with m-sand in terms of strength and workability. Research works on many chemical admixtures are underway to enhance the desired properties of concrete made up of m-sand or stone dust. While promoting the m-sand as an appropriate alternative to river sand, the other undesirable effects of m-sand are overlooked. There may be larger commercial aspects of m-sand over such effects, but these may be dealt efficiently and resolved. While the primary focus is on strength & workability, the durability is one such aspect that needs more attention if locally available m-sand is being used in concrete.

The drying shrinkage which is one such unwarranted effect of most of the m-sands based concrete, needs to be addressed efficiently. While steel fibers are rapidly gaining attention to be used as primary or secondary steel reinforcement, most of the research work concentrates on improvement of tensile & flexural strength of concrete made up by OPC type cement and river sand. Therefore, more research work is required to study the behavior of PPC based concrete aggregated with locally available m-sand, which is subjected to drying shrinkage. There is very less work to know the effect of steel fibers to check the drying shrinkage in relatively thinner sections such as pavement top and unsupported floor slabs.

REFERENCES:

ACI. (1997). American Concrete Institute. ACI 209R-92 [Reapproved]. *Prediction of Creep, Shrinkage, and Temperature Effects in Concrete Structures.*, 47, 92.

Bhoopathy, V., Selvan, S. S., & Vijaya, B. (2018). Corrosion effects on the flexural strength behaviour of concrete containing Manufactured Sand Corrosion effects on the flexural strength behaviour of concrete containing Manufactured Sand View project Experimental and Analytical Investigation on Medium strength concrete containing Manufactured sand View project CORROSION EFFECTS ON THE FLEXURAL STRENGTH BEHAVIOUR OF CONCRETE CONTAINING MANUFACTURED SAND. *Article in Journal of Engineering and Applied Sciences, 13*(5). www.arpnjournals.com

Harle, S. M. (2019). *Network Security and Communication Experimental Investigation on Steel Fibers in the Concrete. 4*, 1–7. www.ijsrnsc.org

Ilangovana, R., Mahendrana, N., & Nagamanib, K. (2008)... *Strength and Durability Properties of Concrete Containing Quarry Rock Dust As Fine Aggregate.*, 3(5), 20–26.

Kumar, A., Liu, Y., & Kumar, R. (Eds.). (2024). *Handbook of Intelligent and Sustainable Manufacturing: Tools, Principles, and Strategies.* CRC Press., DOI: 10.1201/9781003405870

Kumar, A., Rajak, D. K., Kumar, P., & Kumar, A. (Eds.). (2024). *Advances in Sustainable Biomaterials: Bioprocessing 4.0, Characterizations, and Applications* (1st ed.). CRC Press., DOI: 10.1201/9781003434313

Maggenti, R., Knapp, C., & Fereira, S. (2013, July). Controlling Shrinkage Cracking. *Concrete International*, (July), 36–41.

Manju, M., & K, A. N. (2016). Durability Studies on Concrete by Replacing Natural Sand with M-Sand-A Review. *International Journal of Emerging Technology and Advanced Engineering*, 6(3), 3–7.

Mishra, R. K., Tripathi, R. K., & Dubey, V. (2016). Early age shrinkage pattern of concrete on replacement of fine aggregate with industrial by-product. *Journal of Radiation Research and Applied Sciences*, 9(4), 386–391. DOI: 10.1016/j.jrras.2016.05.003

Ng, T. S., & Nyan, T. S. H. (2017). Structural Application of Steel Fibres Reinforced Concrete With and Without Conventional Reinforcement. *The New Zealand Concrete Industry, 3101*(2006), 2–15.

Revilla-Cuesta, V., Evangelista, L., de Brito, J., Skaf, M., & Ortega-López, V. (2022). Mechanical performance and autogenous and drying shrinkage of MgO-based recycled aggregate high-performance concrete. *Construction and Building Materials, 314*(November 2021). DOI: 10.1016/j.conbuildmat.2021.125726

Sethuraman & Kala. (2013). Shrinkage Properties of HPC using Granite Powder as Fine Aggregate. *International Journal of Engineering and Advanced Technology*, 2(3), 637–643.

Shanmuga Priya, T. (2017). Experimental investigation on high performance RC column with manufactured sand and silica fume. *IOP Conference Series: Materials Science and Engineering, 263*(3), 0–8. DOI: 10.1088/1757-899X/263/3/032021

Suriya, D., Chandar, S. P., & Ravichandran, P. T. (2023). Impact of M-Sand on Rheological, Mechanical, and Microstructural Properties of Self-Compacting Concrete. *Buildings*, 13(5), 1126. Advance online publication. DOI: 10.3390/buildings13051126

Umamaheswaran, V., Sudha, C., Ravichandran, P. T., & Kannan Rajkumar, P. R. (2015). Use of M Sand in High Strength and High Performance Concrete. *Indian Journal of Science and Technology*, 8(28). Advance online publication. DOI: 10.17485/ijst/2015/v8i28/84018

Vijaya, B., & Senthil Selvan, S. (2015). Comparative study on the strength and durability properties of concrete with manufactured sand. *Indian Journal of Science and Technology*, 8(36). Advance online publication. DOI: 10.17485/ijst/2015/v8i36/88614

Zaid, O., Mukhtar, F. M., M-García, R., El Sherbiny, M. G., & Mohamed, A. M. (2021, November). M-García, R., El Sherbiny, M. G., & Mohamed, A. M. (2022). Characteristics of high-performance steel fiber reinforced recycled aggregate concrete utilizing mineral filler. *Case Studies in Construction Materials*, 16, e00939. DOI: 10.1016/j.cscm.2022.e00939

Zhao, S., Li, C., Zhao, M., & Zhang, X. (2016). Experimental Study on Autogenous and Drying Shrinkage of Steel Fiber Reinforced Lightweight-Aggregate Concrete. *Advances in Materials Science and Engineering*, 2016, 1–9. Advance online publication. DOI: 10.1155/2016/2589383

Chapter 10

Experimental Investigations on Thermo-Hydraulic Performance of TIO2/ Water Nanofluids in a Flat Tube

Anish Sihag
Guru Jambheshwar University of Science and Technology, India

Neeti Arora
Guru Jambheshwar University of Science and Technology, India

Munish Gupta
Guru Jambheshwar University of Science and Technology, India

Puneet Katyal
Guru Jambheshwar University of Science and Technology, India

ABSTRACT

The investigation aims to experimentally determine the thermophysical properties and heat transfer characteristics of TiO2 and water nanofluids flowing across a flat tube while characterizing the TiO2 nanoparticles. Thus, the paper contributes by experimentally assessing TiO2/water nanofluids in a flat tube, potentially offering improved heat transfer efficiency and insights into nanoparticle characterization, benefiting engineering and sustainability. The characterization of nanoparticles is

DOI: 10.4018/979-8-3693-5513-8.ch010

performed using FESEM, XRD, and FT-IR. The experiment was conducted under a constant heat flux in a laminar regime. The nanofluid exhibited a maximum increase of 14.39% in heat transfer coefficient (HTC), while the highest pressure drop occurred at a 0.04% concentration by weight and a flow rate of 0.5 liters per minute (lpm) when compared to the base fluid. However, at lower concentrations, there was only a minor increase in HTC. It was noted that the pressure drop is directly proportional to the nanoparticle concentration.

I. INTRODUCTION

Heat transfer, a process of utmost importance, in many applications like energy conversion etc. Wide range of applications make it a hot topic for research so there's been a good number of research in the field of heat transfer. Researchers had extensively studied transfer of heat both numerically and experimentally. This manuscript studies on forced convection and various heat conditions, different Reynold's number and through different tubes. The sections ahead give an overview of such discoveries to keep the current situation perspective.

The usage of nanoparticles-based nanofluid in heat transfer was first made known by K.V. Liu et al. (1988) at Argone National Laboratory, which lead towards a vital increase in thermal conductivity. M.M. Heyhat et al. (2012) found that heat transfer coefficient (HTC) using alumina oxide particles in water in round tube with sustained divider temperature underneath turbulent flow was increased by 23%. M. Gupta et al. (2016) investigate the heat transfer characteristics of Multi walled carbon nanotube and water basefluid with a constant heat flux condition. It was found that inclusion of a little MWCNT nanoparticle by weight at a specific flow rate decreases the Reynolds Number, and HTC keeps on increasing as weight fraction of nanoparticle increases and flow velocity increases. M. Modak et al. (2015) experimentally calculated the increase in HTC by Alumina oxide nanoparticle and water base fluid through a circular tube in turbulent regime (Re 5000-12000) and 0.15 and 0.6 percent volume concentration. There was an increase of 12.8% and 32.6% than pure water with volume concentration of 0.15 and 0.6 percent respectively. G.S Sokhal et al. (2018) conducted experiment to determine the HTC and PD inside a flat tube using Copper oxide and H_2O nanofluid at 0.1 and 0.5 percent concentration by weight of nanofluid. It was seen that owing to increase in weight concentration, Reynold no., and temperature of the nanofluid, the Nusselt number for flat tube rises. C. Qi et al. (2018) experimentally studied the heat transfer and hydraulic effects of TiO_2/H_2O nanofluid in elliptical tube of 0.5, 0.3, and 0.1 weight percent. They found that there was an improvement of 27.9% in Nusselt number when compared against circular tube. M.Kh. Abdolbaqi et al. (2017) conducted experimental analysis of HTC using Bioglycol and water based Titanium dioxide nanofluid through the flat tube with turbulent condition with 50 nm average particle size with 0.5 to 2.0% concentration by volume. It was found that the Nusselt number is greater by 28.2% than the H_2O and it improved with increasing Reynolds number and operating temperature, and the frictional resistance at 1.0% volume was more by 6.1% than the base fluid. Using TiO_2 and Silicon dioxide nanoparticles in H_2O/EG (Ethylene Glycol) base fluid the HTC was observed to be increased by 9.02% by K. Abdul Hamid et al. (2018). The nanofluid contains 50:50 Titanium dioxide and SiO_2 at 1.0% concentration by volume. In experimental study of thermo-hydraulic performance of TiO_2/H_2O nanofluid flowing in triangular cross section tube with tapes (twisted) at nanoparticle mass concentration of 0.1%, 0.3%, and 0.5% and Re 400 to 9000 by Cong Qi (2018). It was observed that in laminar and turbulent regime there was improvement of 52.5% and 34.7%. Orhan et al. (2019) experimentally and numerically studied the TiO_2/water nanofluid in turbulent region at 0.5, 0.75 and 1.0% weight fraction of nanoparticle. The maximum Nusselt number observed was 133.61. M. Vakili et al. (2013) in an experiment with TiO_2/ethylene glycol & water nanofluid shows a fair increase in HTC in contrast of only water as basefluid. A general correlation for finding heat transfer using nanofluids is developed for convective heat transfer by Jahar Sarkar (2011). Yanjun Li (2015) experimentally find the PD and the heat transfer in convection using ZnO and EG – water nanofluids which displayed an increase of 30% in heat transfer. Kin Yuen Leonga (2017) observed using copper/TiO_2 nanoparticle in EG base the thermal conductivity was increasing with increasing weight concentration.

Many papers have concluded that researchers have used various methods to increase the heat transfer coefficient of thermal fluids using nanoparticles. These methods include the use of different materials for nanoparticles, various sizes, the matrix of surfactants to enhance stability, the use of different tube cross-sections, the addition of various types of tapes in tubes to increase surface area, changing the nanoparticle concentration in the base fluid, and mixing two or more nano sized particles or thermal fluids, such as the base fluid.

While there has been a significant amount of research on TiO_2 and water nanofluids, but there is limited data available for low concentrations of nanoparticles (0.01%, 0.02%, 0.03%, and 0.04% weight fraction) in laminar flow conditions inside a flat tube. To address this research gap, the current study and experiment are being conducted. The flat tube offers a greater mean effective diameter (i.e., hydraulic diameter). This study has the potential to benefit various applications, such as automobile cross-flow heat exchangers that require flow through narrow flat tubes with low power consumption and minimal impact on the engine case due to the low mass of the narrow particles, resulting in low impact force. Additionally, nanoparticles like TiO_2 and base fluids like water could find application in the biomedical field, offering a favourable cost-to-effectiveness ratio. These materials provide excellent thermohydraulic performance at a low cost, are corrosion-resistant, and do not react with the human body when used in biomedical applications like in dialysis machine. Furthermore, due to their small size, they also have a high surface area to volume ratio, resulting in a significantly increased surface area.Top of Form

II. CHARACTERISATION OF TiO_2 NANOPARTICLES

The nanoparticles are characterized using XRD, FESEM, EDAX, and FT-IR. The FESEM magnified image shows the shape and size of TiO_2 nanoparticles (Behl, V et al. 2022; Kumar, A. 2023). The average size (35 nm) and spherical shape of the TiO_2 nanoparticles can be seen in Figure 1, obtained using FESEM at a magnification of x80,000. The XRD analysis as shown in Figure 2 indicates that the TiO_2 has an anatase crystal structure (K. Manikandan et al. 2017). The functional group of the nanoparticle is determined using FT-IR (Andrea León et al. 2017). The TiO_2 functional groups are visible in Figure 3. The three bands of TiO_2 nanoparticles are shown in the FT-IR spectrum. The band at 3436.26 cm^{-1} corresponds to the O-H (hydroxyl group) stretching vibration of the TiO_2 nanoparticles. The second band at 1634.06 cm^{-1} represents the bending mode of water Ti-OH. The peaks at 463.02 and from 505.43 to 700 cm^{-1} indicate the bending and stretching modes of Ti-O-Ti, confirming the presence of TiO_2.

Figure 1. FESEM scans at 80000 times magnification

Figure 2. XRD peaks of TiO_2 nanoparticles

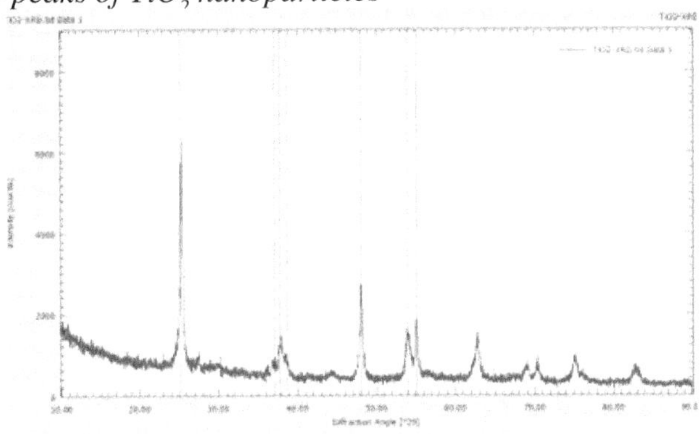

Figure 3. FT-IR spectrum of TiO_2 nanoparticles

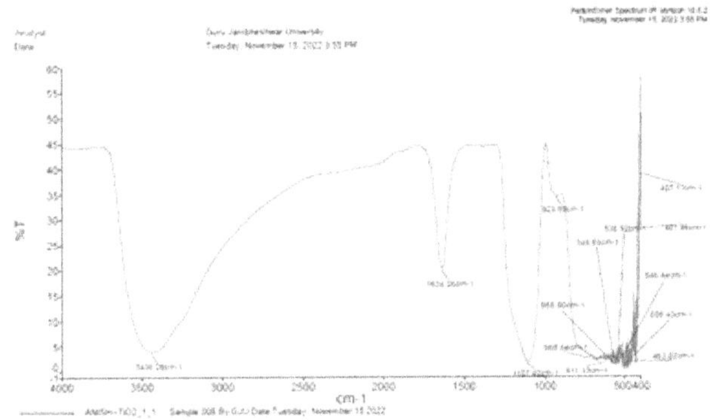

III. NANOFLUIDS PREPARATION

The nanofluids are developed by two-step method with TiO_2 and H_2O as shown in Figure 4 (Singh, V. et. al. 2021). They have a thermal conductivity of 8.4 W/mK and 0.62 W/mK respectively. The nanoparticle has average particle size of 35 nm, and a density of 4230 kg/m^3. The experiment is conducted with four different nanoparticle concentrations: 0.01%, 0.02%, 0.03%, and 0.04% by weight, calculated using Equation (1)

$$w\% = \frac{w_p}{w_p + w_w} \tag{1}$$

To establish proper dispersion of the nanoparticles in the matrix, they are initially stirred using a magnetic stirrer. Ultrasonication is employed to prevent early nanoparticle agglomeration and to break up any pre-existing agglomerated particles. The prepared nanofluids can be observed in Figure 5.

Figure 4. Steps in nanofluids preparation

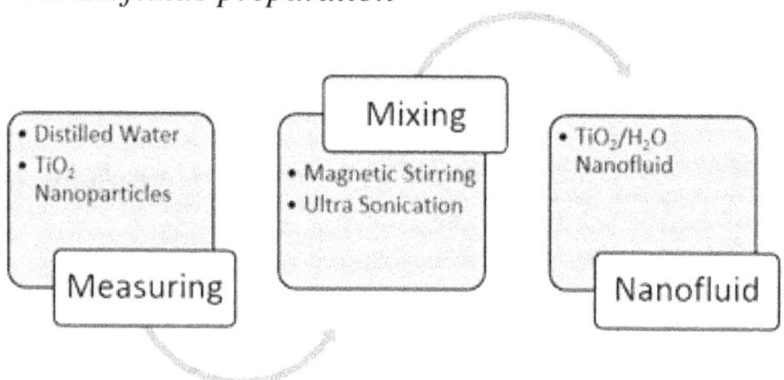

Figure 5. Final prepared nanofluids

IV. EXPERIMENTAL SETUP

The delineative diagram of the experimental rig can be observed in Figure 6 and a photographic image can be seen in Figure 7. It comprises a copper flat tube with a nichrome heater, a refrigeration unit, a peristaltic pump with flow regulation, a pressure gauge, a storage tank for the nanofluid, thermocouples, stop valves, and a data acquisition system. The two-meter-long flat tube used has a major diameter of 17.5mm and a minor diameter of 5.5mm. It is wrapped with a nichrome heating wire and properly insulated. Eight K-type thermocouples are placed with a constant distance along the tube. The wall thickness of the flat tube is 1mm. The controlled flow peristaltic pump is connected to the nanofluid storage tank to circulate the nanofluid with laminar flow. The pump can adjust the discharge within a range of 0.01 lpm to 3 lpm, with a step size of 0.01 lpm. A constant heat flux is maintained by the refrigeration unit. The pressure drop across the flat tube is measured employing a Bourdon tube pressure gauge. Readings can be recorded automatically using the data acquisition system.

Figure 6. Experimental setup schematic diagram

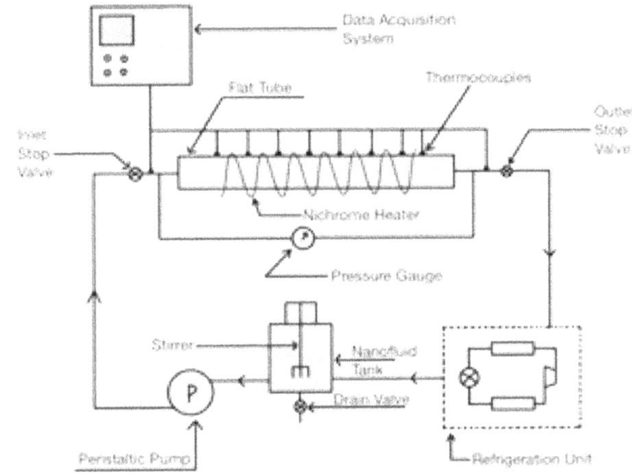

Figure 7. Experimental setup photographic image

V. DATA ANALYSIS

To calculate the HTC and Nusselt number the following variable will have to be first calculated using the setup specification and the data from the experiment.

Hydraulic diameter (D_h) of flat tube is calculated using equation 2 (R. Ravisankar et al. 2017).

$$D_h = 4*\frac{\frac{\pi d^2}{4} + (D - d)*D}{\pi * d + 2*(D - d)} \tag{2}$$

d = minor diameter

D = major diameter

Wetted Surface area, A_s,

$$A_s = \pi * d + 2*(D - d)*L \tag{3}$$

Heat supplied by Nichrome Heater,

$$Q_1 = V * I \tag{4}$$

Heat absorbed by nanofluid in the tube

$$Q_2 = mc_p(T_{out} - T_{in}) \tag{5}$$

Total heat, Q

$$Q = \frac{Q_1 + Q_2}{2} \tag{6}$$

Specific heat of nanofluid is calculated using equation 7 (K. Khanafer and K. Vafai 2011)

$$c_{pnf} = (1 - \alpha)c_{pbf} + \alpha c_{pp} \tag{7}$$

The density is obtained using Pak and Cho Equation (B.C. Pak and Y.I. Cho 1998) as follows,

$$\rho_{nf} = (1 - \alpha)\rho_{bf} + \alpha\rho_{np} \tag{8}$$

Where,

ρ_{nf} = Density of nanofluid
ρ_{bf} = Density of base fluid
α = mass concentration of nanoparticle
ρ_{np} = Density of nanoparticle
Thermal Conductivity of nanofluid using Reynolds equation (J.C. Maxwell 1881)

$$\frac{K_{nf}}{K_b} = \frac{K_p + 2K_p - 2(K_b - K_p)\alpha}{K_p + 2K_p + (K_b - K_p)\alpha} \tag{9}$$

Here;
K_{nf} = Thermal conductivity of nanofluid
K_p = Thermal conductivity of nanoparticle
K_b = Thermal conductivity of base fluid
α = mass concentration of nanoparticles
To calculate the viscosity of nanofluid equation 10 is used (D.A. Drew and S.L. Passman 1998),

$$\mu_{nf} = (1 + 2.5\alpha)\mu_{bf} \tag{10}$$

Where,
μ_{nf} = nanofluid viscosity
μ_{bf} = viscosity of base fluid
α = weight concentration
Constant heat flux, q

$$\dot{q} = \frac{Q}{A_s} \tag{11}$$

Experimental Convective heat transfer coefficient is calculate as (B. Raei 2021)

$$h_{exp} = \frac{\dot{q}}{T_s - T_m} \tag{12}$$

$$T_m = \frac{T_{out} + T_{in}}{2} \tag{13}$$

$$T_s = \frac{\left[\sum_{i=1}^{8} T_s(i)\right]}{8} \tag{14}$$

Finally, Nusselt number can be calculated using

$$Nu_{exp} = \frac{h_{exp} * D_h}{K_{nf}} \tag{15}$$

To calculate friction factor (f),

$$f = \frac{1}{2}\frac{\Delta P}{L}\frac{D_h}{\rho V^2} \quad (16)$$

VI. RESULTS AND DISCUSSION

The study uses four distinct weight concentration (0.01%, 0.02%, 0.03% & 0.04%) of TiO_2 nanoparticles in distilled water. It focuses on effect of different concentrations and flow rate, which is varied from 0.2 lpm to 0.5lpm i.e., laminar regime, on the convective HTC of the nanofluid.

A. Experimental Setup Validation

The experimental setup can be validated using Shah Equation, as given below (R.K. Shah 1975)

$$Nu(x) = = \begin{cases} 1.953\left(Re.Pr.\frac{D_h}{X}\right)^{1/3} \; if\left(Re.Pr.\frac{D_h}{X}\right) \geq 33.3 \\ 4.364 + 0.0722 Re.Pr.\frac{D_h}{X} \; if\left(Re.Pr.\frac{D_h}{X}\right) < 33.3 \end{cases} \quad (17)$$

As shown in the graph in Figure 8, it was seen that the Nusselt number obtained from experiment is close to the one obtained from Shah equation. It tells that the setup has pretty good accuracy. The maximum deviation observed was of 2.65 percent.

Figure 8. Comparison between experimental Nusselt number and Nusselt number calculated using Shah equation

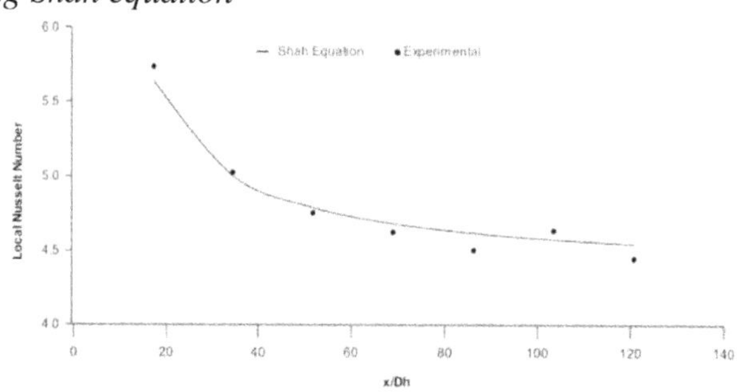

B. **Heat** *Transfer Coefficient of Nanofluid*

Figure 9 illustrates the impact of flow rate HTC at different weight concentrations of TiO_2 nanoparticles in H_2O. The use of nanofluid enhances heat transfer in contrast to the base fluid, and this improvement increases with higher nanoparticle weight concentrations. TiO_2/H_2O nanofluid enhances heat transfer by up to 14.39%, with the most significant enhancement observed at 0.04% weight concentration and 0.5 lpm flow rate. Higher flow rates increase Brownian motion and, consequently, heat transfer. For instance, at 0.5 lpm, there is 5.6% increase with 0.01% weight concentration and 14.39% increase with 0.04% weight concentration. In contrast, lower flow rates result in less Brownian motion and lower heat transfer. At 0.2 lpm, there is only a 0.045% increase with 0.01% weight concentration and a 5.62% increase with 0.04% weight concentration. The overall increase in heat transfer with higher weight concentrations is due to enhanced energy transfer between particles and improved fluid intermixing resulting from particle movement.

Figure 9. Comparison of HTC of TiO_2/water nanofluids

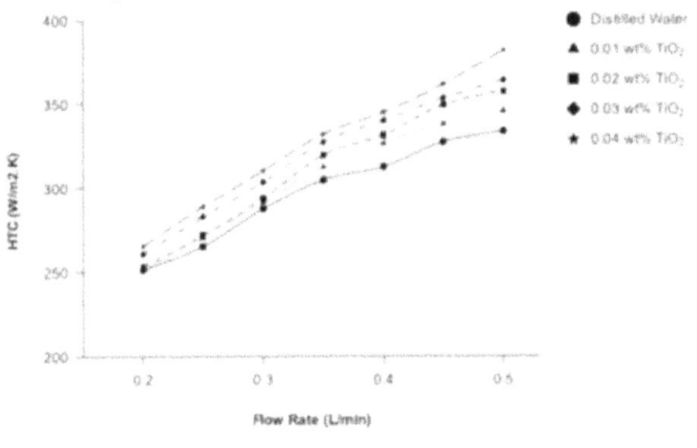

C. Pressure Drop of Nanofluid

The experiment investigates the drop in pressure inside the flat tube at various concentrations and flow rates. As shown in Figure 10, increase in concentration and flow rate worsens the pressure drop within the flat tube. The highest observed pressure drop, reaching 3.4%, was recorded at a wt. concentration of 0.04% and a flow rate of 0.5 lpm. This increase in pressure drop can be due to rising velocity and mass density of the nanofluid

Figure 10. Pressure drop across flat tube at varying weight concentration and with varying flow rate

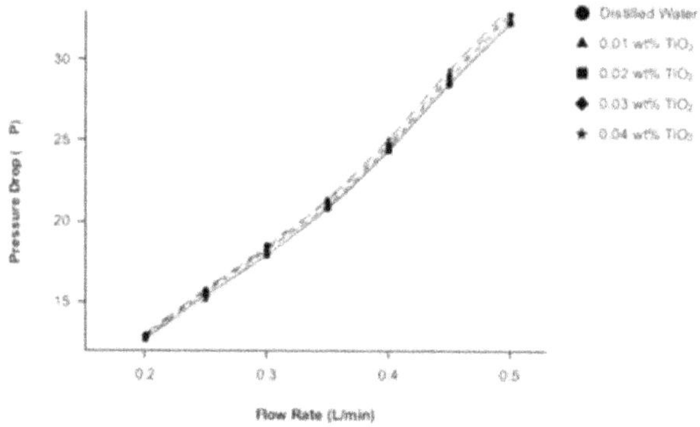

D. Friction Factor of Nanofluid

The frictional factor is determined by using the pressure drop data obtained from the experiment. As the flow rate increases, there is a decrease in the friction factor. Even though the pressure drop increases with higher flow rates, it remains relatively lower. Figure 11 illustrates the correlation between the friction factor and the flow rate of the nanofluid at various weight concentrations. The highest increase in the frictional factor is seen at a flow rate of 0.2 lpm with a 0.04% weight concentration. The maximum percentage increase, reaching 16.89%, is observed at a discharge of 0.5 lpm with a 0.04% weight concentration against the base fluid.

Figure 11. The friction coefficient in a flat tube under different levels of weight concentration and across various flow rates

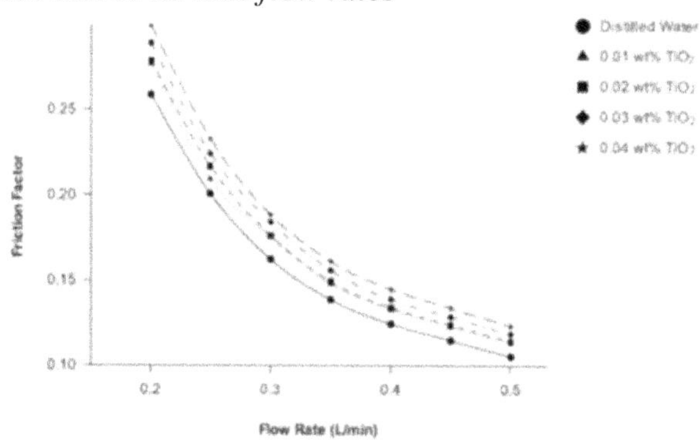

VII. CONCLUSION

The experiment's conclusions encompass the convective HTC and pressure drop of nanofluids composed of titanium dioxide and distilled water flowing within a flat tube, while maintaining a constant heat flux. The study also involves the characterisation of the nanoparticles through techniques such as FT-IR, EDAX, XRD, and FESEM. The nanofluids are prepared with four weight concentrations (0.01%, 0.02%, 0.03%, and 0.04%) using a two-step method.

The concluded parameters can be summarised as follows:

- FESEM was employed to examine the surface morphology of nanoparticles, revealing an average diameter of 35 nm and a spherical shape. The XRD pattern exhibited a tetragonal crystal structure. FT-IR analysis was utilized to determine the functional groups of the TiO_2. The band at 3436.26 cm^{-1} corresponded to the stretching vibration of the O-H (hydroxyl group) in the TiO_2 nanoparticles. The second band at 1634.06 cm^{-1} indicated the bending mode of the Ti-OH in water. The peaks at 463.02cm^{-1} and within the range of 505.43 to 700 cm^{-1} represented the stretching and bending modes of Ti-O-Ti.
- The maximum increase in HTC was found to be 14.39% using 0.04% concentration nanofluids at 0.5 lpm compared to the base fluid (water) at 0.5 lpm. When using 0.01% weight concentration the maximum HTC improvement was 3.62%.
- The pressure drop across flat tube increases with the concentration and flow rate. There was a highest increase of 3.4% with 0.04% concentration compared to the base fluid.
- The highest increase of frictional factor is observed with 0.04% weight concentration at 0.2 lpm flow rate. The maximum percentage increase of 16.89% is observed at 0.5 lpm with 0.04% weight concentration compared to the base fluid.

The agglomeration, dispersion stability and sometimes the surface area to volume ratio are some of the issues faced by the researchers in nanofluids. These are some of the issues which limit the nanofluids usage in certain application. Also, the flat tube provide better hydraulic diameter. One more issue which make the nanofluids unsuitable for an application is its cost. So, it is necessary to develop a nanofluid which can omit some the drawbacks and can be effectively used in applications like commercial heat exchangers, biomedical usage etc and also have long term stability with less cost.

REFERENCES

Abdolbaqi, M. Kh. (2017). Rizalman Mamat, Nor Azwadi Che Sidik, W.H. Azmi, P. Selvakumar Experimental investigation and development of new correlations for heat transfer enhancement and friction factor of BioGlycol/water based TiO_2 nanofluids in flat tubes, International Journal of Heat and Mass Transfer Volume 108. *Part A*, (May), 1026–1035.

Abdul Hamid, K., Azmi, W. H., & Nabil, M. F. (2018, March). Rizalman Mamat (2017), "Experiment investigation of nanoparticle mixture ratios on TiO_2-SiO_2 nanofluids heat transfer performance under turbulent flow. *International Journal of Heat and Mass Transfer*, 118, 617–627. DOI: 10.1016/j.ijheatmasstransfer.2017.11.036

León, A., Reuquen, P., Garín, C., Segura, R., Vargas, P., Zapata, P., & Orihuela, P. A. (2017). FTIR and Raman characterization of TiO2 nanoparticles coated with polyethylene glycol as carrier for 2-methoxyestradiol. *Applied Sciences (Basel, Switzerland)*, 7(1), 49.

Behl, V., Singh, V., Dahiya, V., & Kumar, A. (2022). Characterization of physicochemical and functional properties of fly ash concrete mix. *Materials Today: Proceedings*, 50, 941–945. DOI: 10.1016/j.matpr.2021.06.353

Drew, D. A., & Passman, S. L. (1999). *Theory of Multi Component Fluids*. Springer. DOI: 10.1007/b97678

GS. (2018, September 4). Sokhal D, Gangacharyulu VK, Bulasara, "Influence of copper oxide nanoparticles on the thermophysical properties and performance of flat tube of vehicle cooling system". *Vacuum*, 2018(157), 268–276.

Gupta, M., Kumar, R., Arora, N., Kumar, S., & Dilbagi, N. (2016). Forced Convective Heat Transfer of MWCNT/Water Nanofluid under Constant Heat Flux: An Experimental Investigation. *Arabian Journal for Science and Engineering*, 41(2), 599–609. DOI: 10.1007/s13369-015-1699-5

Heyhat, M. M., Kowsary, F., Rashidi, A. M., Esfehani, S. A. V., & Amrollahi, A. (2012). Experimental investigation of turbulent flow and convective heat transfer characteristics of alumina water nanofluids in fully developed flow regime. *International Communications in Heat and Mass Transfer*, 39(8), 1272–1278. DOI: 10.1016/j.icheatmasstransfer.2012.06.024

Keklikcioglu, O., Dagdevir, T., & Ozceyhan, V. (2019, November). Heat transfer and pressure drop investigation of graphene nanoplatelet-water and titanium oxide-water nanofluids in a horizontal tube. *Applied Thermal Engineering*, 162(114256), 5.

Khanafer, K., & Vafai, K. (2011). A critical synthesis of thermophysical characteristics of Nanofluids. *International Journal of Heat and Mass Transfer*, 54(19-20), 4410–4428. DOI: 10.1016/j.ijheatmasstransfer.2011.04.048

Kin Yuen Leonga. (2017, November 20). Idayu Razalia, K.Z. Ku Ahmada, Hwai Chyuan Ongb, M.J. Ghazalic, Mohd Rosdzimin Abdul Rahmana, "Thermal conductivity of an ethylene glycol/water-based nanofluid with copper-titanium dioxide nanoparticles: An experimental approach". *International Communications in Heat and Mass Transfer*, 90, 34–34.

Kumar, A., Kumar, P., Srivastava, A. K., & Goyat, V. (Eds.). (2023). *Modeling, Characterization, and Processing of Smart Materials*. IGI Global., DOI: 10.4018/978-1-6684-9224-6

Li, Y., Fernández-Seara, J., Du, K., Pardiñas, Á. Á., Latas, L. L., & Jiang, W. (2016). Experimental investigation on heat transfer and pressure drop of ZnO/ethylene glycol-water nanofluids in transition flow. *Applied Thermal Engineering*, 93, 537–548.

Liu, K. V., Choi, U. S., & Kasza, K. E. (1988). *Measurements of pressure drop and heat transfer in turbulent pipe flows of particulate slurries* (No. ANL-88-15). Argonne National Lab.(ANL), Argonne, IL (United States).

Manikandan, K., JafarAhamed, A., & Brahmanandhan, G. M. (2017). Synthesis, Structural and Optical Characterization of TiO2 Nanoparticles and its Assessment to Cytotoxicity Activity. *Journal of Environmental Nanotechnology*, 6(3), 94–102.

Maxwell, J. C. (1881). *A Treatise on Electricity and Magnetism*. Clarendon Press.

Modak, M., Srinivasan, S., Garg, K., Chougule, S. S., Agarwal, M. K., & Sahu, S. K. (2015). Experimental investigation of heat transfer characteristics of the hot surface using Al_2O_3 -water nanofluids. *Chemical Engineering and Processing*, 91, 104–113. Advance online publication. DOI: 10.1016/j.cep.2015.03.006

Pak, B. C., & Cho, Y. I. (1998). Hydrodynamic and heat transfer study of dispersed fluids with submicron metallic oxide particles. *Experimental Heat Transfer*, 11(2), 151–170. DOI: 10.1080/08916159808946559

Qi, C., Liu, M., Luo, T., Pan, Y., & Rao, Z. (2018, December). Effects of twisted tape structures on thermo-hydraulic performances of nanofluids in a triangular tube. *International Journal of Heat and Mass Transfer*, 127(Part C), 146–159. DOI: 10.1016/j.ijheatmasstransfer.2018.08.017

Qi, C., Yang, L., Chen, T., & Rao, Z. (2018). Experimental study on thermo-hydraulic performances of TiO_2-H_2O nanofluids in a horizontal elliptical tube. *Applied Thermal Engineering*, 129, 1315–1324. DOI: 10.1016/j.applthermaleng.2017.10.137

Raei, B. (2021). Forced Convective Heat Transfer of MgO/Water Nanofluid under Constant Heat Flux: Experimental and Statistical Investigation. *Chall. Nano Micro Scale Sci. Tech.*, 9(1), 69–80.

Ravisankar, R., Venkatachalapathy, V. S. K., & Alagumurthy, N. (2017). "Thermal performance improvement of tractor radiator using CuO/Water nanofluid", Heat Tran. *Asian Res.*, 46(1), 61–74.

Sarkar, J. (2011). A critical review on convective heat transfer correlations of nanofluids. *Renewable & Sustainable Energy Reviews*, 15(6), 3271–3277.

Shah, R. K. (1975, December). Thermal entry length solutions for the circular tube and parallel plates. In *Proceedings of 3rd national heat and mass transfer conference* (Vol. 1, pp. 11-75). Delhi: Indian Institute of Technology Bombay.

Singh, V., Kumar, A., Kumar, K., & Gupta, M. (2021). Synthesis and Stability of Al_2O_3/Water Nanofluids. In Advances in Materials and Mechanical Engineering: Select Proceedings of ICFTMME 2020 (pp. 3-11). Springer Singapore.

Vakili, M., Mohebbi, A., & Hashemipour, H. (2013). Experimental study on convective heat transfer of TiO 2 nanofluids. *Heat and Mass Transfer*, 49, 1159–1165.

Chapter 11
Fracture Analysis of Porous FGM Plate Using Finite Element Method

Kanishk Sharma
Department of Mechanical Engineering, School of Engineering and Technology, JECRC University, India

Rhythm Ajmani
Department of Mechanical Engineering, School of Engineering and Technology, JECRC University, India

Vikash Choudhary
Department of Mechanical Engineering, School of Engineering and Technology, JECRC University, India

Ajay Kumar
https://orcid.org/0000-0001-7306-1902
Department of Mechanical Engineering, School of Engineering and Technology, JECRC University, India

Parveen Kumar
Department of Mechanical Engineering, Rawal Institute of Engineering and Technology, Faridabad, India

ABSTRACT

The present study uses Finite Element Method to calculate the stress intensity factor of a porous FGM plate. The First Order Shear Deformation Theory is applied to investigate the effect of different material and geometrical parameters on stress intensity factor under tensile loading conditions. The FGM plate with continuous

DOI: 10.4018/979-8-3693-5513-8.ch011

variation of material properties across the thickness is modelled as laminate consisting of multiple perfectly bonded layers of homogenous material. The material properties are calculated using power law and the rule of mixtures. It is assumed that the Poisson's ratio of FGM plate across the thickness is constant. An APDL script is written to carry out the fracture analysis of the FGM plate using general-purpose FEA software ANSYS. It is found that the stress intensity factor of the FGM plate is considerably affected by the power law index and crack size.

1. INTRODUCTION

Functionally Graded Material is characterized by a distinct genre of composite material in which the composition and structure moderately vary over volume based on a particular function, developing corresponding changes in the material properties (Sharma & Kumar, 2016; Ahmad et al., 2023; Kumar et al., 2024).

FGMs are superior to traditional composite materials as their properties vary throughout the material volume, achieving the desired thermal, mechanical, chemical, and electrical properties (Sharma & Kumar, 2017, Waigh et al., 2024).

The most widely utilized and accepted FGM are particle-reinforced FGMs and their respective composition is based on their position. FGMs are preferred to avoid drawbacks associated with the presence of an interface between two materials with stress distinctiveness due to distinct elastic or thermal properties and poor adhesion. The Ceramic-Metal FGMs can be used to compensate for high thermal stresses by taking advantage of the thermal resistant properties of ceramics and to introduce properties like toughness, machinability, and strength from metals.

Alongwith its promising advantages in terms of engineering applications, due to the complex micromaterial structure, FGMs also pose challenges in terms of its material characterization, designing under critical conditions and modeling of fracture behavior. With the limited knowledge on the critical conditions leading to crack nucleation and crack propagation constraints the use of FGMs in wide range of engineering applications (Butcher et al., 1998, Srivastava et al., 2023, Kumar et al, 2024).

It is evident that the presence of microcracks in the ceramic region of the Ceramic-Metal FGMs leads to failure through crack propagation (Anlas et al., 2000; Srivastava et al., 2024). To investigate the effect of existing cracks and crack size on the load-carrying capacity of FGM it is necessary to perform a fracture mechanics analysis of FGMs. In literature various analytical, numerical and experimental techniques have been used to analyze the fracture behavior of FGM structures (Bhandari & Purohit, 2022a; Singh et al., 2023; Kumar et al., 2024). For instance, Palladino et al. (Palladino et al., 2022) developed a semi-analytical method to compute the stress intensity factor of 2-D and 3-D FGM plates. Dorduncu et al. (Dorduncu et al., 2022; Kumar et al., 2024) used peri-Dynamic fracture mechanics approach to analyze the condition of crack propagation in two-dimensional FGM plates subjected to dynamic and quasi-static loading conditions. Fracture behavior of functionally graded materials using the FEM was investigated by Iqbal et al. (Iqbal et al., 2022; Jaglan et al., 2023; Kumar et al., 2024). An extensive review on investigation of fracture behavior of FGM plate using numerous mathematical models is presented by Bhandari and Purohit (Bhandari & Purohit, 2022b).

Porosity in FGMs refers to the presence of voids or empty spaces within the material that can be induced due to manufacturing and /or processing defects (Chaabani et al., 2023). By inclusion of porosity the FGMs can exhibit specific mechanical, thermal, or electrical properties (Ramteke & Panda, 2023).

Present study investigates the fracture behavior of porous FGM plate with even type porosity. Various numerical studies are conducted to investigate the effect of material inhomogeneity, loading condition, aspect ratio thickness and porosity on fracture behavior of FGM plate. The result in the form of stress intensity factor is presented.

Figure 1. Rectangular porous FGM plate.

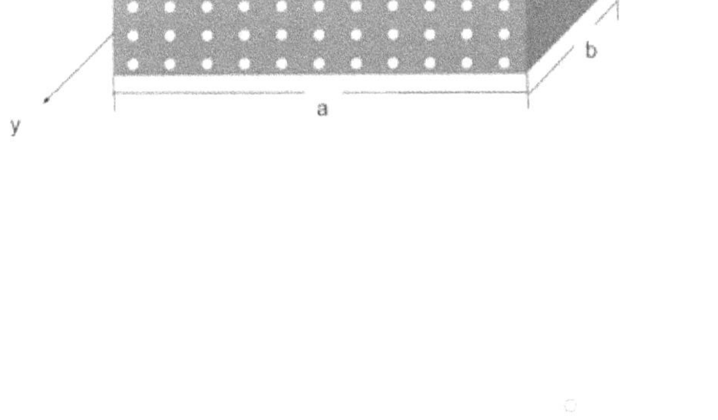

2. MATHEMATICAL MODEL OF POROUS FGM PLATE

In the present study an FGM plate having length a, width b and thickness h is considered and shown in Figure 1. The even type porosity distribution across the thickness of FGM plate is considered. The FGM plate is assumed to be made of ceramic and metal wherein the material composition varies from pure metal at bottom to pure ceramic at top of FGM plate. In between top and bottom of FGM plate the material composition is varied as per power law. Therefore, by assuming even type of porosity with porosity volume fraction α (i.e., $0 \leq \alpha < 1$) and the volume fractions of ceramic and metal, i.e., V_c and V_m respectively, the expression to calculate the Young's modulus of FGM plate in accordance of power law takes following form (Nogata, 1997):

$$E(z) = (E_c - E_m)\left(\frac{2z+h}{2h}\right)^n + E_m - \frac{\alpha}{2}(E_c + E_m)$$

,

$$n \geq 0 \qquad (1)$$

where, the subscripts 'c' and 'm' stand, respectively, for ceramic and metal; E_c and E_m represent a material property at the top (i.e., of pure ceramic) and bottom (i.e., of pure metal) surfaces of FGM plate, respectively, and n is power law exponent.

Figure 2. Modelling of actual non-homogeneous FGM plate into a laminate composed of multiple perfectly-bonded homogeneous layers.

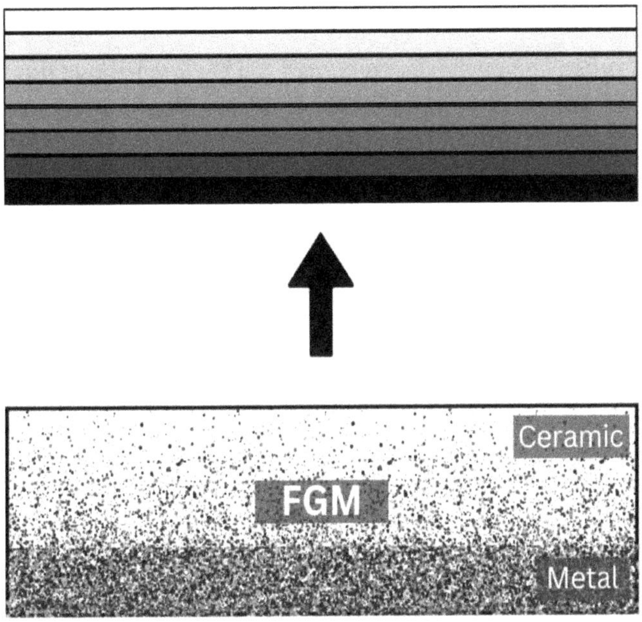

3. NUMERICAL RESULTS AND DISCUSSION

Various numerical studies are conducted to investigate the effect of material inhomogeneity, loading condition, aspect ratio thickness and porosity on fracture behaviour of FGM plate. The result in the form of stress intensity factor is presented. It is to be mentioned here that until and unless stated the power law index for all studies is considered to be unity. Further the porosity parameter is taken as 0.1 for all studies except wherein the effect of porosity is investigated. The effect of various material geometrical and loading parameters on stress intensity factor is investigated and discussed in following paragraphs.

It is assumed that the top surface of FGM plate is of pure Alumina (Al_2O_3) whereas the bottom surface is made of Aluminum (Al) and the material properties are graded across the thickness of plate as per the power law given by Eq. (1). The value of Young's modulus for parent material i.e., for Alumina and Aluminum are taken as 380 GPa and 70 GPa respectively, while the poison's ratio for both material is considered as 0.30. The length and width of FGM plate is assumed to be 1 m whereas, until unless specified the slenderness ratio is taken as 0.10.

Figure 3. (a) Finite element mesh of a typical FGM plate with a crack, (b) Enlarged view of the crack tip node of the crack, (c) Configuration of Eight-Nodded Shell 281 Element.

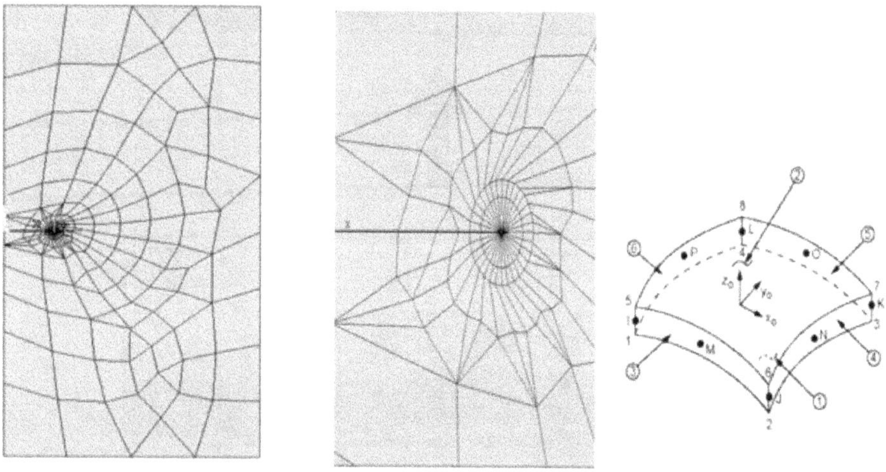

(a) (b) (c)

An APDL script has been written to carry out the fracture analysis of porous FGM plate. The present study uses Eight-nodded Shell 281 Element to model the porous FGM plate. The meshed structure of porous FGM plate and enlarged view of crack are shown in Figs. 3(a-b). The configuration and node numbering scheme of FGM plate of Eight-nodded Shell 281 Element is depicted in Fig. 3(c). The various steps associated to calculate the stress intensity factors of porous cracked FGM plate are as:

Step 1. Click on Preferences > Structural > OK

Step 2. Click on Pre-Processor > Element Type > Add/Edit/Delete > Add > Shell > 8 Node 281 > OK > Options > Change K8 to All Layers

Step 3. Click on Material Props > Material Models > Structural > Linear > Elastic > Isotropic > Enter the values for EX & PRXY > OK (Repeat the procedure 20 times for all 20 layers by adding new material model from the upper ribbon, in this study, we've used Power Law in MATLAB to calculate the values of Ex)

Step 4. Click on Sections > Shell > Lay-up > Add/Edit > Enter the layer thickness for each layer > Tools > Add Multiple copies > Enter number copies (No of layers -1) > Change the Material id for each layer > OK

Step 5. Click on Modelling > Create > Keypoints > In active CS > Enter the keypoint number and its coordinates > Apply > Repeat the process for all the coordinates.

Step 6. Click on Modelling > Create > Lines > Lines > In active CS > Choose the points to b joined > Apply > Repeat the procedure for all the points > OK

Step 7. Click on Meshing > Mesh > Size Cntrl > Manual Size > Lines > Picked Lines > Select L6,7,8,9 > OK > Enter num of div = 4 > Apply > Select L5,10 > Enter num of div = 6 & spacing ratio = 0.2 > OK

Step 8. Click on Concentrate KPs > Choose KP 3 > Enter radius = 0.0025, num of elements = 32 > Select ¼ > OK

Step 9. Click on Modelling > Create > Areas > Arbitrary > By Lines > Select Loop > Select any Line > OK

Step 10. Click on Loads > Define Loads > Apply > Structural > Displacement > Symmetry BC > … With Area > Select L5 > Apply > Area > Apply > L10 > Apply > Area > OK

Step 11. Click on keypoints > Select KP 8 > Constrain UY & UY > OK

Step 12. Click on Pressure > On Lines > Select L6,9 > OK > Add the value as '-100' for tensile force > OK

Step 13. Click on Meshing > Mesh > Areas > Free > Select the Area > OK (A warning appears, Click on OK)

Step 14. Click on Solution > Analysis Type > New Analysis > Static > OK

Step 15. Click on Solve > Current LS > OK > Close

Step 16. Click on General Postproc > Path Operation > Define Path > By Nodes > Enter the nodes on the crack path > OK

Step 17. Click on Work Plane > Local Coordinate System > Create Local CS > By 3 Nodes > Choose the crack tip and the x & y coordinate around the tip.

Step 18. Click on Nodal Calcs > Stress intensity factor > Select Plain Strain > Choose the desired Material num > Select Full Crack Model

Step 19. A dialogue box appears with values of K1, K2 & K3.

3.1. Effect of Material Inhomogeneity

In order to analyse the effect of material gradation profile on fracture behaviour of porous FGM plate various value of power law index i.e., n = 0,1,2,5, and 10, are considered. The value of Young's modulus of FGM plate for particular value of n is calculated using Eqs. 1-2. Variation of stress intensity factor over thickness of FGM plate for different values of power law index is shown in Fig. 4. It can be divulged from figure that the value of stress intensity factor increases monotonically across the thickness the FGM plate and maximum value of stress intensity factor is obtained at the top of FGM plate. Further it is also evident from Fig. 5 that the stress intensity factor increases considerably with an increment in power law index.

Figure 4. Variation of Stress Intensity Factor (K_I) across the thickness of a porous FGM plate (α = 0.10) for different power law index.

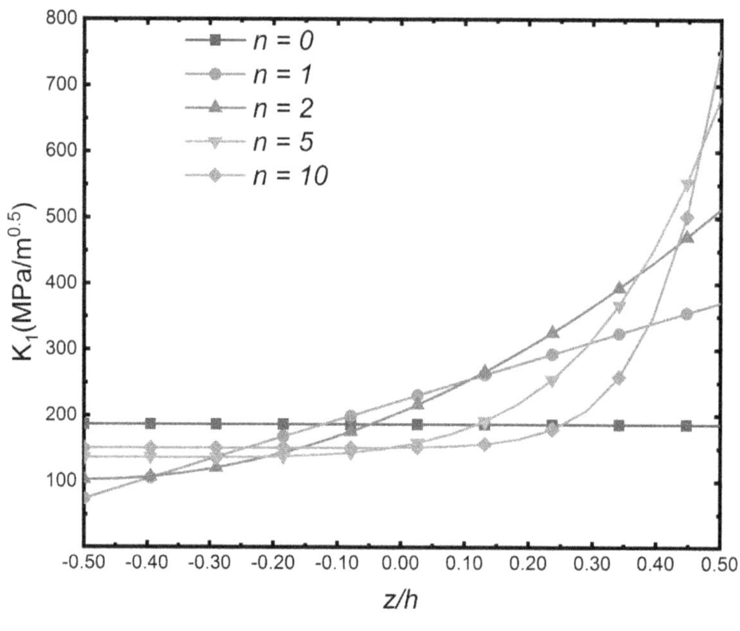

The increase in the stress intensity factor with power law index can be attributed to increased value of material rigidity due to an increase in proportion of ceramic for higher value of power law index.

3.2 Effect of Aspect Ratio of the Plate

The effect of aspect ratio of rectangular FGM plate on stress intensity factor is also analysed and the results are depicted in Fig. 5. From Fig. 5 it can be observed that for a/b = 0.5, 1, 2, and 3 the variation of stress intensity factor over thickness of FGM plate is not affected, magnificently however for a/b = 4 the value of stress intensity factor at particular value of thickness of FGM plate is much lower as compared to other values of aspect ratios.

Figure 5. Variation of Stress Intensity Factor (K_I) across the thickness of FGM plate for different aspect ratio of the plate.

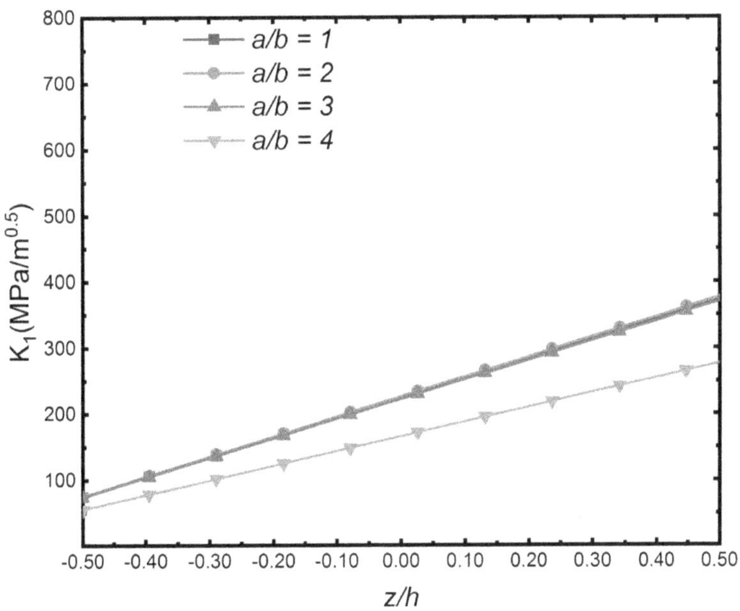

3.3. Effect of Crack Size

It is evident that the fracture behaviour of FGM plate is significantly affected with the crack size. In order to investigate the effect of crack length on stress intensity factor of FGM plate a study is conducted by considering $a/a_0 = 2,3,5$ and 10. The variation of stress intensity factor across the thickness for different values of crack size is depicted in Fig. 6. As shown in Fig. 6, the stress intensity factor of FGM plate is significantly affected by crack length and the value of stress intensity factor increases monotonically with crack length. The results are in accordance with the conventional wisdom as the plate with greater crack size is more susceptible towards the fracture failure.

Figure 6. Variation Stress Intensity Factor (K_I) across an FGM plate for different crack size on the plate.

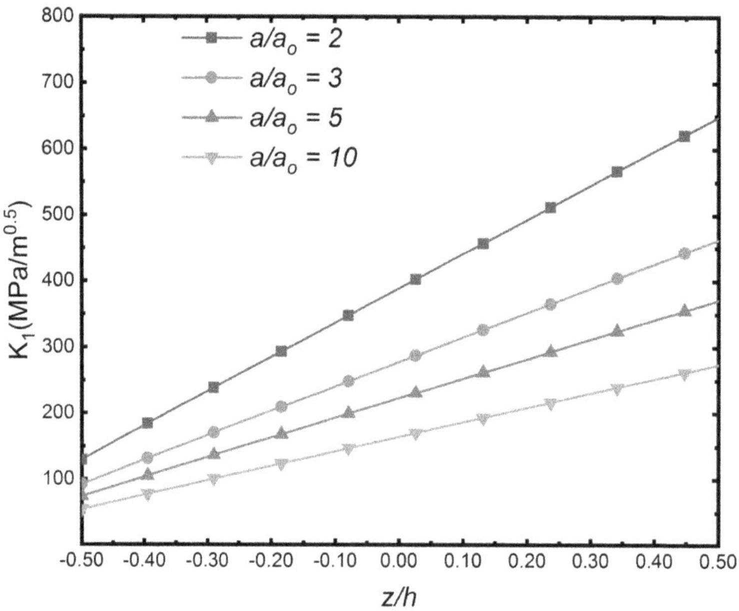

3.4. Effect of Load Intensity

The effect of load intensity on stress intensity factor is also evaluated and presented in Fig. 7. To investigate the effect of load magnitude on stress intensity factor the magnitude of tensile load is considered to be F= 50, 100 and 200 kN. As shown in Fig. 7 the stress intensity is affected by magnitude of tensile loading conditions considerably and with the increase in value of tensile load the stress intensity increases magnificently. The highest value of stress intensity factor over the thickness of FGM plate is obtained for 200 kN.

Figure 7. Variation of Stress Intensity Factor (K_I) across the thickness of a porous FGM plate for different load intensity on the plate.

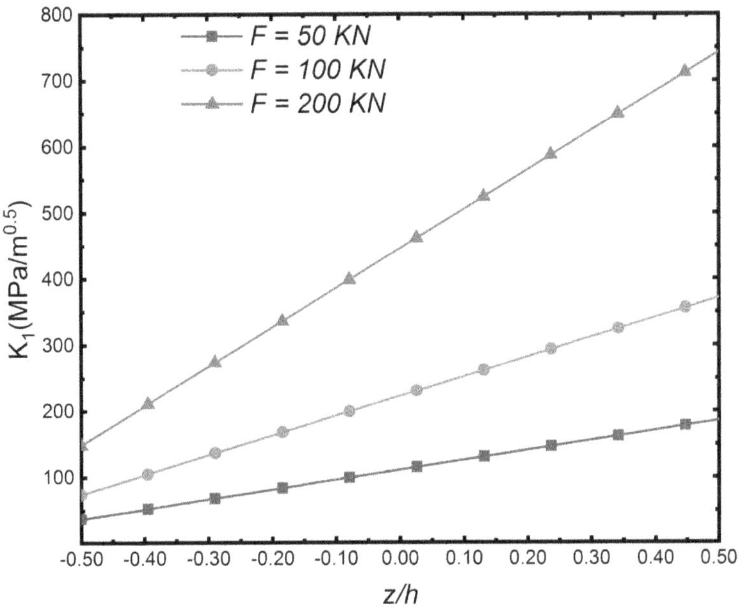

6. CONCLUSION

A study on fracture behavior of cracked an aluminum-zirconia FGM plate with a central crack is conducted under tensile mechanical loadings condition. The formulation was based on first-order shear deformation theory. Finite element numerical technique is employed by writing an ANSYS script. Numerical results are presented to show the parametric effect of various parameters, viz. material gradation exponent, geometric variables, crack size and loading conditions, on the stress intensity factor of the FGM plate. Based on the present study, the following important conclusions can be drawn:

- The material gradation exponent has significant effect on the stress intensity factor of FGM plate.

- It is observed that up to a certain value of aspect ratio (i.e., $a/b = 3$) the value of stress intensity factor remains unaffected form increment in aspect ratio however for the FGM plate with a/b = 4, the stress intensity factor decreased significantly.
- In accordance with the conventional wisdom, it is observed that the FGM plate with larger crack size depicted higher stress intensity factor and is more susceptible towards the fracture.

REFERENCES

Ahmad, S., Sharma, J., & Gambhir, V. (Eds.). (2023). *Handbook of Sustainable Materials: Modelling, Characterization, and Optimization*. CRC Press.

Anlas, G., Santare, M. H., & Lambros, J. (2000).. . *Numerical Calculation of Stress Intensity Factors in Functionally Graded Materials.*, 1994, 131–143.

Bhandari, M., & Purohit, K. (2022a). Dynamic Fracture Analysis of Functionally Graded Material Structures – A Critical Review. *Composites Part C: Open Access*, 7, 100227. DOI: 10.1016/j.jcomc.2021.100227

Bhandari, M., & Purohit, K. (2022b). Mathematical methods to investigate the fracture behaviour of functionally graded material (FGMs)—A critical review. Advances in Mechanical and Industrial Engineering, 102–108.

Butcher, R. J., Rousseau, C. E., & Tippur, H. V. (1998). A functionally graded particulate composite: Preparation, measurements and failure analysis. *Acta Materialia*, 47(1), 259–268. DOI: 10.1016/S1359-6454(98)00305-X

Chaabani, H., Mesmoudi, S., Boutahar, L., & El Bikri, K. (2023). A high-order finite element continuation for buckling analysis of porous FGM plates. *Engineering Structures*, 279, 115597. DOI: 10.1016/j.engstruct.2023.115597

Dorduncu, M., Olmus, I., & Rabczuk, T. (2022). A peridynamic approach for modeling of two dimensional functionally graded plates. *Composite Structures*, 279, 114743. DOI: 10.1016/j.compstruct.2021.114743

Iqbal, M. D., Birk, C., Ooi, E. T., Pramod, A. L. N., Natarajan, S., Gravenkamp, H., & Song, C. (2022). Thermoelastic fracture analysis of functionally graded materials using the scaled boundary finite element method. *Engineering Fracture Mechanics*, 264, 108305. DOI: 10.1016/j.engfracmech.2022.108305

Jaglan, A., Dogra, N., Kumar, A., & Kumar, P. (2023). Finite Element Simulation Approach in Incremental Sheet Forming Process. Handbook of Flexible and Smart Sheet Forming Techniques: Industry 4.0 Approaches, 129-139. DOI: 10.1002/9781119986454.ch7

Kumar, A., Kumar, P., & Liu, Y. (2024). *Industry 4.0 Driven Manufacturing Technologies*. Springer Nature., DOI: 10.1007/978-3-031-68271-1

Kumar, A., Kumar, P., Sharma, N., & Srivastava, A. K. (Eds.). (2024). *3D Printing Technologies: Digital Manufacturing, Artificial Intelligence, Industry 4.0*. Walter de Gruyter GmbH & Co KG. DOI: 10.1515/9783111215112

Kumar, A., Liu, Y., & Kumar, R. (Eds.). (2024). *Handbook of Intelligent and Sustainable Manufacturing: Tools, Principles, and Strategies*. CRC Press., DOI: 10.1201/9781003405870

Kumar, A., Rajak, D. K., Kumar, P., & Kumar, A. (Eds.). (2024). *Advances in Sustainable Biomaterials: Bioprocessing 4.0, Characterizations, and Applications* (1st ed.). CRC Press., DOI: 10.1201/9781003434313

Kumar, P., Hussain, S. S., Kumar, A., Srivastava, A. K., Hussain, M., & Singh, P. K. (2024). 10 Finite element method investigation on delamination of 3D printed hybrid composites during the drilling operation. 3D Printing Technologies: Digital Manufacturing, Artificial Intelligence, Industry 4.0, 223.

Nogata, F. (1997). Learning about design concepts from natural functionally graded materials. *ASME International Mechanical Engineering Congress and Exposition*, 18343, 11–18. DOI: 10.1115/IMECE1997-0646

Palladino, S., Minutolo, V., & Esposito, L. (2022). Hybrid semi-analytical calculation of the stress intensity factor for heterogeneous and functionally graded plates. *Engineering Fracture Mechanics*, 274, 108763. DOI: 10.1016/j.engfracmech.2022.108763

Ramteke, P. M., & Panda, S. K. (2023). Computational Modelling and Experimental Challenges of Linear and Nonlinear Analysis of Porous Graded Structure: A Comprehensive Review. *Archives of Computational Methods in Engineering*, 30(5), 3437–3452. DOI: 10.1007/s11831-023-09908-x

Sharma, K., & Kumar, D. (2016). Thermo-mechanical analysis of Functionally Graded Material plate. *Advanced Science Letters*, 22(11), 3813–3816. DOI: 10.1166/asl.2016.8046

Sharma, K., & Kumar, D. (2017). Elastoplastic stability and failure analysis of FGM plate with temperature dependent material properties under thermomechanical loading. *Latin American Journal of Solids and Structures*, 14(7), 1361–1386. DOI: 10.1590/1679-78253747

Singh, A., Srivastava, A. K., Kumar, A., Bajaj, R., & Singh, G. K. (2023). Design and development of plentiful fly ash-based glass powder-reinforced plastic composite bricks for low water absorption and high compressive and flexural strength. [IJIDeM]. *International Journal on Interactive Design and Manufacturing*, •••, 1–10. DOI: 10.1007/s12008-023-01580-6

Srivastava, A. K., Kumar, A., Kumar, P., Gautam, P., & Dogra, N. (2023). Research Progress in metal additive manufacturing: Challenges and Opportunities. [IJIDeM]. *International Journal on Interactive Design and Manufacturing*, •••, 1–17. DOI: 10.1007/s12008-023-01661-6

Srivastava, A. K., Tiwari, S., Pachauri, P., Gupta, N., Sunil, B., & Kumar, A. (2024). Bonding strength and microstructural features of Al5083-AZ31B alloys laminated sheet through friction stir additive manufacturing. *Journal of Adhesion Science and Technology*, 38(4), 583–596. DOI: 10.1080/01694243.2023.2240637

Wagih, A., Junaedi, H., Mahmoud, H. A., Lubineau, G., Kumar, A., & Sebaey, T. A. (2024). Enhanced damage tolerance and fracture toughness of lightweight carbon-Kevlar fiber hybrid laminate. *Journal of Composite Materials*, 58(9), 00219983241235853. DOI: 10.1177/00219983241235853

Chapter 12
Use of Waste Marble to Produce Sustainable Mortar and Concrete:
A Review

Ram Vilas Meena
JECRC University, India

Ankit Singh Beniwal
https://orcid.org/0000-0001-8655-903X
JECRC University, India

Ghanshyam Gupta
https://orcid.org/0009-0008-0580-9614
JECRC University, India

Mohmmad Akaram
https://orcid.org/0009-0008-2189-9509
JECRC University, India

Ajay Kumar
https://orcid.org/0000-0001-7306-1902
Department of Mechanical Engineering, School of Engineering and Technology, JECRC University, Jaipur, India

ABSTRACT

Marble stones are natural rocks that demonstrate special texture, unique patterns and vibrant colour. They are durable and tough, and, hence can be simply formed into several sizes and shape; to be utilized as floor decorative and cladding purpose.

DOI: 10.4018/979-8-3693-5513-8.ch012

An enormous amount of non-biodegradable trash is produced during the mining and processing of dimensional stones like marble. Since it disrupts the ecology in so many ways, this has become a serious issue over time. Finding a method to securely dispose of or reuse marble waste is thus important. There are studies showing that this discarded can be utilized in replacement of conventional building materials. The present review paper investigates the influence of waste marble stone on workability, mechanical and durability characteristics when utilized as partial to full substitution in type of mortar and concrete mixtures. The concluded review of the literature, waste marble stone partial replacement of cement, and partial and full substitution of fine and coarse aggregate has been used to produce mortar and concrete.

1. INTRODUCTION

For the development and revolution of the dumping or waste material as flawless exploitation without affecting the environment or useful resources, concrete has been a great asset to the concrete construction sector (Vilas Meena, Singh Chouhan, et al., 2021). There are defined as sustainable development for the world and future generation, use the valuable resources on the earth without any losses of natural minerals (Vilas, Kumar, & Singh, 2022). India should also use the waste from the industry to create new resources so that the remaining resources in the country can be conserved and protected. When heavy manufacturing industry such as cement, coal, petroleum etc. deforestation and burning of fossil fuels are highly emitting of the carbon dioxide(CO_2) mainly the global cement industry produce as 10% of greenhouse gas (such as CO_2, CH_4, N_2O, ozone, water vapor etc.) emission to the atmosphere (Yoro & Daramola, 2022). Due to which we are getting to see the problem like global warming in nature. Nearly 70% of the weight of concrete is made up of aggregates, which constitute its main component (Vilas, Kumar, Singh, et al., 2022). The majority of aggregates are found in river beds or natural quarries (Meena, Beniwal, & Gupta, 2023). Together with the growth in construction activity, the need for natural aggregates is also rising, which encourages an even greater exploitation of the environment's resources (Kumawat et al., 2023). For instance, natural river sand, considered to be the best fine aggregate, is frequently employed to satisfy the continually increasing demand for concrete manufacturing. So, save the environmental from the ill effects of cement and to prevent highly congestion of natural resources by using industrial waste instead of these resources, we can avoid harming the environment (Meena et al., 2021).

Marble is a type of the metamorphic rock and its transform their geological property in pure limestone (sedimentary rock). The utilization of marble gained from the industry's waste generated during marble cutting. It is possible to reuse leftover marble. Waste in the form of powder and solid particles are produced in massive quantities by the marble cutting business. While cutting, dressing, polishing marble produce the large amount of slurry and solid are formed which replace cement, fine and coarse aggregate in mix design of concrete. 30% to 40% of each marble industry's annual production in India is lost during the quarrying and cutting stages as marble waste (Singh et al., 2017). The thickness of the gang saw blades directly correlates with the volume of slurry generated during marble processing; for example, using a 5 mm thick blade to cut a 25 mm block into a 20 mm block will result in 20% of the marble being wasted. 70% of the water in this slurry waste is made up of marble particles. Through a settling tank, the water is recycled and the collected, damp marble dust is then moved for disposal. About 40% of the particles in marble dust, an extremely fine powder, are less than 75 micrometers in diameter, and about 30% of them are smaller than 25 micrometers. Marble slurry has a specific gravity of 2.70–3.00 (Singhal et al., 2020) and a water content of 5–20%. Concrete's quality can be improved by using the remaining marble powder. The discarded marble slurry, which resembles mud, is thrown there, and any additional water is drained off and recycled (Kore et al., 2020). The very fine-grained marble sludge makes up 20–30% of the weight of the treated marble. It has largely been used to substitute cement in concrete due of its incredibly tiny particle size. A combination of drying and load treatments can be used to further process the marble sludge, which has an exceptionally fine grain, into marble powder. For usage in a variety of industrial applications, including the plastics, glass, paper, and dyeing industries, this powder is periodically refined even more. Waste product management has become a critical challenge in developing countries like India (Vilas Meena, Kumar Jain, et al., 2021). The disposal of such waste material is another issue which a nation is facing in the modern world & the quantity of such material is huge due to the huge population in each country. Therefore a large area required for such disposal. By substituting waste from the industrial sector for current natural resources, we can lessen the ongoing consumption of those resources (Meena, Beniwal, Jain, et al., 2023). So, that the coming generation can benefit from the healthy development of the eco-friendly environment.

2. LITERATURE SURVEY

(Hebhoub et al., 2011) tested the used of marble waste to substitute natural aggregate in concrete. In their mixes, up to 100% natural aggregate was replaced by waste marble, at 0%, 25%, 50%, 75% and 100% replacements. In order to determine workability, compressive strength, density, split tensile strength experiments were conducted. For substitutions at 25%, 50%, and 75%, the increase in resistance is very large. (Lezzerini et al., 2022) used mortar manufacturing with marble waste powder, as the swap of binder content. Waste marble powder was added in various proportions between (0% - 25%, 5% interval) by weight of binder. In overall, increasing percentage of waste marble powder in mortar as binder leads to a decrease of mechanical strength of mortar, with still acceptable values for various applications up to a replacement of less than 25%. (Vardhan et al., 2019) reveled the feasibility of waste marble (WM) as a swap of natural river sand in concrete mix. The WM was incorporation in different proportions between 10 and 60% by weight. According to the test results, WM can potentially replace with fine aggregate up to 40%, however concrete workability reduce with increment of WM in concrete. The increase in mechanical strength was attributed to better filler effect occurring due to the presence of WM. (Kore Sudarshan & Vyas, 2019) conducted an experimental study on concrete with replacing the natural coarse aggregate by 75% with the Stone marble waste. The results shown that the concrete with 75% Coarse Aggregate of stone marble waste resists temperature as high as 400°C and the results were found satisfactorily. Ultrasonic pulse velocity values, compressive strength, water absorption, and mass loss were found to be essentially the same between the experimental and control mixtures in terms of their attributes, but if the temperature rises over 400°C, concrete begins to degrade. (Alyamaç & Aydin, 2015) study were conducted on concrete containing waste marble dust by swap of (10%, 20%, 30%, 40%, 50% and 90%) natural sand. The experimental results are drawn and shown that the concrete containing up to 40% marble dust waste provide the required characteristics of concrete, and also the marble powder used in this study is inexpensive and environment friendly. (Singh et al., 2019) study were conducted using marble dust as a partial replacement of binder in concrete. Experimental studies were carried out to study the effect of marble dust power on hydration process, compressive strength of concrete. In the study cement was swapped by marble dust as a percentage of 0%, 10%, 15%, 20% and 25% by weight of cement. The mechanical strength of concrete of quarry dust added to concrete was predicted using regression model and was based on strength ratios as opposed to specimen form or kind. Study concluded that substitute of 10% binder by marble dust reduces the CO_2 emission by 5.2%. The higher compressive strength was observed at 15% substitute of cement. (Gameiro et al., 2014) this research has presented the utilization of WM found from marble quarrying industry in concrete industry as a substitute of natural river sand. In order to assess the mechanical performance of the concrete, different proportion (0, 20, 50 and 100%) of WM was consider as a fine aggregate of the total volume of aggregate. The experimental results of concrete made with marble powder shows that water absorption decreases for 20% and 50% replacement level, and also workability decreases with increase replacement ratio. (Bostanci, 2020) this study's objective was to analyze the characteristics of concretes with similar 28 day design compressive strengths

(45 MPa) constructed using raw aggregates, Portland cement, & substitutions of 5%, 10%, and 20% marble dust for Portland cement and natural sand, respectively, in the Portland cement mix. The sustainability of laboratory mixes in terms of the environment and the economy was then evaluated in further detail. The outcomes demonstrated that using waste materials offered comparable fresh attributes to conventional mixture. The 28- day compressive block strength reduced by 10%–19% for mixtures that additionally included marble particles and reused glass sand. Table 1 show past research finding on using marble waste.

Table 1. Research from the past on using marble waste with mortar and concrete.

S.NO.	Type of mortar and Concrete	Type of marble wastes	Percentage of substitution	Properties Examined	Outcomes	References.
1	Self-compacting Mortar	Waste marble	Natural sand (0%, 15%, 30%, 75% and 100%)	Workability, compressive strength, flexural strength, SEM and EDX	30% waste marble was used in self-compacting mortar, the 90 day, compressive and flexural strength increased by 8% and 17% respectively of conventional mix.	(Bentlemsan et al., 2023)
2	Concrete	Marble waste	Fine aggregate (10-60%)	Compressive Strength, split tensile strength, strength permeation, SEM,XRD	The Better Results was observed at 40% replacement of Natural fine aggregate with Waste marble.	(Vardhan et al., 2019)
3	Concrete	Waste marble	Coarse aggregate (75%)	Compressive strength, water holding capacity, velocity of ultrasonic pulses values, and mass loss.	Concrete with 75% of Marble waste and 25% of conventional CA are shown better result when the temperatures rises to 400°C.	(Kore Sudarshan & Vyas, 2019)
4	Concrete	Marble Powder	Natural sand (10%, 20%, 30%, 40%, 50% and 90%)	Compressive strength, Water Absorption, Split Tensile Strength, and Abrasion Resistance value.	Results have shown that 40% of swap of river sand with marble dust gives better outcomes.	(Alyamaç & Aydin, 2015)
5	Concrete	Waste Marble dust	Cement (0%, 5%, 10% and 20%)	Slump values, Split tensile strength, Compressive strength, Flexural strength	Compressive strength decreases after by 10% replacement of marble dust powder as binder content in concrete.	(Kumar & Kumar, 2018)
6	Self-compacting concrete	Marble waste	Fine aggregate (0, 20 and 40%)	Bending strength, Compressive strength, Water absorption, Impact resistance test	The compressive strength of the concrete manufactured by using fine aggregate in place of waste marble was more than that of the reference specimen.	(Ahmadi et al., 2024)

3. CONCRETE PROPERTIES

3.1 Workability

The standard method of determining concrete's workability is to measure the slump value. (Kabeer & Vyas, 2018), utilized marble powder as substitution of sand from (0% to 100%) with a 20% interval. It was noted that slump value slightly increased up to 20% replacement, and then after increase replacement level slump value was reduced. (Ahmad et al., 2021), evaluated the aptness of using marble slurry as a partial substitute of binder at (0% to 30%, 5% interval). It was reported the slump value reduced with the increased quantity of marble slurry in concrete. (singh & S Bansal, 2015) examined the use of marble waste powder with the replace of binder content up to 20% in concrete mixes. Slump value was investigated and it was found that the slump value decreased up to 20% replacement level. (Kore & Vyas, 2016) used marble waste as a substitute (by weight) of coarse aggregate from (0% to 100%) with a 20% gap. It was detected that workability of concrete increased with the increase of WM compared to the control mixtures. It was found that the workability of concrete blends incorporating WM was 14% more than that of conventional concrete.

3.2. Compressive Strength

The mechanical characteristic that gives a complete picture of the quality of concrete is its strength, which is an essential mechanical parameter. (Alyamaç & Aydin, 2015) calculated concrete incorporating of waste marble powder as sand substitution. Using (10%, 20%, 30%, 40%, 50% and 90%) waste marble powder to replace sand increased compressive strength for 20% substitution as related to reference concrete. (singh & S Bansal, 2015) studied compressive strength assessment of concrete make with WM powder replacing binder content up to 20% Outcomes displayed a rise in compressive strength value for 12% substitution. (Topçu et al., 2009) observed compressive strength of self-compacting concrete with 0, 50, 100, 150, 200, 250 and 300 kg/m^3 substitution of binder contain with marble dust. It was detected that compressive strength continue reduced with increasing replacement ratio of waste marble dust in self-compacting concrete. (Vaidevi, 2013) conducted the compressive strength assessment of concrete having WM in replace to cement up to 20%. The outcomes indicated that the best compressive strength for 10% substitution was detected as compared to the reference mix.

3.3 Flexural Strength

Memon et al., (2019) performed a study to calculate the flexural strength of concrete mixes containing waste marble dust as sand replacement from 0 to 100% at 25% interval. It was detected that the 50% replacement mix showed highest flexural strength as compared to conventional mix. (Vigneshpandian et al., 2017) observed that replacing 25%, 50% and 100% waste marble in concrete as fine aggregate. It was detected that the 50% replacement ratio showed the highest flexural strength. (Singhal et al., 2020) stated that increasing the percentage of waste marble slurry with 30% fly ash in self-compacting concrete decreased the flexural strength of concrete. Bentlemsan et al., (2023) determined that the 30% waste marble was used in self-compacting mortar, the 90 day flexural strength increased by 17% respectively of conventional mix.

3.4 Split Tensile Strength

(Vardhan et al., 2019) performed the spilt tensile strength experiment on concrete having waste marble as a fine aggregate replacement. It was found the increases spilt tensile strength values as an increase waste marble powder ratio in the concrete mixtures up to 40%. In an identical study, (Alyamaç & Aydin, 2015) displayed that when 20% marble powder substituted sand, the 28 days spilt tensile strength was more than the reference concrete mixture. (Ahmad et al., 2021), evaluated the suitability of using marble slurry as a partial substitute of binder at (0% to 30%, 5% interval). It was observed highest split tensile strength for 20% substitution.

4. CONCLUSION

In this review paper study, the probability of recycling marble waste as aggregate and cement replacement in the mortar and concrete manufacturing processes was investigated. From the study, the following outcomes may be drawn:

- The findings showed that using marble waste in the production of mortar and concrete is advantageous in terms of producing mortar and concrete that is affordable and durable and resolving environmental issues.
- Use of 20-30% marble slurry as cement substitution can result in good performance of mortar and concrete blends, depending on the particle size distribution of cement and WM.

- Comparing the usage of WM to ordinary mixes, the mechanical qualities of mortar and concrete are maintained when fine and coarse aggregate is partially replaced by it up to 50%.

In addition to this, the use of marble waste helps to create an environment that is both viable and sustainable while also reducing the use of non-renewable resources in artificial resources.

Abbreviations

EDX: Energy-Dispersive X-ray analysis or spectroscopy
SEM: Scanning Electron Microscope
WM: Waste Marble
XRD: X-ray diffraction

REFERENCES

Ahmad, J., Zaid, O., Shahzaib, M., Abdullah, M. U., Ullah, A., & Ullah, R. (2021). Mechanical properties of sustainable concrete modified by adding marble slurry as cement substitution. *AIMS Materials Science*, 8(3), 343–358. DOI: 10.3934/matersci.2021022

Ahmadi, M., Abdollahzadeh, E., & Kioumarsi, M. (2024). Materials Today : Proceedings Using marble waste as a partial aggregate replacement in the development of sustainable self-compacting concrete. *Materials Today: Proceedings*, xxxx. Advance online publication. DOI: 10.1016/j.matpr.2023.04.103

Alyamaç, K. E., & Aydin, A. B. (2015). Concrete properties containing fine aggregate marble powder. *KSCE Journal of Civil Engineering*, 19(7), 2208–2216. DOI: 10.1007/s12205-015-0327-y

Bentlemsan, N., Yahiaoui, W., & Kenai, S. (2023). Case Studies in Construction Materials Strength and durability of self-compacting mortar with waste marble as sand substitution. *Case Studies in Construction Materials*, 19(June), e02331. DOI: 10.1016/j.cscm.2023.e02331

Bostanci, S. C. (2020). Use of waste marble dust and recycled glass for sustainable concrete production. *Journal of Cleaner Production*, 251, 119785. DOI: 10.1016/j.jclepro.2019.119785

Gameiro, F., De Brito, J., & Correia da Silva, D. (2014). Durability performance of structural concrete containing fine aggregates from waste generated by marble quarrying industry. *Engineering Structures*, 59, 654–662. DOI: 10.1016/j.engstruct.2013.11.026

Hebhoub, H., Aoun, H., Belachia, M., Houari, H., & Ghorbel, E. (2011). Use of waste marble aggregates in concrete. *Construction & Building Materials*, 25(3), 1167–1171. DOI: 10.1016/j.conbuildmat.2010.09.037

Kabeer, K. I. S. A., & Vyas, A. K. (2018). Utilization of marble powder as fine aggregate in mortar mixes. *Construction & Building Materials*, 165, 321–332. DOI: 10.1016/j.conbuildmat.2018.01.061

Kore, S. D., & Vyas, A. K. (2016). Impact of marble waste as coarse aggregate on properties of lean cement concrete. *Case Studies in Construction Materials*, 4, 85–92. DOI: 10.1016/j.cscm.2016.01.002

Kore, S. D., Vyas, A. K., & Syed, S. A. (2020). A brief review on sustainable utilisation of marble waste in concrete. *International Journal of Sustainable Engineering*, 13(4), 264–279. DOI: 10.1080/19397038.2019.1703151

Kore Sudarshan, D., & Vyas, A. K. (2019). Impact of fire on mechanical properties of concrete containing marble waste. *Journal of King Saud University. Engineering Sciences*, 31(1), 42–51. DOI: 10.1016/j.jksues.2017.03.007

Kumar, R., & Kumar, S. K. (2018). *" Partial Replacement of Cement with Marble Dust Powder " " Partial Replacement of Cement with Marble Dust Powder ."* *5*(August 2015), 106–114.

Kumawat, V., Meena, R. V., Beniwal, A. S., & Gupta, G. (2023). *Materials Today : Proceedings '' Utilization of Kota stone slurry along with demolition and construction waste in concrete ". xxxx*. DOI: 10.1016/j.matpr.2023.06.103

Lezzerini, M., Luti, L., Aquino, A., Gallello, G., & Pagnotta, S. (2022). Effect of Marble Waste Powder as a Binder Replacement on the Mechanical Resistance of Cement Mortars. *Applied Sciences (Basel, Switzerland)*, 12(9), 4481. Advance online publication. DOI: 10.3390/app12094481

Meena, R. V., Beniwal, A. S., & Gupta, G. (2023). Materials Today : Proceedings '' Fresh characteristics and compressive strength evaluation of self- compacting concrete prepared with ceramic waste tile and fly ash. *Materials Today: Proceedings*, xxxx. Advance online publication. DOI: 10.1016/j.matpr.2023.05.687

Meena, R. V., Beniwal, A. S., Jain, A., Choudhary, R., & Mandolia, R. (2023). Evaluating resistance of ceramic waste tile self-compacting concrete to sulphuric acid attack. *Construction & Building Materials*, 393(March), 132042. DOI: 10.1016/j.conbuildmat.2023.132042

Meena, R. V., Jain, J. K., Chouhan, H. S., & Beniwal, A. S. (2021). Mechanical and durability performance of self-compacting concrete with waste ceramic tile as a replacement for natural river sand. *Materials Today: Proceedings*, xxxx. Advance online publication. DOI: 10.1016/j.matpr.2021.12.303

Memon, M. J., Jhatial, A. A., Rid, Z. A., Rind, T. A., & Sandhu, A. R. (2019). Marble Powder As Fine Aggregates in Concrete. *Engineering, Technology &. Applied Scientific Research*, 9(3), 4105–4107. DOI: 10.48084/etasr.2698

Singh, M., Choudhary, K., Srivastava, A., Singh Sangwan, K., & Bhunia, D. (2017). A study on environmental and economic impacts of using waste marble powder in concrete. *Journal of Building Engineering*, 13, 87–95. DOI: 10.1016/j.jobe.2017.07.009

Singh, M., Srivastava, A., & Bhunia, D. (2019). Analytical and Experimental Investigations on Using Waste Marble Powder in Concrete. *Journal of Materials in Civil Engineering*, 31(4), 1–13. DOI: 10.1061/(ASCE)MT.1943-5533.0002631

singh, J., & S Bansal, E. R. (2015). Partial Replacement of Cement With Waste Marble Powder With M25 Grade. *International Journal of Technical Research and Applications, 3*(2), 202–205. www.ijtra.com

Singhal, V., Nagar, R., & Agrawal, V. (2020). Sustainable use of fly ash and waste marble slurry powder in concrete. *Materials Today: Proceedings, 32*(xxxx), 975–981. DOI: 10.1016/j.matpr.2020.05.650

Topçu, I. B., Bilir, T., & Uyguno lu, T. (2009). Effect of waste marble dust content as filler on properties of self-compacting concrete. *Construction & Building Materials*, 23(5), 1947–1953. DOI: 10.1016/j.conbuildmat.2008.09.007

Vaidevi, C. (2013). Study on marble dust as partial replacement of cement in concrete. *Indian Journal of Engineering*, 4(9), 14–16.

Vardhan, K., Siddique, R., & Goyal, S. (2019). Strength, permeation and micro-structural characteristics of concrete incorporating waste marble. *Construction & Building Materials*, 203, 45–55. DOI: 10.1016/j.conbuildmat.2019.01.079

Vigneshpandian, G. V., Shruthi, E. A., Venkatasubramanian, C., & Muthu, D. (2017). Utilisation of Waste Marble Dust as Fine Aggregate in Concrete. *IOP Conference Series. Earth and Environmental Science*, 80(1), 012007. Advance online publication. DOI: 10.1088/1755-1315/80/1/012007

Vilas, R., Kumar, J., & Singh, A. (2022). *" Sustainable self-compacting concrete containing waste ceramic tile aggregates : Fresh, mechanical, durability, and microstructural properties ." 57*(March).

Vilas, R., Kumar, J., Singh, H., & Singh, A. (2022). Use of waste ceramics to produce sustainable concrete : A review. *Cleaner Materials*, 100085, 100085. Advance online publication. DOI: 10.1016/j.clema.2022.100085

Vilas Meena, R., Kumar Jain, J., & Singh Chouhan, H., mandolia, R., & Singh Beniwal, A. (. (2021). Impact of waste ceramic tile on resistance to fire and abrasion of self-compacting concrete. *Materials Today: Proceedings*, xxxx. Advance online publication. DOI: 10.1016/j.matpr.2021.12.287

Vilas Meena, R., Singh Chouhan, H., Kumar Jain, J., & Kumar Satankar, R. (2021). Construction and Demolition Waste as an Alternative of Rigid Pavement Material: A Review. *IOP Conference Series. Earth and Environmental Science*, 795(1), 012019. DOI: 10.1088/1755-1315/795/1/012019

Yoro, K. O., & Daramola, M. O. (2022). *Engineering a Sustainable Carbon Future.*

Chapter 13
Utilization of Waste Marble Slurry Along With Construction and Demolition Waste in Concrete

Ram Vilas Meena
JECRC University, India

Swikrit Gupta
https://orcid.org/0009-0009-6882-5192
JECRC University, India

Ankit Singh Beniwal
https://orcid.org/0000-0001-8655-903X
JECRC University, India

Ajay Kumar
https://orcid.org/0000-0001-7306-1902
Department of Mechanical Engineering, School of Engineering and Technology, JECRC University, Jaipur, India

ABSTRACT

In this current study investigates the possibility of construction demolish waste (CDW) and waste marble slurry (WMS) as a coarse aggregate (20mm) and cement in the making of concrete. CDW added in (0% to 30% at interval of 10%), and WMS used in (0% to 30% at interval of 10%) to make the concrete blends. Total of four blends was cast including control mixture, and three mixtures for combined use

DOI: 10.4018/979-8-3693-5513-8.ch013

of WMS and CDW. Workability, water absorption, and mechanical strength tests were performed to judge the performance of concrete prepared with WMS and CDW. With WMS and CDW, the workability of concrete was decreased, although within acceptable limits as per the Indian standard code. Concrete blend prepared with combined used of 10% WMS and 10% CDW (i.e. M10C10 blend) displayed maximum compressive and split tensile strength (STS). The results suggested the optimum use of CDW up to 10% and WMS up to 10% as a coarse aggregate and cement replacement in concrete.

1. INTRODUCTION

Concrete is most widely used globally in construction sector and demand also increases as we live in developing countries further which impact on environment (Vilas, Kumar, & Singh, 2022). Concrete is mixture of water, cement, sand and aggregate theses are main materials of concrete (Meena et al., 2021). In which sand and aggregate are obtained from disintegration of rocks. The major difference between sand and aggregate was their particle size. If the particle size is greater than 4.75 mm called coarse aggregate and less than 4.75 mm called fine aggregate. Day by day natural resources are decreasing(Vilas, Kumar, Singh, et al., 2022). As we live in developing countries, natural resources are scarce and we must balance them with our future needs. After aggregate cement is the major use in concrete and mortar (Vilas Meena, Kumar Jain, et al., 2021). Now, day by day consumption of cement increases in mortar and concrete (Meena et al., 2023). Cement production reached 329 million tonnes (MT) in finance year 2020 and is predict to reach 381MT by finance year 2022. From 2015 to 2020, emissions of carbon dioxide from cement production increased by 1.8% per year. Parallel to this, much waste material creates global environmental problems (Ajay et al, 2024). In this study, the main focus is on waste marble slurry and construction and demolition waste. There are now a number of viable cement replacements, including metakaolin, fly ash (FA), granite waste, silica fume (SF), waste marble slurry (WMS), and rice husk ash (Choudhary, Gupta, Nagar, et al., 2021). India generate annually 16MT marble waste (Choudhary, Gupta, Alomayri, et al., 2021). Marble is a type of the metamorphic rock and its transform their geological property in pure limestone (sedimentary rock). The utilization of marble gained from the industry's waste generated during marble cutting. It is possible to reuse leftover marble. Waste in the form of powder and solid particles are produced in massive quantities by the marble cutting business. While cutting, dressing, polishing marble produce the large amount of slurry and solid are formed which replace cement, fine and coarse aggregate in mix design of concrete (Kumar et al. 2024). Construction and demolition waste are generated from various construction and demolition activities like renovation, buildings, roads, repair, bridges, flyover etc (Vilas Meena, Singh Chouhan, et al., 2021; Kumar et al., 2024). Construction and demolition garbage is typically improperly disposed of in a disorganized way in developing nations like India, which contributes to environmental problems. It contains most of the non-biodegradable material such as steel, wood, concrete, plaster, metal, bricks, tiles, and hazardous materials also. These wastes are present in bulky, heavy with high density and engaged with road side, bare land and dump yard. Utilizing these wastes in concrete reduces the pollution in environment, conserves natural resources, and promotes economic development.

2. LITERATURE REVIEW

(Shah et al. 2021) studied about useful properties of recycled coarse aggregate. The replacement of 100% natural aggregate by recycled coarse aggregate can decrease the cost of up to 16%. (Kumawat et al.2023) reported the influence of CDW and kota stone waste slurry as a substitution of fine aggregate and cement in sustainable concrete. Substitutions were done by weight in variations of 0%, 10%, 20% and 30% both type waste materials. The concrete blend was designed for a 28-day compressive strength of 30 MPa. On testing concrete mixture samples attained a peak 28 day compressive and split tensile strength for 10% substitution for both type waste materials(CDW and kota stone waste slurry) combined used. (Kumar, 2017) assessed the concrete's compressive and flexural strength, abrasion resistance, fresh density, and slump. In this investigation, when it came to laying concrete, RCA completely replaced NCA. (Silva et al., 2019) employed construction and demolition waste materials in new construction applications, such as bitumen-bound, hydraulically-bound, and unbound concrete, as well as non-structural concrete in building and road construction. (Sudarshan and Vyas) conducted a study on concrete with replacing the Natural Coarse Aggregate by 75% with the Stone marble waste. The results shown that the concrete with 75% Coarse Aggregate of stone marble waste resist temperature as high as 400°C and the results were found satisfactorily. Ultrasonic pulse velocity values, compressive strength, water absorption, and mass loss were found to be essentially the same between the experimental and control mixtures in terms of their attributes, but if the temperature rises over 400°C, concrete begins to degrade. (Hebhoub et al.) examined the utilization of waste marble to replace natural aggregate in concrete. In their mixes, up to 100% natural aggregate was replaced by waste marble, at 0%, 25%, 50%, 75% and 100% replacements. In order to determine workability, compressive strength, STS, density tests were conducted. For substitutions at 25%, 50%, and 75%, the increase in resistance is very large (Hebhoub et al., 2011). (Alyamaç and Aydin) study were conducted on concrete containing waste marble dust by replacement of (10%, 20%, 30%, 40%, 50% and 90%) natural sand. The experimental results are drawn and shown that the concrete containing up to 40% waste marble dust provide the required properties of concrete, and also the marble powder used in this study is inexpensive and environment friendly. (Singh et al., 2019) study was conducted using marble dust as a partial replacement of binder in concrete. Experimental studies were carried out to study the effect of marble dust power on hydration process, compressive strength of concrete. In the study cement was swapped by marble dust as a percentage of 0%, 10%, 15%, 20% and 25% by weight of cement. The mechanical strength of concrete of quarry dust added to concrete was predicted using regression model and was based on strength ratios as opposed to specimen form or kind. Study concluded that substitute of 10% binder by marble dust reduces the CO_2 emission by 5.2%.

3. RAW MATERIALS

Waste WMS was collected from the VKI industrial area Jaipur, Rajasthan, India. By sieving, the WMS materials size was evaluated. The size of the WMS materials was observed to be less than 90μ. WMS is disposed of by trucks or tankers, which is a laborious process that keeps the industry's area unoccupied. It is therefore free of charge. When the WMS was collected, it was damp. The wet WMS was left out in the sun to dry. The WMS was utilized in the experimental program once it had fully dried and transformed into a dry powder condition. Ordinary Portland Cement 43 conforming to BIS: 8112 (BIS, 2013) was utilized as a cement in the research work. Banas river sand was used as fine aggregate, zone conforming to the IS 383: 2016 (Standard, 2016). The basalt type of rock stones was crushed using a cone crusher to produce coarse aggregate. The nominal size of crushed gravel coarse aggregates utilized was maximum 20 mm. CDW obtained from the near JECRC University Jaipur, India was initially screened for unwanted materials such as litters, plastic, wood, steel etc. In these wastes materials contain demolish RCC concrete, brick, stone waste, and mortar concrete etc. CDW material was then crushed into smaller sizes manually and further with the help of a hammer. In this investigation BASF superplasticizer was utilized as water reducers admixture. Incorporating admixture in concrete mix to improve the workability and mechanical properties of concrete. Concrete was prepared with tab water that conformed with IS: 456 (IS 456, 2000) and was also utilized for curing. Sieve analyses of CDW and coarse aggregate are displayed in Figure 1. Table 1 shown physical properties of raw materials.

Table 1. Physical properties of raw materials

Sr. No.	*Properties*	*WMS*	*Cement*	*Fine aggregates*	*Coarse aggregate*	*CDW*
1	Fineness (%)	2.3	1.9	2.46		
2	Specific Gravity	2.67	3.15	2.70	2.78	2.93
3	Normal Consistency		30			
4	Compressive strength of (28 days)		43 N/ mm^2			
5	Silt Content			6.11		
6	Water Absorption (%)			0.40	0.06	3.47

Figure 1. Gradation of coarse aggregate and CDW

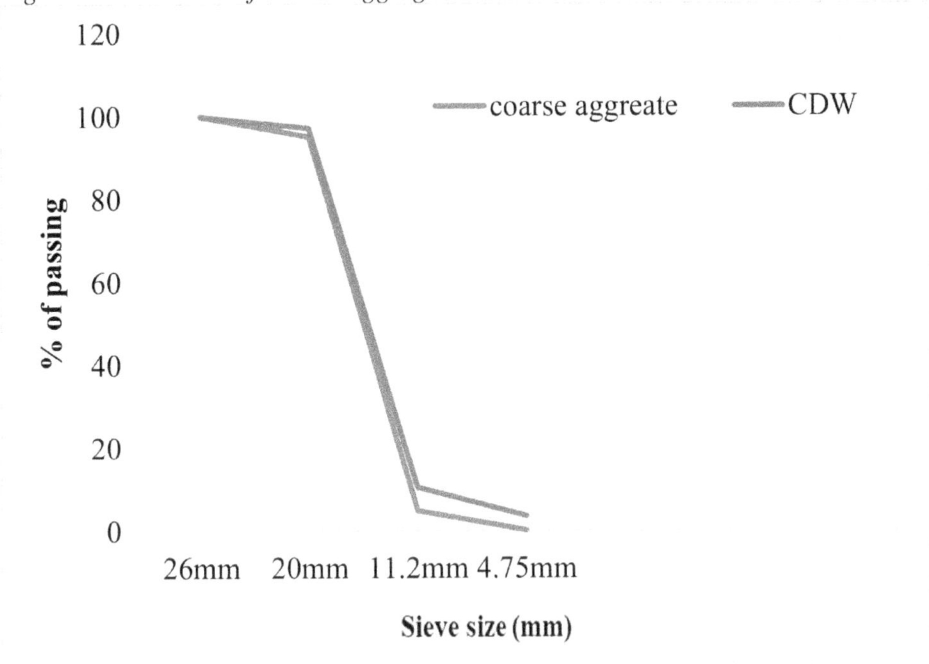

4. MATERIALS AND METHODS

The Indian standard process was followed for mixing and preparing the concrete specimens. Concrete mix design was carried out as per the guideline provided in IS: 416:2000 (IS 456, 2000) and IS: 10262:2019 (BIS, 2019). In the current research total four mixtures of concrete were ready of which one reference blend without WMS and CDW other three mixtures with WMS and CDW in % of (10 to 30% at 10% interval) by weight of coarse aggregate and cement. Study mixtures IDs along with blend proportions are depicted in Table 2, for example $M_{10}C_{10}$ stands for 10% swap of sand and cement with WMS and CDW. Newly prepared concrete mixtures were tested for compaction factor and slump values in order to judge their workability. All concrete mixtures under research work were examined for their initial compaction factor and slump in accordance with IS code. For every concrete mix 150 mm cubes were cast for determined STS, compressive strength and water absorption. The samples were cured in a curing tank until the time of testing, after which they were demolded after 24 hours. Compressive strength testing was carried out as per IS: 516 (IS 516:2014, 2004) for measuring of their compressive strength at 7 and 28 days of water curing. Water absorption and STS of concrete mixtures were examined at 28 days of water curing as per ASTM C642 (ASTM C642, 1997) and IS 5816 [11].

Table 2. Mixture proportion (Kg/m^3)

S.No.	Mix	Cement	Coarse aggregate		Fine aggregate	Admixture	Water	WMS	CDW
			10mm (40%)	20mm (60%)					
1.	**Control mix (CM)**	**440**	**529.2**	**793.8**	**641**	**4.4**	**144**		
2.	$M_{10}C_{10}$	396	529.2	714.42	641	4.4	144	44	79.38
3.	$M_{20}C_{20}$	352	529.2	635.04	641	4.4	144	88	158.76
4.	$M_{30}C_{30}$	308	529.2	555.66	641	4.4	144	132	238.14

5. EXPERIMENTAL RESULTS

5.1 Slump Test

To understand the effects of replacing cement and natural aggregates with WMS and CDW on the concrete's fresh characteristics, the slump of concrete mixtures was measured. The slump values detected for various mixtures were observed to be within specified limits for concrete (39 and 48) mm as per IS code 1199:1959 (IS : 1199- 1959, 2012). The admixtures dosage was similar for the all mixtures. Incorporations of CDW and WMS did reduce the slump values of concrete mixture as visible from trends shown in Figure 2. This decrease for CDW and WMS blend is well expected because of the shape of CDW materials and more water absorption might be the reason for lowering workability. (singh & S Bansal, 2015) examined the use of marble waste powder with the replace of binder content up to 20% in concrete mixes. Slump value was investigated and it was found that the slump value decreased up to 20% replacement level.

Figure 2. Slump value trends for all concrete mixtures

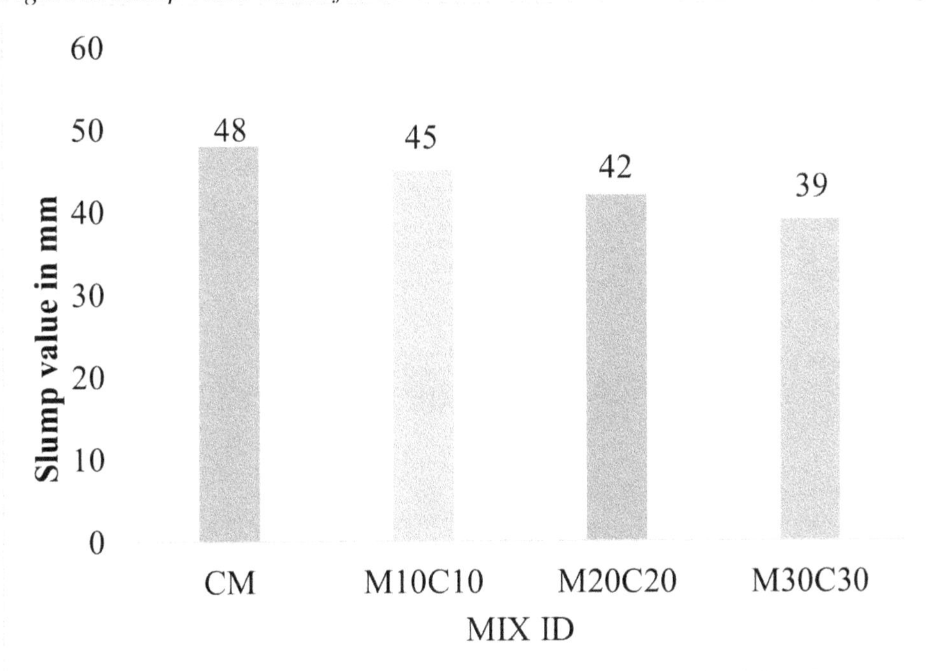

5.2 Compaction Factor

To understand the effects of replacing cement and natural aggregates with WMS and CDW on the concrete's fresh characteristics, the compaction factor of concrete mixtures was measured. The compaction factor values detected for various mixtures were observed to be within specified limits for concrete (0.9 and 0.95) as per IS code IS 1199:1959 (IS : 1199- 1959, 2012). Incorporations of CDW and WMS did reduce the compaction factor values of concrete mixture as visible from trends shown in Figure 3. This decrease for CDW and WMS blend is well expected because of the shape of CDW materials and more water absorption might be the reason for lowering workability.

Figure 3. Compaction factor value trends for all concrete mixtures

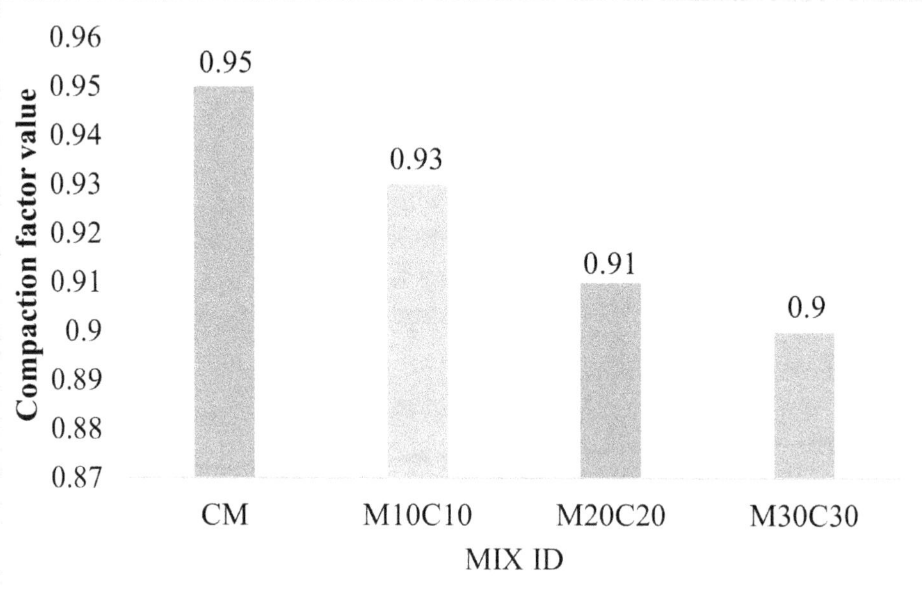

5.3 Compressive Strength

Figure 4 displayed the compressive strength of the all concrete mixtures after 7 and 28 days of curing. The experiment outcomes of the compressive strength experiment presented that with increase in % of the CDW and WMS the compressive strength of the concrete increased up to 10% replacement and above this % the compressive strength starts decreasing.

Figure 4. Compressive strength experiment outcomes

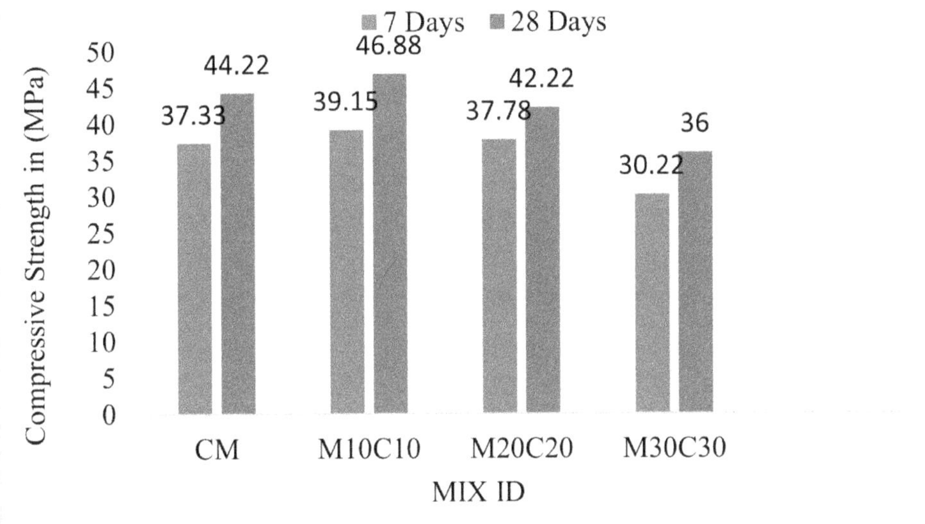

For CDW and WMS integrating concrete, $M_{10}C_{10}$ mixture at all curing times had the maximum compressive strength and $M_{30}C_{30}$ mix had the minimum. The rough surface and irregular shape of CDW materials, along with WMS acting as filler for the pores in the concrete, influence be the reason for increase in compressive strength of $M_{10}C_{10}$ construction and demolition waste and marble stone slurry modified concrete mixes.

5.4 Split Tensile Strength

Figure 5 displayed the STS of the all concrete mixtures after 28 days of curing. The concrete blends' STS at 28-days were 3.33 to 4.51 MPa. The experiment outcomes of the STS experiment presented that with increase in % of the CDW and WMS the split tensile strength of the concrete increased up to 10% replacement and above this % the STS starts decreasing. For CDW and WMS integrating concrete, $M_{10}C_{10}$ mixture at all curing times had the maximum STS and $M_{30}C_{30}$ mix had the minimum. The rough surface and irregular shape of CDW materials, along with WMS acting as filler for the pores in the concrete, influence be the reason for increase in STS of $M_{10}C_{10}$ construction and demolition waste and marble stone slurry modified concrete mixes. Similarly a study by (Anurag and Singh, 2021), found improvement in STS of self-compacting concrete with steel fibers and recycled waste demolition aggregates.

Figure 5. Split tensile strength experiment outcomes

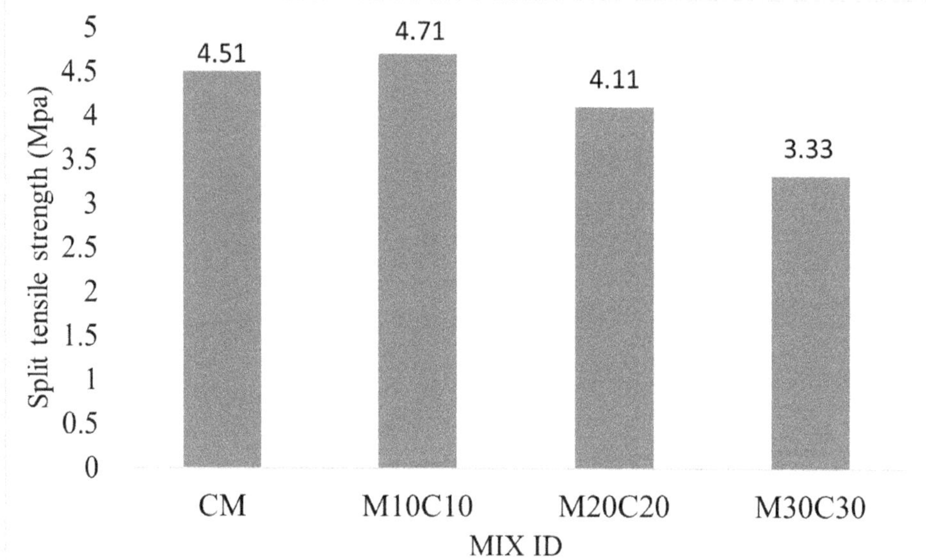

5.5 Water Absorption

The water absorption results (display in Figure 6) of CDW and WMS mixes was found to be lower than that of control mix up to 10% replacement, then increases replacement ratio increases water absorption. The maximum water absorption % was observed of $M_{30}C_{30}$ mixtures.

Figure 6. Water absorption trend for all concrete mixtures

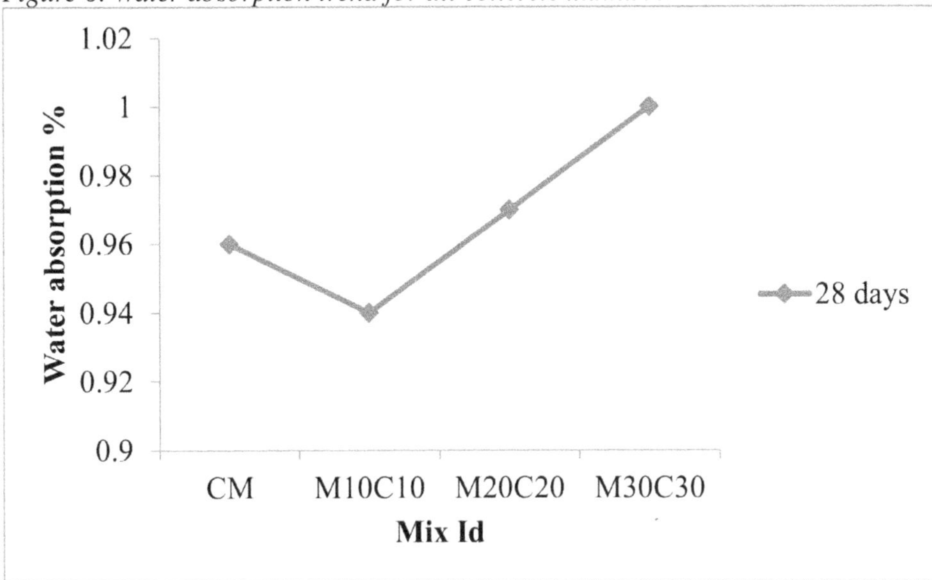

CONCLUSION

This research work on the recycled concrete in which cement is replaced by waste marble slurry and coarse aggregate is replaced by C and D waste. Both are mixed in simultaneously in ratio of 10%, 20% and 30%. Physical and mechanical properties were evaluated. Bases on result and discussion following conclusion were drawn from this study of grade M-40 concrete.

- The WMS and CDW decrease the slump and compaction factor of the concrete mix when it replaces the cement and coarse aggregate. The workability of decline due to high angular particles presence in the concrete and rate of water soaking also high that is slump and compaction factor decreases continuously when increases both materials quantities.
- 10% replacement of cement by WMS and coarse aggregate (20mm) by CDW increase the compressive and split tensile strength, further replacement of cement and coarse aggregates by WMS and CDW reduces the both strength.
- The use of CDW and WMS up to 10% reduced the water absorption of the concrete mixes. The water absorption further higher with the integration of CDW and WMS concrete mixes.

From the above results, it may be inferred that 20% of coarse aggregate and cement may be replaced by CDW and WMS.

REFERENCES

Ajay, P. Kumar, A., Mittal, R.K., & Goel, R. (Eds.). (2023). Waste Recovery and Management: An Approach Toward Sustainable Development Goals (1st ed.). CRC Press. DOI: 10.1201/9781003359784

BIS. (2013). ORDINARY PORTLAND CEMENT, 43 GRADE — SPECIFICATION. In *Indian Standards*.

BIS. (2019). IS 10262:2019, Indian standard, Concrete mix Proportioning – Guidelines (Second Revision). -. *Bis, January*. https://www.google.com/search?q=IS+10262%3A2019%2C+Indian+standard%2C+Concrete+mix+Proportioning+–+Guidelines+(Second+Revision).&rlz=1C1CHBF_enIN984IN984&oq=IS+10262%3A2019%2C+Indian+standard%2C+Concrete+mix+Proportioning+–+Guidelines+(Second+Revision).&a

Choudhary, R., Gupta, R., Alomayri, T., Jain, A., & Nagar, R. (2021). Permeation, corrosion, and drying shrinkage assessment of self-compacting high strength concrete comprising waste marble slurry and fly ash, with silica fume. *Structures*, 33(May), 971–985. DOI: 10.1016/j.istruc.2021.05.008

Choudhary, R., Gupta, R., Nagar, R., & Jain, A. (2021). Mechanical and abrasion resistance performance of silica fume, marble slurry powder, and fly ash amalgamated high strength self-consolidating concrete. *Construction and Building Materials, 269*(xxxx), 121282. DOI: 10.1016/j.conbuildmat.2020.121282

Hebhoub, H., Aoun, H., Belachia, M., Houari, H., & Ghorbel, E. (2011). Use of waste marble aggregates in concrete. *Construction & Building Materials*, 25(3), 1167–1171. DOI: 10.1016/j.conbuildmat.2010.09.037

IS : 1199- 1959. (. (2012). Indian Standard METHODS OF SAMPLING AND ANALYSIS OF CONCRETE. *Recent Patents on Nanotechnology*, 7(1), 13–25. DOI: 10.2174/18722105130103

IS 456. (2000). Plain Concrete and Reinforced. *Bureau of Indian Standards,New Dehli*, 1–114.

IS 516:2014. (2004). Method of Tests for Strength of Concrete. *IS : 516 - 1959 (Reaffirmed 2004)*, New Delhi,India.

IS 5816 : 1999. (1999). Indian standard Splitting tensile strength of concrete- method of test (first revision). *Bureau of Indian Standards, New Delhi, (reaffirme*, 1–14. Kumar, R. (2017). Influence of recycled coarse aggregate derived from construction and demolition waste (CDW) on abrasion resistance of pavement concrete. *Construction and Building Materials, 142*, 248–255. DOI: 10.1016/j.conbuildmat.2017.03.077

Kumar, A., Liu, Y., & Kumar, R. (Eds.). (2024). *Handbook of Intelligent and Sustainable Manufacturing: Tools, Principles, and Strategies*. CRC Press., DOI: 10.1201/9781003405870

Kumar, A., Rajak, D. K., Kumar, P., & Kumar, A. (Eds.). (2024). *Advances in Sustainable Biomaterials: Bioprocessing 4.0, Characterizations, and Applications* (1st ed.). CRC Press., DOI: 10.1201/9781003434313

Meena, R. V., Beniwal, A. S., Jain, A., Choudhary, R., & Mandolia, R. (2023). Evaluating resistance of ceramic waste tile self-compacting concrete to sulphuric acid attack. *Construction & Building Materials*, 393(March), 132042. DOI: 10.1016/j.conbuildmat.2023.132042

Meena, R. V., Jain, J. K., Chouhan, H. S., & Beniwal, A. S. (2021). Mechanical and durability performance of self-compacting concrete with waste ceramic tile as a replacement for natural river sand. *Materials Today: Proceedings*, xxxx. Advance online publication. DOI: 10.1016/j.matpr.2021.12.303

Silva, R. V., de Brito, J., & Dhir, R. K. (2019). Use of recycled aggregates arising from construction and demolition waste in new construction applications. *Journal of Cleaner Production*, 236, 117629. DOI: 10.1016/j.jclepro.2019.117629

Singh, M., Srivastava, A., & Bhunia, D. (2019). Analytical and Experimental Investigations on Using Waste Marble Powder in Concrete. *Journal of Materials in Civil Engineering*, 31(4), 1–13. DOI: 10.1061/(ASCE)MT.1943-5533.0002631

singh, J., & S Bansal, E. R. (2015). Partial Replacement of Cement With Waste Marble Powder With M25 Grade. *International Journal of Technical Research and Applications, 3*(2), 202–205. www.ijtra.com

Standard, I. (2016). *Coarse and Fine Aggregate for Concrete — Specification. 19*(January).

Vilas, R., Kumar, J., & Singh, A. (2022). " *Sustainable self-compacting concrete containing waste ceramic tile aggregates : Fresh, mechanical, durability, and microstructural properties* ." *57*(March).

Vilas, R., Kumar, J., Singh, H., & Singh, A. (2022). Use of waste ceramics to produce sustainable concrete : A review. *Cleaner Materials*, 100085, 100085. Advance online publication. DOI: 10.1016/j.clema.2022.100085

Vilas Meena, R., Kumar Jain, J., & Singh Chouhan, H., mandolia, R., & Singh Beniwal, A. (. (2021). Impact of waste ceramic tile on resistance to fire and abrasion of self-compacting concrete. *Materials Today: Proceedings*, xxxx. Advance online publication. DOI: 10.1016/j.matpr.2021.12.287

Vilas Meena, R., Singh Chouhan, H., Kumar Jain, J., & Kumar Satankar, R. (2021). Construction and Demolition Waste as an Alternative of Rigid Pavement Material: A Review. *IOP Conference Series. Earth and Environmental Science*, 795(1), 012019. DOI: 10.1088/1755-1315/795/1/012019

Chapter 14
Renewable Energy Outlook in India:
An Analysis

N. Sivakumar
Varuvan Vadivelan Institute of Technology, India

R. Arulmurugan
https://orcid.org/0000-0002-8893-2894
Annasaheb Dange College of Engineering and Technology, India

S. Mahendran
Er. Perumal Manimekalai College of Engineering, India

ABSTRACT

Significant rises in the renewable energy generation in all over the world. The solar energy plays a particular role to generate electricity in recent times all over the world and in an India. Though solar energy transition is very much important for environment prospects. Even some challenges were encountered while increasing the renewable penetration in the modern grid system. This article discuss various scheme initiated by government and trends of Indian energy transition. During the last couple of decades, there has been a significant increase in the installation of renewable energy source worldwide. According to international renewable energy agency India is ranked fifth in the list with installed capacity of renewable energy 117.955 GW power. In this article discussed grid connected rooftop scheme, analysis of solar PV system, and production linked incentive scheme, etc.

DOI: 10.4018/979-8-3693-5513-8.ch014

I. INTRODUCTION

The renewable solar scheme, known as the Pradhan Mantra Kisan Urja Suraksha evam Utthan Mahabhiyan (PM KUSUM) was introduced in the year 2019 (Rai, 2022). In addition with the financial year 2020-21 was planned to extend the installation of solar plant capital was 30.8GW by 2022, up for the initial goal value of 25.7 GW power (Gupta et al., 2014). The centralized financial support was allocated under the Pradhan mantra kisan scheme was 34,035 cores, with proper plane and implementing the domestic content requirement and supporting domestic manufacturers can significantly enhance their visibility and competitiveness and potential exposure are vast opportunities in the Indian market due to its significant demand (Wafa hayder et al., 2020).

The schemes is further classified into the three distinct components.

- Component Scheme one is decentralized ground based solar energy or other renewable energy based plant setting up of 10GW (Guiza dhaouadi et al, 2021) .
- Component Scheme two is standalone solar agriculture pumps 2 million installation (Ankit Gupta et al., 2014).
- The third component of the Scheme focuses on the solarization of 1.5 million existing grid connected agriculture pumps.

Utilizing based on domestic content requirement modules and cell is mandatory for component scheme two and three (Rajeshree patil, 2017). As of November 2021 only 20 megawatts had been installed under component scheme one, out of the total sanctioned amount. Under component scheme two, 75,000 standalone solar pumps have been installed across various states (Patil & Anantwar, 2017; Kumar et al., 2024). over one million grid connected solarization pumps have been assigned across two variations of components three (Arulmurugan, et.al, 2013). The variants are one is individual pump and another one is feeder level. Out of the allocated only 1000 units had been installed (Arulmurugan & Anushalini, 2018). The smart manufacturing and Industry 4.0 book explained the recent trends adopted on the industry and state of the art review included smart manufacturing and forecasting the future industry (Arulmurugan & Chandramouli, 2018; Kumar et al., 2024). The book modelling processing and characterization of the materials and incremental of sheet forming technologies on recent era was descripted (Arulmurugan, et al., 2013; Kumar et al., 2024). Everywhere the waste management play the big roles, government took lot of the efforts about sustainable development, artificial intelligence technique utilization for industry manufacturing sector was detailed in (Arulmurugan &Venkatesan, 2015).

India is shifting towards cleaner energy sources by boosting renewable energy and improving energy efficiency. This includes increasing investments in solar and wind power and modernizing the energy grid to better integrate these resources. India's energy sector relies heavily on coal, with increasing use of solar and wind energy (Arulmurugan,et al, 2014). However, the grid infrastructure is outdated and struggles to integrate renewable sources effectively. The Indian government has rolled out several energy policies, focusing on increasing renewable energy and improving efficiency (Kumar et al., 2024). Yet, these policies are often not thoroughly detailed, making their implementation and effectiveness. Recent progress in renewable energy research includes innovations in solar technology, such as advanced photovoltaic cells and new materials like perovskite, which improve energy efficiency(Kumar & Sharma, 2020). Wind energy research has led to the development of larger, more efficient turbines that perform better in low-wind conditions. Advances in energy storage include the creation of high-capacity and solid-state batteries. Research into green hydrogen is advancing, providing potential for cleaner fuel options. Additionally, smart grid technologies are evolving to better integrate renewable energy and enhance grid stability. These developments are driving more efficient and sustainable renewable energy systems (Kumar et al., 2023; Kumar et al., 2024).

The Indian government has introduced several key policies to expand solar energy capacity. The National Solar Mission aims to achieve 100 GW of solar power by 2030, while the Solar Park Scheme helps develop large-scale projects by providing necessary infrastructure. Financial incentives and subsidies support both residential and commercial solar installations, making them more affordable. Policies on grid connectivity and net metering enable solar producers to sell excess power back to the grid, enhancing the financial benefits. Recent initiatives, including import duties, are intended to strengthen domestic solar manufacturing. These measures have significantly increased solar energy capacity, spurred economic growth, and supported environmental goals by cutting greenhouse gas emissions(Laturkar & Laturkar, 2023).

Figure.1 shows the as of November 2021 PMKUSUM allocated capacity under various components

Figure 1. November 2021 PMKUSUM allocated capacity under various components

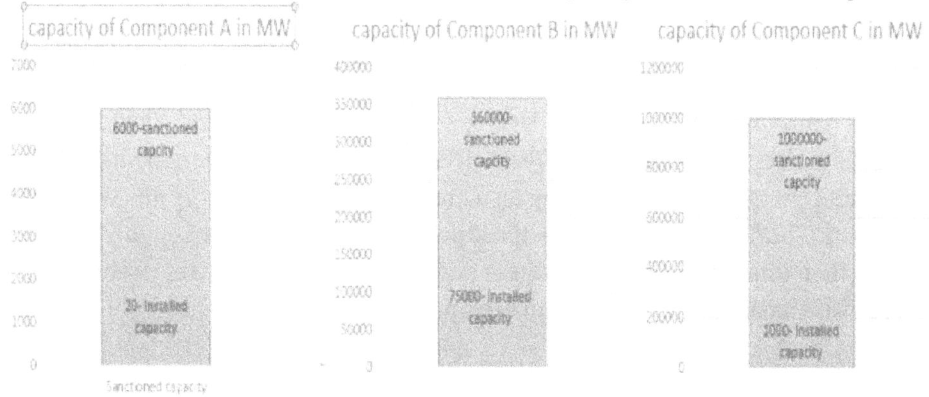

Source; JMK Research

II. ROOFTOP SOLAR SYSTEM INTEGRATED TO THE GRID

In December 2015, they started phase one of the program to encourage homes, social areas and institution with incentives. Additional government segments and achievement links were introduced.

Phase two was accepted in the year 2019 February, aiming for a cumulative capacity of 40GW from rooftop solar projects by 2022. In phase two, incentives in the form of central financial assistance were tailored based on the capacity of rooftop solar PV installations.

- The rooftop solar plant system with up to 3KW capacity offers 40% by central financial assistance.
- The rooftop solar plant system beyond 3KW to 10KW capacity offer 20% by central financial assistance.
- In addition, the limits of 500Kw capacity for the supply of power to common facilities of group housing societies and welfare associations offer 20% by central financial assistance. Figure.1 shows the allocated capacity under the PM KUSUM scheme as of November 2021. Moreover, for the supply of the power to common facilities of the group housing societies and welfare associations, there is a capacity limit of 500KW, with 20 percentage covered by central financial assistance. Figure.1 illustrate the allocated capacity under the PM KUSUM scheme of the November 2021.

The various union territories and states are implementing the rooftop solar system program. The domestic content requirement is applicable to the residential sector of the program, aiming to stimulate the growth of domestic photovoltaic manufacturing. Phase two sets a target of 4 Giga watts specifically for the residential sector. Out of the allocated 3.4 Giga watt to union territories and states, only 1.07 Giga watt had been installed as of November 2021. Figure.2 illustrated the allocated capacity of roof top solar system integrated to the grid(Kothari et al., 2021).

Figure 2. Roof top solar system integrated to the grid status

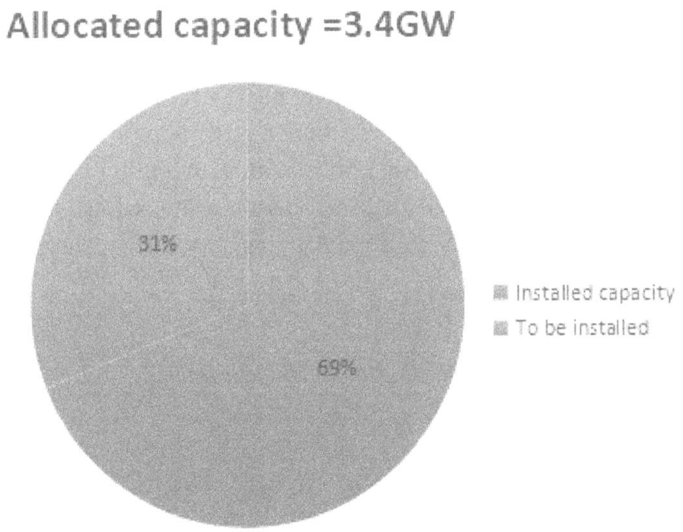

The parliamentary panel observed that achieving the set target for solar roof top plants would require proper implementation of net metering or gross meter. They emphasized the importance of maintaining uniformity in regulations, model operating procedures, and unified portals(Majid, 2020).

Recent updates on various energy sources show advancements in wind and hydropower, with improved turbine technologies and efficient hydroelectric projects enhancing their output. Nuclear energy is being considered for its low-emission benefits, though issues related to safety and waste management persist. Fossil fuels continue to play a major role in energy production but are under growing pressure due to environmental impacts, driving a transition towards cleaner energy options (Bekun, 2022).

Fossil fuels are dependable and efficient but are major sources of pollution and contribute to climate change. Solar energy is renewable and environmentally friendly, though its production varies with sunlight and needs energy storage. Wind energy is sustainable and produces minimal emissions but relies on wind conditions and can impact wildlife. Hydropower generates steady power with low emissions but can alter ecosystems and requires significant infrastructure. Nuclear energy offers low carbon emissions and high output but comes with safety risks and waste disposal issues. A balanced approach is necessary to address the strengths and weaknesses of each energy source for a sustainable future(Eren, et al., 2019).

To undersand the theoretical content more relatable, with few of the examples as follows, the Tata Power Station in Mumbai exemplifies the use of fossil fuels for large-scale energy production. The Bhadla Solar Park in Rajasthan illustrates the effectiveness and scope of solar energy through its extensive installations. Suzlon Energy showcases the practical application of wind power with its advanced turbine technology. The Bhakra Nangal Dam highlights how hydropower provides consistent electricity and manages water resources. Lastly, the Kudankulam Nuclear Power Plant in Tamil Nadu demonstrates how nuclear energy can offer a low-carbon energy solution.

Figure 3. extent of manufacturing Integration

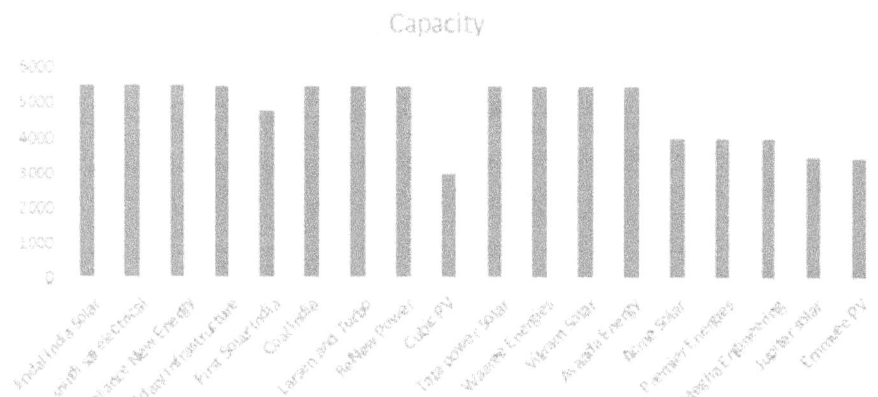

(Y axis is capacity in MW, X axis is manufacturer)

III. PRODUCTION LINKED INCENTIVES SCHEMES BY GOVERNMENT

The production-lined incentives schemes were introduced government of India. The approval of the national program on high-efficiency solar photovoltaic modules came in April 2021. The scheme with a financial outlay of Rs. 4,500, aimed to support the establishment of integrated solar photovotlaic manufacturing plats with the capacity of 10 Giga watt. In addition, the production lined incentive scheme foresaw both direct and indirect benefits related to this initiative. The direct investment of about Rs.17,200 core in solar photovoltaic manufacturing projects. Over the five years, the scheme anticipates a demand of Rs. 17,500 core for balance of materials. It is expected to create direct employment for about 30,000 people and indirect employment for approximately 1,20,000 individuals, while also achieving import substitution of about Rs.1,500 crore annually. Further more, the scheme aims to provide a boost to research and development, resulting in higher efficiency in solar PV modules.

In 2021 May, the Ministry of New and Renewable Energy, in collaboration with the Indian Renewable Energy Development agency initiated the bidding process under the production linked incentive schemes for integrated solar manufacturing units. The total of 54.8GW worth of bids were received, resulting in more than a fourfold over subscription. Bidders were shortlisted based on two parameters. Extern of connection of manufacturing capacity and manufacturing stage. Shiradi sai electrical, Reliance New Energy Solar, and Jindal India Solar Energy following were chosen as the production-linked incentives beneficiaries for establishing the first four categories. Such as ingot-wafer, polysilicon, cell and module integrated manufacturing. Shirdi Sai and Jindal were each awarded 4 Giga watt, while reliance secured 2,483 Mega watt. Figure.3 shows the extent of manufacturing integration. Stage one contains the five manufacturers. Stage two contains four manufacturers. Stage three contains nine manufacturers.

India's heave reliance on imported high efficiency solar PV products renders it highly vulnerable to supply chain disruptions and market price fluctuations. The production-like incentives schemes can help mitigate the photovoltaic supply chain risks(Eren, et al., 2019).

The energy landscape of the Indian subcontinent is diverse, with coal as the primary source, accounting for about 70% of electricity production. Oil and natural gas are essential for transportation and industrial processes, while renewable sources like solar and wind are expanding, contributing roughly 4-5% and 10% of the energy mix, respectively. Hydropower provides around 12-14%, and nuclear energy adds about 2-3%. However, reliance on coal and other fossil fuels results in significant environmental challenges, including air pollution, greenhouse gas emissions, and associated health issues. While renewables offer cleaner alternatives, they encounter issues such as variability and land use. Hydropower can affect ecosystems and communities, and nuclear energy, although low in emissions, presents concerns about radioactive waste and safety. To address these problems, a transition to cleaner energy sources and improved sustainability practices is essential.

Integrating renewable energy sources into the existing energy infrastructure poses several challenges. The variable nature of solar and wind energy can disrupt grid stability, which was designed for steady outputs from fossil fuels. Additionally, current grid systems may struggle to accommodate the decentralized and fluctuating nature of renewable sources, requiring significant upgrades. Effective energy storage is essential for balancing supply and demand, but existing technologies can be costly and may not meet large-scale needs. High costs associated with developing infrastructure for renewable projects, such as land and installation, and inconsistent policies can further complicate the transition. To address these challenges and alleviate energy deficits, authorities and industry stakeholders should invest in modernizing the grid, advancing energy storage solutions, and establishing supportive and consistent policies. Hybrid systems combining renewable sources with traditional energy can offer reliability, while public-private partnerships can help share financial and logistical responsibilities. Additionally, boosting research and development can lead to innovations that reduce costs and improve performance, making the integration of renewable energy more feasible.

Existing energy sources, such as coal, oil, and natural gas, have significant environmental drawbacks. Coal-fired power plants are major sources of carbon dioxide and particulate matter, contributing to air pollution, climate change, and health problems like respiratory diseases. While oil and natural gas are somewhat cleaner alternatives, they still produce greenhouse gases and carry risks such as oil spills and gas leaks. Renewable energy sources like solar and wind offer cleaner options but come with their own challenges. Solar and wind energy are intermittent, which can affect grid stability and require costly advancements in storage and grid management. The development of renewable infrastructure, including solar panels and wind turbines, involves high initial costs and can impact land use. Hydropower, though low in emissions, can disrupt ecosystems and displace communities. Nuclear energy produces minimal emissions but raises concerns about managing radioactive waste and safety risks. To address these issues, a comprehensive approach is needed, incorporating technological innovation, supportive policies, and sustainable practices to effectively integrate renewable energy while minimizing adverse effects.

The current energy infrastructure faces several critical issues. Much of it is outdated, including aging power plants and transmission networks, which lead to inefficiencies and frequent failures. The traditional grid was designed for centralized fossil fuel power generation, making it challenging to incorporate decentralized and variable renewable sources. This mismatch results in reliability problems and capacity constraints, as many systems struggle to accommodate increased demands or new energy inputs. Energy losses during transmission and distribution further reduce efficiency, and the intermittent nature of renewable sources like solar and wind necessitates advanced storage and grid management solutions. Upgrading or replacing this infrastructure involves significant financial investment, creating challenges for both governments and utilities. Additionally, outdated regulations may impede the rapid adoption of new technologies, while increased digitalization raises the risk of cyber-attacks, complicating the management and security of the energy supply.

Adopting renewable energy sources brings several benefits, including a substantial reduction in greenhouse gas emissions, which helps combat climate change and enhance air quality. Solar and wind energy are abundant and sustainable, reducing dependence on finite fossil fuels and improving energy security. Additionally, renewables support job creation in new sectors and offer long-term economic advantages as technologies advance. Despite these benefits, there are challenges to overcome. The variable nature of solar and wind energy can affect grid reliability, requiring advanced storage solutions and better grid management. High initial investment costs and the need for significant upgrades to existing infrastructure can be financial obstacles. Furthermore, developing renewable projects may impact land use and local ecosystems. Addressing these challenges while leveraging the advantages is essential for a successful transition to a more sustainable energy system.

The future of renewable energy promises significant advancements driven by technological innovation and research. Improvements in the efficiency of solar panels and wind turbines, coupled with breakthroughs in energy storage technologies like advanced batteries and hydrogen storage, will facilitate better integration of renewables into existing energy systems. The evolution of smart grid technologies will enhance energy management and grid reliability, while decentralized energy systems and microgrids will offer increased resilience and local control. With growing investments and supportive policies, the adoption of renewable energy is set to accelerate. Additionally, expanding renewable energy use in sectors such as transportation and heating will further reduce carbon emissions. These developments will position renewable energy as a cornerstone of a sustainable and adaptable energy future.

A deeper exploration of renewable energy reveals promising advancements and future directions. Recent innovations include the development of perovskite solar cells, which offer greater efficiency and lower production costs compared to traditional photovoltaic technologies. In wind energy, floating wind turbines are extending the reach of wind farms to deeper waters with stronger winds. Energy storage is also evolving, with solid-state batteries and large-scale storage solutions improving the management of intermittent energy sources. The advancement of smart grids and artificial intelligence is enhancing grid management and facilitating the integration of renewables. Looking ahead, efforts will focus on scaling these technologies, integrating them into existing infrastructure, and expanding their use into sectors like electric vehicles and heating. Research into hybrid systems that combine multiple renewable sources and advanced storage solutions will contribute to a more stable and resilient energy network. Supportive policies and increased investment in research and development will be crucial for accelerating the transition to a sustainable energy future.

IV. RESULTS AND DISCUSSION

An Analysis considered 300Watt solar panel with simulation and real-time data were carried out. The waveform shown in Figure.4 to Figure.8. Figure.4 shows the voltage versus current and voltage versus power of the panel. Similarly, Figure.5 shows the constant irradiance and variable temperature. The change in temperature was 0, 18, 23,26 were considered in the simulation. Table.I observed the rating of photovoltaic panels, which consists of short circuit current, open circuit voltage, maximum current, maximum voltage, maximum power, and efficiency. Figure.6 shows the change in irradiance of different sample days. The days were November 01 and 30th of 2021.

Table 1. Observed rating of PV panel on 30.11.21

G-30.11.21	T-30.11.21	Isc-2	Voc-2	R-2	Im-2	Pm-2	Vm-2	efficiency-2
22.6	19.87	0.1916	37.18	11.46	0.1801	5.415	30.07	1.80
186.92	20.59	1.588	38.83	1.444	1.492	46.85	31.4	15.616
395.62	21.87	3.372	40.78	0.7144	3.168	104.5	32.98	34.833
576.48	23.48	4.933	42.33	0.5068	4.636	158.7	34.23	52.9
703.25	24.83	6.039	43.33	0.4239	5.675	198.9	35.05	66.3
758.2	25.64	6.523	43.73	0.396	6.13	216.8	35.37	72.26
747.75	26.2	6.443	43.57	0.3994	6.055	213.3	35.24	71.1
663.42	26.45	5.719	42.77	0.4417	5.375	185.9	34.59	61.96
513.98	26.41	4.431	41.38	0.5517	4.164	139.4	33.47	46.46
319.12	26	2.748	39.55	0.8501	2.582	82.6	31.99	27.53
118.88	25.02	1.02	37.64	2.177	0.9598	29.22	30.44	9.74
5.67	23.15	0.04874	36.65	44.42	0.0458	1.358	29.64	0.45
5.67	22.29	0.04863	36.74	44.63	0.0457	1.358	29.72	0.452

Figure 4. Change in temperature at constant irradiance of 300watt PV module

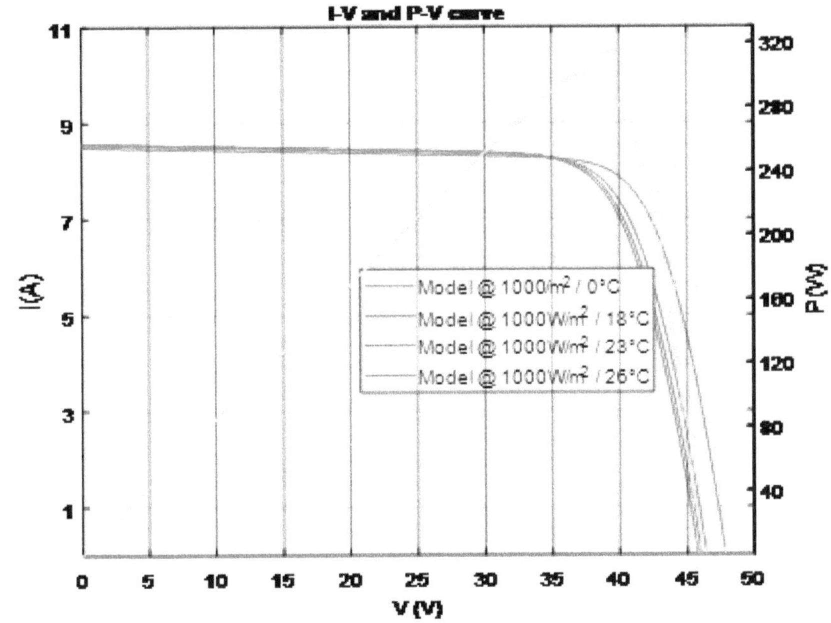

Figure 5. Observed irradiance of two different sample data

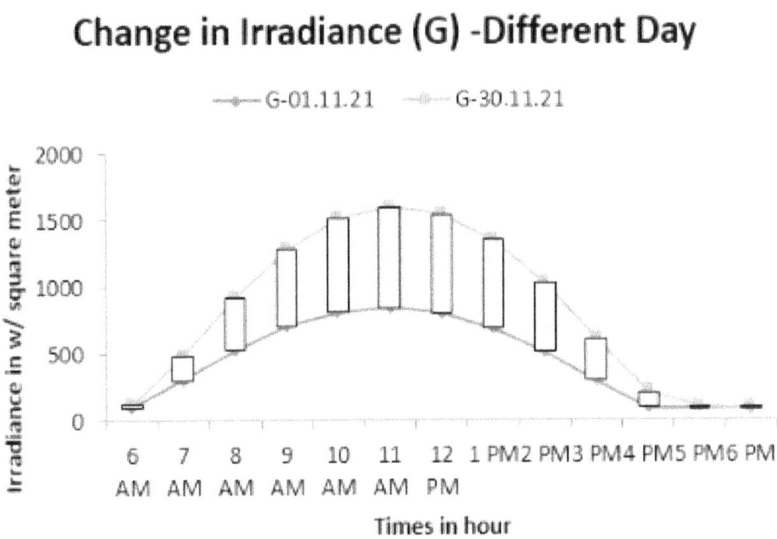

Figure 6. Observed temperature of two different sample data

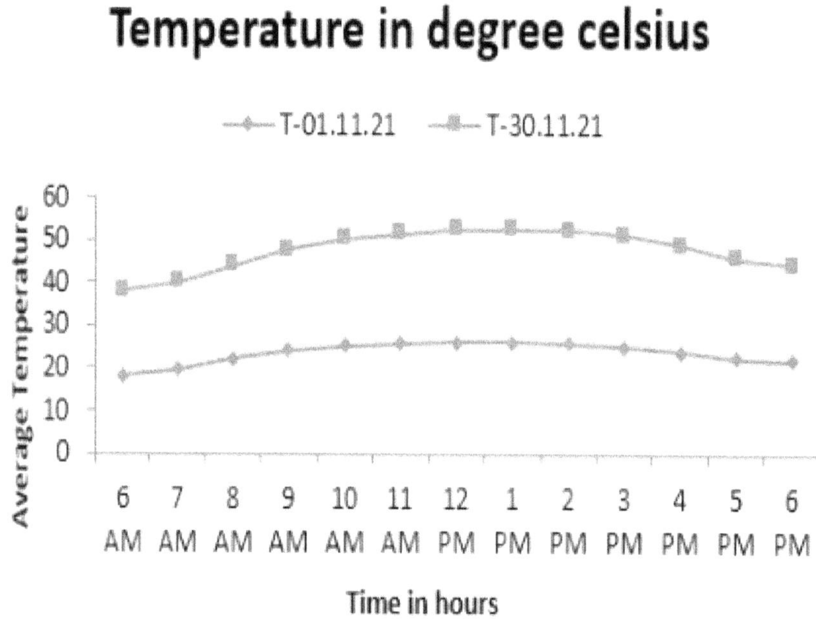

Figure.7 displays the open circuit voltage of the photo voltiac panel for various sample days. Day two sample are higher than day one sample. Figure.8 shows the short circuit different sample day.

Figure 7. Observed open circuit voltage of two different sample data

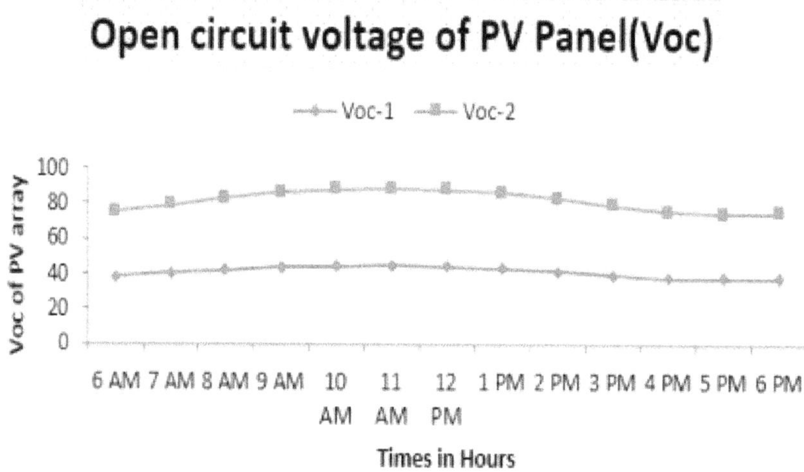

Figure 8. Observed short circuit current of two different sample data

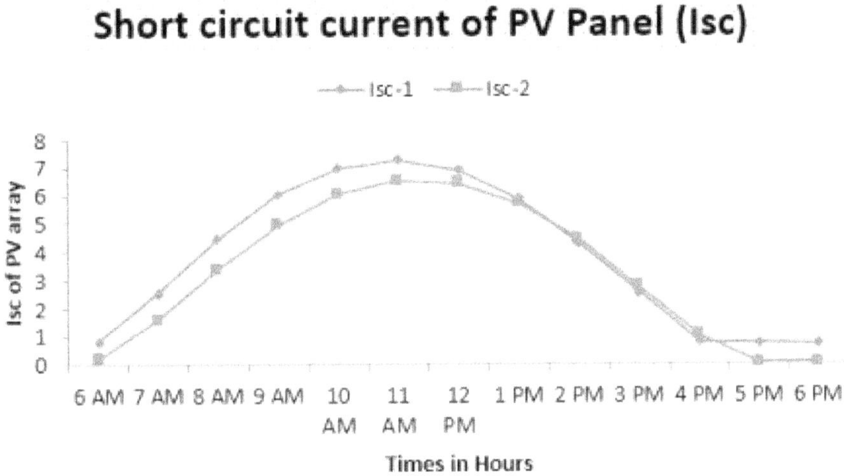

V. CONCLUSION

The analysis shows that India has an enormous potential for renewable and the entire load requirement of the county may be by the proper utilization of these energy sources. Although solar energy sources have become more popular in recent days, the article also carried various incentives and power production of energy from solar and renewable systems. Ladakh Renewable Initiative is a scheme implemented by the Ministry of New Renewable Energy in 2010 June, which insists and focuses on the reduction of the dependence on kerosene and diesel in the Ladakh region. Therefore several policy were carried out throughout the globe to minimize the carbon emission and size and cost of solar panel production.

REFERENCES

Arulmurugan, R.. (2013). Intelligent fuzzy MPPT controller using analysis of DC to DC novel buck converter for photovoltaic energy system applications. In *2013 International Conference on Pattern Recognition, Informatics and Mobile Engineering* (pp. 1-6). IEEE. https://doi.org/DOI: 10.1109/ICPRIME.2013.6496477

Arulmurugan, R.. (2013). Numerical analysis of incremental conductance with cyclic measurement of the MPPT algorithm for PV power generation. In *IET Chennai Fourth International Conference on Sustainable Energy and Intelligent Systems (SEISCON 2013)* (pp. 1-6). IET. https://doi.org/DOI: 10.1049/ic.2013.0297

Arulmurugan, R.. (2014). An intelligent novel FLC based FOINC MPPT technique for PV applications. In *2014 International Conference on Advances in Electrical Engineering (ICAEE)* (pp. 1-6). IEEE. DOI: 10.1109/ICAEE.2014.6838541

Arulmurugan, R., & Anushalini, T. (2018). Implementation of fuzzy controlling structure of PV-Fe for a grid connected system. In *2018 4th International Conference on Electrical Energy Systems (ICEES)*. IEEE. DOI: 10.1109/ICEES.2018.8443202

Arulmurugan, R., & Chandramouli, A. (2018). Design and implementation of a nine-level inverter with shunt active filter for distributed system. In *2018 4th International Conference on Electrical Energy Systems (ICEES)*. IEEE. DOI: 10.1109/ICEES.2018.8444145

Arulmurugan, R., & Venkatesan, T. (2015). Performance calculation and probability lifecycle price investigation of standalone photovoltaic power supply for residential applications. In *2015 Online International Conference on Green Engineering and Technologies (IC-GET)*. IEEE. DOI: 10.1109/GET.2015.7453785

Bekun, F. V. (2022). Mitigating emissions in India: Accounting for the role of real income, renewable energy consumption and investment in energy. *International Journal of Energy Economics and Policy*, 12(1), 188–192. DOI: 10.32479/ijeep.12652

Dhaouadi, G., ... (2021). Fuzzy logic controller based MPPT for a photovoltaic system. In *2021 IEEE 1st International Maghreb Meeting of the Conference on Sciences and Techniques of Automatic Control and Computer Engineering (MI-STA)* (p. 160). IEEE. DOI: 10.1109/MI-STA52233.2021.9464439

Eren, B. M., Taspinar, N., & Gokmenoglu, K. K. (2019). The impact of financial development and economic growth on renewable energy consumption: Empirical analysis of India. *The Science of the Total Environment*, 663, 189–197. DOI: 10.1016/j.scitotenv.2019.01.323 PMID: 30711585

Gupta, A.. (2014). Performance analysis of neural network and fuzzy logic based MPPT techniques for solar PV systems. In *International Conference on Energy, Communication, Data Analytics and Soft Computing (ICECDS)* (p. 90). IEEE. DOI: 10.1109/POWERI.2014.7117722

Gupta, A., Kumar, P., Pachauri, R. K., & Chauhan, Y. K. (2014). Performance analysis of neural network and fuzzy logic based MPPT techniques for solar PV systems. In *2014 6th IEEE Power India International Conference (PIICON)* (p. 66). IEEE. DOI: 10.1109/POWERI.2014.7117722

Hayder, W., ... (2020). Intelligent MPPT algorithm for PV system based on fuzzy logic. In *2020 17th International Multi-Conference on Systems, Signals & Devices (SSD)* (p. 70). IEEE. DOI: 10.1109/SSD49366.2020.9364195

Kothari, D. P., Ranjan, R., & Singal, K. C. (2021). *Renewable energy sources and emerging technologies. Springer. Kumar, A., Kumar, P., & Liu, Y. Industry 4.0 Driven Manufacturing Technologies*. Springer Nature., DOI: 10.1007/978-3-031-68271-1

Kumar, A., Liu, Y., & Kumar, R. (Eds.). (2024). *Handbook of Intelligent and Sustainable Manufacturing: Tools, Principles, and Strategies*. CRC Press., DOI: 10.1201/9781003405870

Kumar, A., Rajak, D. K., Kumar, P., & Kumar, A. (Eds.). (2024). *Advances in Sustainable Biomaterials: Bioprocessing 4.0, Characterizations, and Applications* (1st ed.). CRC Press., DOI: 10.1201/9781003434313

Kumar, A., Rajak, D. K., Kumar, P., & Rathee, S. (Eds.). (2024). *Smart Electric and Hybrid Vehicles: Fundamentals, Strategies and Applications* (1st ed.). CRC Press., DOI: 10.1201/9781003495574

Kumar, A., Rajak, D. K., Kumar, P., & Rathee, S. (Eds.). (2024). *Smart Electric and Hybrid Vehicles: Design, Modeling, and Assessment by Industry 4.0 Approaches* (1st ed.). CRC Press., DOI: 10.1201/9781003502470

Kumar, A., Singh, H., Kumar, P., & AlMangour, B. (2023). *Handbook of smart manufacturing: Forecasting the future of Industry 4.0*. CRC Press., DOI: 10.1201/9781003333760

Kumar, L., & Sharma, R. K. (2020). Smart manufacturing and Industry 4.0: State-of-the-art review. In *Handbook of Smart Manufacturing* (pp. 1–28). CRC Press., DOI: 10.1201/9781003333760-1

Laturkar, K., & Laturkar, K. (2023). Environmental sustainability measures in the oil and gas industry. In *Multidisciplinary Approaches to Sustainable Human Development* (pp. 176–205). IGI Global. DOI: 10.4018/978-1-6684-8223-0.ch008

Majid, M. (2020). Renewable energy for sustainable development in India: Current status, future prospects, challenges, employment, and investment opportunities. *Energy, Sustainability and Society*, 10(1), 1–36. DOI: 10.1186/s13705-020-00249-2

Patil, R. (2017). Fuzzy based MPPT technique for PV system. In *2017 International Conference on Energy, Communication, Data Analytics and Soft Computing (ICECDS)* (pp. 120). IEEE. DOI: 10.1109/ICECDS.2017.8389686

Patil, R., & Anantwar, H. (2017). Comparative analysis of fuzzy based MPPT for buck and boost converter topologies for PV application. In *2017 International Conference on Smart Technologies for Smart Nation (SmartTechCon)* (pp. 1-6). IEEE. DOI: 10.1109/SmartTechCon.2017.8358610

Rai, R. K. (2022). Fuzzy logic based control technique using MPPT for solar PV system. In *2022 First International Conference on Electrical, Electronics, Information and Communication Technologies (ICEEICT)* (pp. 56-65). IEEE. DOI: 10.1109/ICEEICT53079.2022.9768650

Compilation of References

Abali, Y., Bayca, S. U., & Targan, S. (2006). Evaluation of blends tincal waste, volcanic tuff, bentonite and fly ash for use as a cement admixture. *Journal of Hazardous Materials*, 131(1–3), 126–130. DOI: 10.1016/j.jhazmat.2005.09.031 PMID: 16314042

Abdolbaqi, M. Kh. (2017). Rizalman Mamat, Nor Azwadi Che Sidik, W.H. Azmi, P. Selvakumar Experimental investigation and development of new correlations for heat transfer enhancement and friction factor of BioGlycol/water based TiO_2 nanofluids in flat tubes, International Journal of Heat and Mass Transfer Volume 108. *Part A*, (May), 1026–1035.

Abdul Hamid, K., Azmi, W. H., & Nabil, M. F. (2018, March). Rizalman Mamat (2017), "Experiment investigation of nanoparticle mixture ratios on TiO_2-SiO_2 nanofluids heat transfer performance under turbulent flow. *International Journal of Heat and Mass Transfer*, 118, 617–627. DOI: 10.1016/j.ijheatmasstransfer.2017.11.036

Abnar, B., Kazeminezhad, M., & Kokabi, A. (2015). The effect of premixed Al-Cu powder on the stir zone in friction stir welding of AA3003-H18. *Journal of Materials Engineering and Performance*, 24(2), 1086–1093. DOI: 10.1007/s11665-014-1373-1

Achiou, B., Elomari, H., Bouazizi, A., Karim, A., Ouammou, M., Albizane, A., Bennazha, J., Alami Younssi, S., & El Amrani, I. E. (2017). Manufacturing of tubular ceramic microfiltration membrane based on natural pozzolan for pretreatment of seawater desalination. *Desalination*, 419(June), 181–187. DOI: 10.1016/j.desal.2017.06.014

ACI. (1997). American Concrete Institute. ACI 209R-92 [Reapproved]. *Prediction of Creep, Shrinkage, and Temperature Effects in Concrete Structures.*, 47, 92.

Adams, C. A., Tuffour, Y. A., & Kwofie, S. (2016). Effects of Soil Properties and Geogrid Placement on CBR Enhancement of Lateritic Soil for Road Pavement Layers "Effects of Soil Properties and Geogrid Placement on CBR Enhancement of Lateritic Soil for Road Pavement Layers. *American Journal of Civil Engineering and Architecture*, 4, 62–66. DOI: 10.12691/ajcea-4-2-4

Agarwal, A., Kumar, P., Kumar, A., & Ghadai, R. K. (2024). *Strategic Design Optimization of Cutting Tools for Enhanced Manufacturing Efficiency Industry 4.0 Driven Manufacturing Technologies*. Springer.

Ahmad, A., & Kumar, P. (2024). Optimization of electric vehicle lithium-ion battery parameter by using Taguchi methodology. In *Smart Electric and Hybrid Vehicles* (pp. 76–94). CRC Press., DOI: 10.1201/9781003495574-5

Ahmadi, M., Abdollahzadeh, E., & Kioumarsi, M. (2024). Materials Today : Proceedings Using marble waste as a partial aggregate replacement in the development of sustainable self-compacting concrete. *Materials Today: Proceedings*, xxxx. Advance online publication. DOI: 10.1016/j.matpr.2023.04.103

Ahmad, J., Zaid, O., Shahzaib, M., Abdullah, M. U., Ullah, A., & Ullah, R. (2021). Mechanical properties of sustainable concrete modified by adding marble slurry as cement substitution. *AIMS Materials Science*, 8(3), 343–358. DOI: 10.3934/matersci.2021022

Ahmad, S., Sharma, J., & Gambhir, V. (Eds.). (2023). *Handbook of Sustainable Materials: Modelling, Characterization, and Optimization*. CRC Press.

Ahmed, M. M. Z., El-Sayed Seleman, M. M., Fydrych, D., & Çam, G. (2023). Review on friction stir welding of dissimilar magnesium and aluminum alloys: Scientometric analysis and strategies for achieving high-quality joints. *Journal of Magnesium and Alloys*, 11(11), 4082–4127. DOI: 10.1016/j.jma.2023.09.039

Ahmed, S., Jannat, F., Khan, M. A. K., & Alim, M. A. (2021). Numerical development of eco-friendly Cs2TiBr6 based perovskite solar cell with all-inorganic charge transport materials via SCAPS-1D. *Optik (Stuttgart)*, 225, 165765. DOI: 10.1016/j.ijleo.2020.165765

Akarsh, P. K., Marathe, S., & Bhat, A. K. (2021). Influence of graphene oxide on properties of concrete in the presence of silica fumes and M-sand. *Construction & Building Materials*, 268, 121093. DOI: 10.1016/j.conbuildmat.2020.121093

AKOLADE, A.S and OLANIYAN, O.S., (2014). Application of Geogrids on the Geotechnical Properties of Subgrade Materials under Soaked Condition 6, 12–19.

Almoussawi, M., Smith, A. J., & Faraji, M. (2018). Wear of Polycrystalline Boron Nitride Tool During the Friction Stir Welding of Steel. *Metallography, Microstructure, and Analysis*, 7(3), 252–267. DOI: 10.1007/s13632-018-0439-0

Al-Salem, S. M., Lettieri, P., & Baeyens, J. (2009). Recycling and recovery routes of plastic solid waste (PSW): A review. *Waste Management (New York, N.Y.)*, 29(10), 2625–2643. DOI: 10.1016/j.wasman.2009.06.004 PMID: 19577459

Al-Sinaidi, A. R., & Ali, A. H. (2006). Improvement in bearing capacity of soil by geogrid -an experimental approach. *Iaeg*, 2006, 1–5.

Al-Swaidani, A., Hammoud, I., & Meziab, A. (2016). Effect of adding natural pozzolana on geotechnical properties of lime-stabilized clayey soil. *Journal of Rock Mechanics and Geotechnical Engineering*, 8(5), 714–725. DOI: 10.1016/j.jrmge.2016.04.002

Alyamaç, K. E., & Aydin, A. B. (2015). Concrete properties containing fine aggregate marble powder. *KSCE Journal of Civil Engineering*, 19(7), 2208–2216. DOI: 10.1007/s12205-015-0327-y

Alzeer, M. I. M., Cheeseman, C., & Kinnunen, P. (2022). New synthetic glass-based supplementary cementitious materials derived from basalt composition. *Journal of Building Engineering*, 46, 103699. DOI: 10.1016/j.jobe.2021.103699

Ambalavanan, R., & Roja, A. (1996). Feasibility studies on utilisation of waste lime and gypsum with flyash. *Indian Concrete Journal*, 70(11), 611–615.

American Society for Testing and Materials. (2003). *Standard Test Methods for Sampling and Testing Fly Ash or Natural Pozzolans for Use in Portland Cement Concrete* (ASTM C 311-02). https://www.astm.org/c0311-00.html

American Society for Testing and Materials. (2018). *Standard Test Method for Compressive Strength of Hydraulic Cement Mortars* (ASTM C109/C109M-16a). https://www.astm.org/c0109_c0109m-23.html

American Society for Testing and Materials. (2019). *Standard Specification for Coal Fly Ash and Raw or Calcined Natural Pozzolan for Use in Concrete* (ASTM C618-19). https://www.astm.org/c0618-23e01.html

Anand, S. Gayatri, Satyarthi, M. K., Bharti, P. S., Kumar, A., & Rathee, S. (2024). Robotic Arm 3D Printing: Technological Advancements and Applications *Industry 4.0 Driven Manufacturing Technologies* (pp. 293-310): Springer.

Anand, S., Satyarthi, M. K., Bharti, P. S., Kumar, P., & Kumar, A. (2024). *Digital Twin Integration for Enhanced Control in FDM 3D Printing Industry 4.0 Driven Manufacturing Technologies*. Springer.

Andwari, A. M., Pesiridis, A., Rajoo, S., Martinez-Botas, R., & Esfahanian, V. (2017). A review of battery electric vehicle technology and readiness levels. *Renewable & Sustainable Energy Reviews*, 78, 414–430. DOI: 10.1016/j.rser.2017.03.138

Anlas, G., Santare, M. H., & Lambros, J. (2000)... *Numerical Calculation of Stress Intensity Factors in Functionally Graded Materials.*, 1994, 131–143.

Annual Report on Solid Waste Management (2020-21), CPCB, Delhi

Armand, M., & Tarascon, J. M. (2008). Building better batteries. *Nature*, 451(7179), 652–657. DOI: 10.1038/451652a PMID: 18256660

Arulmurugan, R., & Anushalini, T. (2018). Implementation of fuzzy controlling structure of PV-Fe for a grid connected system. In *2018 4th International Conference on Electrical Energy Systems (ICEES)*. IEEE. DOI: 10.1109/ICEES.2018.8443202

Arulmurugan, R., & Chandramouli, A. (2018). Design and implementation of a nine-level inverter with shunt active filter for distributed system. In *2018 4th International Conference on Electrical Energy Systems (ICEES)*. IEEE. DOI: 10.1109/ICEES.2018.8444145

Arulmurugan, R.. (2013). Intelligent fuzzy MPPT controller using analysis of DC to DC novel buck converter for photovoltaic energy system applications. In *2013 International Conference on Pattern Recognition, Informatics and Mobile Engineering* (pp. 1-6). IEEE. https://doi.org/DOI: 10.1109/ICPRIME.2013.6496477

Arulmurugan, R.. (2013). Numerical analysis of incremental conductance with cyclic measurement of the MPPT algorithm for PV power generation. In *IET Chennai Fourth International Conference on Sustainable Energy and Intelligent Systems (SEISCON 2013)* (pp. 1-6). IET. https://doi.org/DOI: 10.1049/ic.2013.0297

Arulmurugan, R.. (2014). An intelligent novel FLC based FOINC MPPT technique for PV applications. In *2014 International Conference on Advances in Electrical Engineering (ICAEE)* (pp. 1-6). IEEE. DOI: 10.1109/ICAEE.2014.6838541

Arulmurugan, R., & Venkatesan, T. (2015). Performance calculation and probability lifecycle price investigation of standalone photovoltaic power supply for residential applications. In *2015 Online International Conference on Green Engineering and Technologies (IC-GET)*. IEEE. DOI: 10.1109/GET.2015.7453785

Asl, N. S., Mirsalehi, S. E., & Dehghani, K. (2019). Effect of TiO2 nanoparticles addition on microstructure and mechanical properties of dissimilar friction stir welded AA6063-T4 aluminum alloy and AZ31B-O magnesium alloy. *Journal of Manufacturing Processes*, 38, 338–354. DOI: 10.1016/j.jmapro.2019.01.023

Atan, M. C. (2015). *Using Cappadocia Tuff As A Natural Pozzolan In the Cement Production* [Unpublished Master's Thesis]. Middle East Technical University.

Ava, T. T., Al Mamun, A., Marsillac, S., & Namkoong, G. (2019). A review: Thermal stability of methylammonium lead halide based perovskite solar cells. *Applied Sciences (Basel, Switzerland)*, 9(1), 188. DOI: 10.3390/app9010188

Aydan, Ö., & Ulusay, R. (2003). Geotechnical and geoenvironmental characteristics of man-made underground structures in Cappadocia, Turkey. *Engineering Geology*, 69(3–4), 245–272. DOI: 10.1016/S0013-7952(02)00285-5

Ayhan, A. (2004). *Geological and Morphological Investigations of the Underground Cities of Cappadocia Using GIS* [Unpublished Master's Thesis]. Middle East Technical University.

Azri, F., Meftah, A., Sengouga, N., & Meftah, A. (2019). Electron and hole transport layers optimization by numerical simulation of a perovskite solar cell. *Solar Energy*, 181, 372–378. DOI: 10.1016/j.solener.2019.02.017

B.M.Patil1, S.B.B., (2017). Improving Soil Stability By Using Polyethylene Geogrid. Int. J. Res. Eng. Technol. 06, 70–72. DOI: 10.15623/ijret.2017.0606012

Bag, A., Radhakrishnan, R., Nekovei, R., & Jeyakumar, R. (2020). Effect of absorber layer, hole transport layer thicknesses, and its doping density on the performance of perovskite solar cells by device simulation. *Solar Energy*, 196, 177–182. DOI: 10.1016/j.solener.2019.12.014

Bagaria, A., & Juneja, D. (2023). Experimental research on influence of marble powder, silica fume and polypropylene fiber on the porous concrete. *Materials Today: Proceedings*. Advance online publication. DOI: 10.1016/j.matpr.2023.03.240

Bahadori, H., Hasheminezhad, A., & Taghizadeh, F. (2019). Experimental Study on Marl Soil Stabilization Using Natural Pozzolans. *Journal of Materials in Civil Engineering*, 31(2), 04018363. Advance online publication. DOI: 10.1061/(ASCE)MT.1943-5533.0002577

Bahrudin, M. S., Abdullah, S. F., Ahmad, I., Zuhdi, A. W. M., & Hasani, A. H., Za'Abar, F., ... & Harif, M. N. (2018, August). J sc and V oc Optimization of Perovskite Solar Cell with Interface Defect Layer Using Taguchi Method. In 2018 IEEE International Conference on Semiconductor Electronics (ICSE) (pp. 192-196). IEEE. DOI: 10.1109/SMELEC.2018.8481203

Baikerikar, A. (2014). A review on green concrete. *Journal of Emerging Technologies and Innovative Research*, 1(6), 472–474.

Bajaj, A., Epps Martin, A., King, G., Glover, C., Kaseer, F., & Arámbula-Mercado, E. (2020). Evaluation and classification of recycling agents for asphalt binders. *Construction & Building Materials*, 260, 119864. DOI: 10.1016/j.conbuildmat.2020.119864

Bajpai, R., Choudhary, K., Srivastava, A., Sangwan, K. S., & Singh, M. (2020). Environmental impact assessment of fly ash and silica fume based geopolymer concrete. *Journal of Cleaner Production*, 254, 120147. DOI: 10.1016/j.jclepro.2020.120147

Baki, V. A., Nayır, S., Erdoğdu, Ş., & Ustabaş, İ. (2020). Determination of the Pozzolanic Activities of Trachyte and Rhyolite and Comparison of the Test Methods Implemented. *International Journal of Civil Engineering*, 18(9), 1053–1066. DOI: 10.1007/s40999-020-00516-5

Balog, A., Cobirzan, N., Aciu, C., & İlutiu-Varvara, D. A. (2014). Valorification of volcanic tuff in constructions and materials manufacturing industry. *Procedia Technology*, 12, 323–328. DOI: 10.1016/j.protcy.2013.12.493

Bandhauer, T. M., Garimella, S., & Fuller, T. F. (2011). A critical review of thermal issues in lithium-ion batteries. *Journal of the Electrochemical Society*, 158(3), R1–R25. DOI: 10.1149/1.3515880

Becerra-Duitama, J. A., & Rojas-Avellaneda, D. (2022). Pozzolans : A review. *Engineering and Applied Science Research, 49*(4), 495–504. https://doi.org/DOI: 10.14456/easr.2022

Behl, V., Singh, V., Dahiya, V., & Kumar, A. (2022). Characterization of physicochemical and functional properties of fly ash concrete mix. *Materials Today: Proceedings*, 50, 941–945. DOI: 10.1016/j.matpr.2021.06.353

Bekun, F. V. (2022). Mitigating emissions in India: Accounting for the role of real income, renewable energy consumption and investment in energy. *International Journal of Energy Economics and Policy*, 12(1), 188–192. DOI: 10.32479/ijeep.12652

Bentlemsan, N., Yahiaoui, W., & Kenai, S. (2023). Case Studies in Construction Materials Strength and durability of self-compacting mortar with waste marble as sand substitution. *Case Studies in Construction Materials*, 19(June), e02331. DOI: 10.1016/j.cscm.2023.e02331

Bhandari, M., & Purohit, K. (2022b). Mathematical methods to investigate the fracture behaviour of functionally graded material (FGMs)—A critical review. Advances in Mechanical and Industrial Engineering, 102–108.

Bhandari, M., & Purohit, K. (2022a). Dynamic Fracture Analysis of Functionally Graded Material Structures – A Critical Review. *Composites Part C: Open Access*, 7, 100227. DOI: 10.1016/j.jcomc.2021.100227

Bhanumathidas, N., & Kalidas, N. (1993). The Renaissance of Mediaeval Age Cement: FaL-G. *Civil Engineering & Construction Review,* 31-33.

Bhanumathidas, N., & Kalidas, N. (1999) The Rationale for Portland Pozzolan Cement Compositions. *Proceedings of International Conference on Waste and By-products as Secondary Resources for Building Materials.*

Bhattacharya, R., Banerjee, R., & Mukherjee, A. (2016). Aerogels: Promising thermal insulators for electric vehicles. *Energy Reports*, 2, 56–63. DOI: 10.1016/j.egyr.2016.04.002

Bhoopathy, V., Selvan, S. S., & Vijaya, B. (2018). Corrosion effects on the flexural strength behaviour of concrete containing Manufactured Sand Corrosion effects on the flexural strength behaviour of concrete containing Manufactured Sand View project Experimental and Analytical Investigation on Medium strength concrete containing Manufactured sand View project CORROSION EFFECTS ON THE FLEXURAL STRENGTH BEHAVIOUR OF CONCRETE CONTAINING MANUFACTURED SAND. *Article in Journal of Engineering and Applied Sciences, 13*(5). www.arpnjournals.com

Bida, S. M., Danraka, M., & Mohammed, J. (2016). Performance of Reclaimed Asphalt Pavement (RAP) as a Replacement of Fine Aggregate in Concrete. [IJSR]. *International Journal of Science and Research (Raipur, India)*, 5(4), 65–67. DOI: 10.21275/v5i4.NOV162395

BIS. (2013). ORDINARY PORTLAND CEMENT, 43 GRADE — SPECIFICATION. In *Indian Standards.*

BIS. (2019). IS 10262:2019, Indian standard, Concrete mix Proportioning – Guidelines (Second Revision). -. *Bis, January*. https://www.google.com/search?q=IS+10262%3A2019%2C+Indian+standard%2C+Concrete+mix+Proportioning+–+Guidelines+(Second+Revision).&rlz=1C1CHBF_enIN984IN984&oq=IS+10262%3A2019%2C+Indian+standard%2C+Concrete+mix+Proportioning+–+Guidelines+(Second+Revision).&a

Black, L. (2016). Low clinker cement as a sustainable construction material. In Jamal M. Khatib (Ed.), *Sustainability of Construction Materials* (Second Edi). Elsevier. DOI: 10.1016/B978-0-08-100370-1.00017-2

Böke, H., Akkurt, S., İpekoğlu, B., & Uğurlu, E. (2006). Characteristics of brick used as aggregate in historic brick-lime mortars and plasters. *Cement and Concrete Research*, 36(6), 1115–1122. DOI: 10.1016/j.cemconres.2006.03.011

Bondar, D. (2009). Alkali Activation of Iranian Natural Pozzolans for Producing Geopolymer Cement and Concrete. In *Doctoral Thesis,The University of Sheffield* (Issue June).

Bostanci, S. C. (2020). Use of waste marble dust and recycled glass for sustainable concrete production. *Journal of Cleaner Production*, 251, 119785. DOI: 10.1016/j.jclepro.2019.119785

Bouzoubaa, N., Zhang, M. H., & Malhotra, V. M. (2001). Mechanical properties and durability of concrete made with high-volume fly ash blended cements using a coarse fly ash. *Cement and Concrete Research*, 31(10), 1393–1402. DOI: 10.1016/S0008-8846(01)00592-0

Bridgwater, A. V. (2012). Review of fast pyrolysis of biomass and product upgrading. *Biomass and Bioenergy*, 38, 68–94. DOI: 10.1016/j.biombioe.2011.01.048

Bundela, V. S., & Patel, D. A. (2022). MSW Management and Best Suitable Option Virtually Waste to Energy Plant for Resource and Energy Recovery for Bhopal (MP) India. *International Journal of Management and Humanities*, 8(11), 1–15.

Butcher, R. J., Rousseau, C. E., & Tippur, H. V. (1998). A functionally graded particulate composite: Preparation, measurements and failure analysis. *Acta Materialia*, 47(1), 259–268. DOI: 10.1016/S1359-6454(98)00305-X

Çam, G. (2011). Friction stir welded structural materials: Beyond Al-alloys. *International Materials Reviews*, 56(1), 1–48. DOI: 10.1179/095066010X12777205875750

Cannon, C., Guido, V., & Wright, L. (2021). *Concrete Solution Guise Mix It Up Supplementary Cementitious Materials (SCMs)*. RMI Innovation Center. http://www.rmi.org/concrete- solutions-guide/

Cardiano, P., Ioppolo, S., De Stefano, C., Pettignano, A., Sergi, S., & Piraino, P. (2004). Study and characterization of the ancient bricks of monastery of " San Filippo di Fragalà " in Frazzanò (Sicily). *Analytica Chimica Acta*, 519(1), 103–111. DOI: 10.1016/j.aca.2004.05.042

Çavdar, A., & Yetgin, Ş. (2007). Availability of tuffs from northeast of Turkey as natural pozzolan on cement, some chemical and mechanical relationships. *Construction & Building Materials*, 21(12), 2066–2077. DOI: 10.1016/j.conbuildmat.2006.05.034

Çevre ve Şehircilik Bakanlığı. (2017, October 18). *Kayadan Oyma Yapıların Tasarım, Hesap ve Yapım Esaslarına Dair Yönetmelik*. Resmi Gazete. https://www.resmigazete.gov.tr/eskiler/2017/10/20171018-7.htm

Chaabani, H., Mesmoudi, S., Boutahar, L., & El Bikri, K. (2023). A high-order finite element continuation for buckling analysis of porous FGM plates. *Engineering Structures*, 279, 115597. DOI: 10.1016/j.engstruct.2023.115597

Chen, D., & Yin, L. (2018). Pyrolysis technologies for municipal solid waste: A review. *Waste Management (New York, N.Y.)*, 79, 277–299.

Cheng, M., Sun, L., Buja, G., & Song, L. (2015). Advanced electrical machines and machine-based system for electric and hybrid vehicles. *Energies*, 8(9), 9541–9564. DOI: 10.3390/en8099541

Cheng, Y., He, P., Yang, H., Wang, H., & Zhang, H. (2019). A review of biomass pyrolysis and hydro-pyrolysis for high-value chemicals. *Fuel Processing Technology*, 196, 106194.

Chen, J., Shi, L., Wu, C., & Jiang, Y. (2021). The effect of tool pin size and taper angle on the thermal process and plastic material flow in friction stir welding. *International Journal of Advanced Manufacturing Technology*, 116(9-10), 2847–2860. DOI: 10.1007/s00170-021-07650-x

Chen, S., Li, Q., & Pan, F. (2020). High-strength and high-conductivity copper alloys for electric vehicle applications. *Materials Science and Engineering R Reports*, 142, 100579. DOI: 10.1016/j.mser.2020.100579

Chen, Y., & Sharma, B. K. (2020). Recent advances in biochar applications in agricultural soils: Benefits and environmental implications. *Clean Technologies and Environmental Policy*, 22(5), 989–1005.

Chen, Z., & Qin, Y. (2020). Next-generation anode materials for high-performance electric vehicle batteries. *Advanced Energy Materials*, 10(2), 1902343. DOI: 10.1002/aenm.201902343

Chinnaraju, K., Subramanian, K., & Senthil Kumar, S. R. R. (2011). Role of Fly Ash and Silica Fume on Durability Characteristics of High Performance Concrete. *International Journal of Advance in Civil Engineering*, 17-27.

Chong, W. T., Poh, S. C., Fazlizan, A., & Bakar, M. H. A. (2015). The assessment of electric vehicle performance on battery sizing and charging infrastructure. *Energy Procedia*, 75, 2223–2228. DOI: 10.1016/j.egypro.2015.07.420

Choudhary, R., Gupta, R., Nagar, R., & Jain, A. (2021). Mechanical and abrasion resistance performance of silica fume, marble slurry powder, and fly ash amalgamated high strength self-consolidating concrete. *Construction and Building Materials, 269*(xxxx), 121282. DOI: 10.1016/j.conbuildmat.2020.121282

Choudhary, R., Gupta, R., Alomayri, T., Jain, A., & Nagar, R. (2021). Permeation, corrosion, and drying shrinkage assessment of self-compacting high strength concrete comprising waste marble slurry and fly ash, with silica fume. *Structures*, 33(May), 971–985. DOI: 10.1016/j.istruc.2021.05.008

Chouhan, H. S., Kalla, P., Nagar, R., & Gautam, P. K. (2019). Gainful utilization of dimensional limestone waste as fine aggregate in cement mortar mixes. *Construction & Building Materials*, 221, 363–374. DOI: 10.1016/j.conbuildmat.2019.06.097

Chowdhury, M. S., Shahahmadi, S. A., Chelvanathan, P., Tiong, S. K., Amin, N., Techato, K. A., Nuthammachot, N., Chowdhury, T., & Suklueng, M. (2020). Effect of deep-level defect density of the absorber layer and n/i interface in perovskite solar cells by SCAPS-1D. *Results in Physics*, 16, 102839. DOI: 10.1016/j.rinp.2019.102839

Christy, J. V., Mourad, A.-H. I., Sherif, M. M., & Shivamurthy, B. (2021). Review of recent trends in friction stir welding process of aluminum alloys and aluminum metal matrix composites. *Transactions of Nonferrous Metals Society of China*, 31(11), 3281–3309. DOI: 10.1016/S1003-6326(21)65730-8

Chung, D. D. L. (2012). Review: Carbon materials for structural self-sensing, electromagnetic shielding and thermal interfacing. *Carbon*, 50(10), 3342–3353. DOI: 10.1016/j.carbon.2012.01.031

D'Orazio, A., Forcellese, A., & Simoncini, M. (2019). Prediction of the vertical force during FSW of AZ31 magnesium alloy sheets using an artificial neural network-based model. *Neural Computing & Applications*, 31(11), 7211–7226. DOI: 10.1007/s00521-018-3562-6

Dabra, V., Dubey, A. K., Kumar, A., & Kumar, P. (2024). Cylindrical lithium-ion battery. Smart Electric and Hybrid Vehicles: Design, Modeling, and Assessment by Industry 4.0 Approaches, 32. https://doi.org/DOI: 10.1201/9781003502470-3

De Los Santos, I. M., Cortina-Marrero, H. J., Ruíz-Sánchez, M. A., Hechavarría-Difur, L., Sánchez-Rodríguez, F. J., Courel, M., & Hu, H. (2020). Optimization of CH3NH3PbI3 perovskite solar cells: A theoretical and experimental study. *Solar Energy*, 199, 198–205. DOI: 10.1016/j.solener.2020.02.026

Demirbas, A. (2004). Pyrolysis of municipal plastic wastes for recovery of gasoline-range hydrocarbons. *Journal of Analytical and Applied Pyrolysis*, 72(1), 97–102. DOI: 10.1016/j.jaap.2004.03.001

Deng, Y., Ye, R., Xu, G., Yang, J., Pan, Q., Peng, B., Cao, X., Duan, Y., Wang, Y., Lu, L., & Yin, Z. (2015). Corrosion behaviour and mechanism of new aerospace Al–Zn–Mg alloy friction stir welded joints and the effects of secondary Al3ScxZr1−x nanoparticles. *Corrosion Science*, 90, 359–374. DOI: 10.1016/j.corsci.2014.10.036

Devi, C., & Mehra, R. (2019). Device simulation of lead-free MASnI 3 solar cell with CuSbS 2 (copper antimony sulfide). *Journal of Materials Science*, 54(7), 5615–5624. DOI: 10.1007/s10853-018-03265-y

Devi, L., Ptasinski, K. J., & Janssen, F. J. (2003). A review of the primary measures for tar elimination in biomass gasification processes. *Biomass and Bioenergy*, 24(2), 125–140. DOI: 10.1016/S0961-9534(02)00102-2

Dewangan, S. K., Banjare, P. N., Tripathi, M. K., & Manoj, M. K. (2023). Effect of vertical and horizontal zinc interlayer on material flow, microstructure, and mechanical properties of dissimilar FSW of Al 7075 and Mg AZ31 alloys. *International Journal of Advanced Manufacturing Technology*, 126(9), 4453–4474. DOI: 10.1007/s00170-023-11348-7

Dey, S. (2016). Study of compressive strength of high volume fly ash concrete with varying proportion of fly ash and silica fume. *International Journal of Civil Engineering*, 3(3), 3–6. DOI: 10.14445/23488352/IJCE-V3I3P102

Dhaouadi, G., ... (2021). Fuzzy logic controller based MPPT for a photovoltaic system. In *2021 IEEE 1st International Maghreb Meeting of the Conference on Sciences and Techniques of Automatic Control and Computer Engineering (MI-STA)* (p. 160). IEEE. DOI: 10.1109/MI-STA52233.2021.9464439

Dhoka, M. C. (2013). Green concrete: Using industrial waste of marble powder, quarry dust and paper pulp. *International Journal of Engineering and Science Invention*, 2(10), 67–70.

Dinç-Şengönül, B., Bayram, M., Oktay, D., & Yüzer, N. (2024). Valorization of Cappadocia waste earth in the production of sustainable lime-based grouts. *Sustainable Chemistry and Pharmacy*, 38, 101503. DOI: 10.1016/j.scp.2024.101503

Diwevdi, S. D. (2014). *Sheeraz Ahmad Wagay, Solid Waste Management InBhanpur Bhopal: Status And Need- A Case Study. Chemical And Process Engineering Research Issn 2224-7467 (Paper) Issn 2225-0913* (Vol. 18). Online.

Diyaley, S., Chakraborty, S., Kumar, A., & Kalita, K. (2024). *Elephant Swarm Water Search Algorithm-Based Optimization of a Laser Beam Machining Process Industry 4.0 Driven Manufacturing Technologies*. Springer.

Donatello, S., Tyrer, M., & Cheeseman, C. R. (2010). Comparison of test methods to assess pozzolanic activity. *Cement and Concrete Composites*, 32(2), 121–127. DOI: 10.1016/j.cemconcomp.2009.10.008

Dorduncu, M., Olmus, I., & Rabczuk, T. (2022). A peridynamic approach for modeling of two dimensional functionally graded plates. *Composite Structures*, 279, 114743. DOI: 10.1016/j.compstruct.2021.114743

Drew, D. A., & Passman, S. L. (1999). *Theory of Multi Component Fluids*. Springer. DOI: 10.1007/b97678

Dumpala, L., Chandra, P. H., & Sriram, N. (2022). Development of bobbin tool and influence of its design parameters on FSW joint. *Materials Today: Proceedings*, 56, 1582–1587. DOI: 10.1016/j.matpr.2022.02.363

Efstathiadis, E. (1978). *Greek concrete of three millenniums* (1st ed.). Public Works Research Centre.

Ekinci, E., Türkmen, İ., Kantarci, F., & Karakoç, M. B. (2019). The improvement of mechanical, physical and durability characteristics of volcanic tuff based geopolymer concrete by using nano silica, micro silica and Styrene-Butadiene Latex additives at different ratios. *Construction & Building Materials*, 201, 257–267. DOI: 10.1016/j.conbuildmat.2018.12.204

Eldera, S. S., Aldawsari, S., & Hamzawy, E. M. A. (2023). Utilization of waste glass with natural pozzolan in the production of self-glazed glass-ceramic materials. *Nanotechnology Reviews*, 12(1), 20220565. DOI: 10.1515/ntrev-2022-0565

El-Maaty, A. E. A., & Elmohr, A. I. (2015). Characterization of recycled asphalt pavement (RAP) for use in flexible pavement. *American Journal of Engineering and Applied Sciences*, 8(2), 233–248. DOI: 10.3844/ajeassp.2015.233.248

El-Sayed, M. M., Shash, A. Y., Abd-Rabou, M., & ElSherbiny, M. G. (2021). Welding and processing of metallic materials by using friction stir technique: A review. *Journal of Advanced Joining Processes*, 3, 100059. DOI: 10.1016/j.jajp.2021.100059

Eren, B. M., Taspinar, N., & Gokmenoglu, K. K. (2019). The impact of financial development and economic growth on renewable energy consumption: Empirical analysis of India. *The Science of the Total Environment*, 663, 189–197. DOI: 10.1016/j.scitotenv.2019.01.323 PMID: 30711585

Esmaeili, M., Naderi, B., Neyestanaki, H. K., & Khodaverdian, A. (2018). Investigating the effect of geogrid on stabilization of high railway embankments. *Soil and Foundation*, 58(2), 319–332. DOI: 10.1016/j.sandf.2018.02.005

Eyvazian, A., Hamouda, A., Tarlochan, F., Derazkola, H. A., & Khodabakhshi, F. (2020). Simulation and experimental study of underwater dissimilar friction-stir welding between aluminium and steel. *Journal of Materials Research and Technology*, 9(3), 3767–3781. DOI: 10.1016/j.jmrt.2020.02.003

Fakhraldin, M. K. (2016). Improvement of Loose Granular Soil By Using Geogrid Reinforcement. *Kufa J. Eng.*, 7(3), 66–79. DOI: 10.30572/2018/KJE/731201

Fawzy, M., & Sanaa, E. (2018). Review on fast pyrolysis of biomass: A potential technology for renewable energy and soil amendment. *Sustainable Environment Research*, 28(6), 283–288.

Frías, M., Villar-Cociña, E., De Rojas, M. I. S., & Valencia-Morales, E. (2005). The effect that different pozzolanic activity methods has on the kinetic constants of the pozzolanic reaction in sugar cane straw-clay ash/lime systems: Application of a kinetic-diffusive model. *Cement and Concrete Research*, 35(11), 2137–2142. DOI: 10.1016/j.cemconres.2005.07.005

Fujii, H., Cui, L., Maeda, M., Sato, Y. S., & Nogi, K. (2006). *Effect of threads on tool in friction stir welding of aluminum alloys.* Paper presented at the Materials science forum. DOI: 10.4028/0-87849-996-2.389

Gameiro, F., De Brito, J., & Correia da Silva, D. (2014). Durability performance of structural concrete containing fine aggregates from waste generated by marble quarrying industry. *Engineering Structures*, 59, 654–662. DOI: 10.1016/j.engstruct.2013.11.026

Gao, H., Zhou, T., Zheng, Y., & Tang, Z. (2017). Performance and life cycle analysis of solid-state batteries for electric vehicles. *Journal of Power Sources*, 359, 126–136. DOI: 10.1016/j.jpowsour.2017.05.073

Ghahari, S. A., Mohammadi, A., & Ramezanianpour, A. A. (2017). Performance assessment of natural pozzolan roller compacted concrete pavements. *Case Studies in Construction Materials*, 7, 82–90. DOI: 10.1016/j.cscm.2017.03.004

Ghais, A., Ahmed, D., Siddig, E., Elsadig, I., & Albager, S. (2014). Performance of concrete with fly ash and kaolin inclusion. *International Journal of Geosciences*, 5(12), 1445–1450. DOI: 10.4236/ijg.2014.512118

Ghani, J. A., Choudhury, I. A., & Hassan, H. H. (2004). Application of Taguchi method in the optimization of end milling parameters. *Journal of Materials Processing Technology*, 145(1), 84–92. DOI: 10.1016/S0924-0136(03)00865-3

Gibaja, C., Marquez, F., & Ruiz, M. C. (2020). A review of thermal management technologies for electric vehicle battery systems: Mathematical models. *Renewable & Sustainable Energy Reviews*, 117, 109504. DOI: 10.1016/j.rser.2019.109504

Gómez Sánchez de Rojas, M. I., & Frías Rojas, M. (2013). Natural pozzolans in eco-efficient concrete. In Pacheco-Torgal, F., Jalali, S., Labrincha, J., & John, V. M. (Eds.), *Eco-Efficient Concrete* (pp. 83–104). Woodhead Publishing Limited., https://www.sciencedirect.com/science/article/abs/pii/B9780857094247500049 DOI: 10.1533/9780857098993.2.83

González-Cortés, S. L., Pérez-Díaz, J. A., & Lozano-Miralles, J. J. (2022). Thermal management in electric vehicles using phase change materials: A review. *Applied Thermal Engineering*, 203, 117924. DOI: 10.1016/j.applthermaleng.2021.117924

González, J. F., Encinar, J. M., & Canito, J. F. (2009). Pyrolysis of different organic waste from agro-industrial activities: Characterization and energy evaluation. *Journal of Chemical Technology and Biotechnology*, 84(8), 1242–1249.

Goodenough, J. B., & Kim, Y. (2010). Challenges for rechargeable Li batteries. *Chemistry of Materials*, 22(3), 587–603. DOI: 10.1021/cm901452z

GS. (2018, September 4). Sokhal D, Gangacharyulu VK, Bulasara, "Influence of copper oxide nanoparticles on the thermophysical properties and performance of flat tube of vehicle cooling system". *Vacuum*, 2018(157), 268–276.

Guo, Y., Zhao, X., & Xu, M. (2019). Magnesium-based batteries for electric vehicles: A review of current challenges and future perspectives. *Journal of Energy Chemistry*, 39, 38–55. DOI: 10.1016/j.jechem.2018.12.003

Gupta, A.. (2014). Performance analysis of neural network and fuzzy logic based MPPT techniques for solar PV systems. In *International Conference on Energy, Communication, Data Analytics and Soft Computing (ICECDS)* (p. 90). IEEE. DOI: 10.1109/POWERI.2014.7117722

Gupta, M., Kumar, R., Arora, N., Kumar, S., & Dilbagi, N. (2016). Forced Convective Heat Transfer of MWCNT/Water Nanofluid under Constant Heat Flux: An Experimental Investigation. *Arabian Journal for Science and Engineering*, 41(2), 599–609. DOI: 10.1007/s13369-015-1699-5

Haider, S. Z., Anwar, H., & Wang, M. (2019). Theoretical device engineering for high-performance perovskite solar cells using CuSCN as hole transport material boost the efficiency above 25%. physica status solidi (a), 216(11), 1900102.

Hariharan, A. R., Santhi, A. S., & Mohan Ganesh, G. (2011). Study on Strength Development of High Strength Concrete Containing Fly Ash and Silica Fume *[IJEST]*. *International Journal of Engineering Science and Technology*, 3, 2955–2961.

Harle, S. M. (2019). *Network Security and Communication Experimental Investigation on Steel Fibers in the Concrete. 4*, 1–7. www.ijsrnsc.org

Harris, S. J., Lu, P., & Manthiram, A. (2017). The role of solid electrolytes in lithium batteries. *Journal of Power Sources*, 352, 66–75. DOI: 10.1016/j.jpowsour.2017.03.028

Hassan, M. A., & Al-Malack, M. H. (2006). A review on oil mist lubrication systems. *International Journal of Machine Tools & Manufacture*, 46(12-13), 1279–1288.

Hattingh, D., Blignault, C., Van Niekerk, T., & James, M. (2008). Characterization of the influences of FSW tool geometry on welding forces and weld tensile strength using an instrumented tool. *Journal of Materials Processing Technology*, 203(1-3), 46–57. DOI: 10.1016/j.jmatprotec.2007.10.028

Hayder, W., ... (2020). Intelligent MPPT algorithm for PV system based on fuzzy logic. In *2020 17th International Multi-Conference on Systems, Signals & Devices (SSD)* (p. 70). IEEE. DOI: 10.1109/SSD49366.2020.9364195

Hebhoub, H., Aoun, H., Belachia, M., Houari, H., & Ghorbel, E. (2011). Use of waste marble aggregates in concrete. *Construction & Building Materials*, 25(3), 1167–1171. DOI: 10.1016/j.conbuildmat.2010.09.037

Heyhat, M. M., Kowsary, F., Rashidi, A. M., Esfehani, S. A. V., & Amrollahi, A. (2012). Experimental investigation of turbulent flow and convective heat transfer characteristics of alumina water nanofluids in fully developed flow regime. *International Communications in Heat and Mass Transfer*, 39(8), 1272–1278. DOI: 10.1016/j.icheatmasstransfer.2012.06.024

Hima, A., & Lakhdar, N. (2020). Enhancement of efficiency and stability of CH3NH3GeI3 solar cells with CuSbS2. *Optical Materials*, 99, 109607. DOI: 10.1016/j.optmat.2019.109607

Hima, A., Lakhdar, N., Benhaoua, B., Saadoune, A., Kemerchou, I., & Rogti, F. (2019). An optimized perovskite solar cell designs for high conversion efficiency. *Superlattices and Microstructures*, 129, 240–246. DOI: 10.1016/j.spmi.2019.04.007

Hirsch, A., Dietrich, F., Berger, S., & Zaeh, M. F. (2019). Manufacturing technologies for lightweight electric vehicle structures. *Journal of Manufacturing Processes*, 38, 377–389. DOI: 10.1016/j.jmapro.2019.01.039

Hossain, J., Moon, M. M. A., Mondal, B. K., & Halim, M. A. (2021). Design guidelines for a highly efficient high-purity germanium (HPGe)-based double-heterojunction solar cell. *Optics & Laser Technology*, 143, 107306. DOI: 10.1016/j.optlastec.2021.107306

Husainat, A., Ali, W., Cofie, P., Attia, J., & Fuller, J. (2019). Simulation and analysis of methylammonium lead iodide (CH3NH3PbI3) perovskite solar cell with Au contact using SCAPS 1D simulator. *American Journal of Optics and Photonics*, 7(2), 33. DOI: 10.11648/j.ajop.20190702.12

Idorn, G. M. (1997). *Concrete progress from antiquity to third millenium*. Thomas Telford. DOI: 10.1680/cpftatttm.26315

Ilangovana, R., Mahendrana, N., & Nagamanib, K. (2008)... *Strength and Durability Properties of Concrete Containing Quarry Rock Dust As Fine Aggregate.*, 3(5), 20–26.

Iqbal, M. D., Birk, C., Ooi, E. T., Pramod, A. L. N., Natarajan, S., Gravenkamp, H., & Song, C. (2022). Thermoelastic fracture analysis of functionally graded materials using the scaled boundary finite element method. *Engineering Fracture Mechanics*, 264, 108305. DOI: 10.1016/j.engfracmech.2022.108305

IS : 1199- 1959. (. (2012). Indian Standard METHODS OF SAMPLING AND ANALYSIS OF CONCRETE. *Recent Patents on Nanotechnology*, 7(1), 13–25. DOI: 10.2174/18722105130103

IS 456. (2000). Plain Concrete and Reinforced. *Bureau of Indian Standards,New Dehli*, 1–114.

IS 516:2014. (2004). Method of Tests for Strength of Concrete. *IS : 516 - 1959 (Reaffirmed 2004)*, New Delhi,India.

IS 5816 : 1999. (1999). Indian standard Splitting tensile strength of concrete- method of test (first revision). *Bureau of Indian Standards, New Delhi, (reaffirme*, 1–14. Kumar, R. (2017). Influence of recycled coarse aggregate derived from construction and demolition waste (CDW) on abrasion resistance of pavement concrete. *Construction and Building Materials, 142*, 248–255. DOI: 10.1016/j.conbuildmat.2017.03.077

Jaglan, A., Dogra, N., Kumar, A., & Kumar, P. (2023). Finite Element Simulation Approach in Incremental Sheet Forming Process. Handbook of Flexible and Smart Sheet Forming Techniques: Industry 4.0 Approaches, 129-139. DOI: 10.1002/9781119986454.ch7

Jeon, N. J., Noh, J. H., Yang, W. S., Kim, Y. C., Ryu, S., Seo, J., & Seok, S. I. (2015). Compositional engineering of perovskite materials for high-performance solar cells. *Nature*, 517(7535), 476–480. DOI: 10.1038/nature14133 PMID: 25561177

Kabeer, K. I. S. A., & Vyas, A. K. (2018). Utilization of marble powder as fine aggregate in mortar mixes. *Construction & Building Materials*, 165, 321–332. DOI: 10.1016/j.conbuildmat.2018.01.061

Kabir, E., Kumar, P., Kumar, S., Adelodun, A. A., & Kim, K. H. (2018). Solar energy: Potential and future prospects. *Renewable & Sustainable Energy Reviews*, 82, 894–900. DOI: 10.1016/j.rser.2017.09.094

Kalemtaş, A. (n.d.). *Ceramic Materials*. https://depo.btu.edu.tr/dosyalar/metalurji/Dosyalar/ Ceramic_Materials_3_Ayse_Kalemtas.pdf

Kang, J., Fu, R., Luan, G., Dong, C., & He, M. (2010). In-situ investigation on the pitting corrosion behavior of friction stir welded joint of AA2024-T3 aluminium alloy. *Corrosion Science*, 52(2), 620–626. DOI: 10.1016/j.corsci.2009.10.027

Kanoun, A. A., Kanoun, M. B., Merad, A. E., & Goumri-Said, S. (2019). Toward development of high-performance perovskite solar cells based on CH3NH3GeI3 using computational approach. *Solar Energy*, 182, 237–244. DOI: 10.1016/j.solener.2019.02.041

Kaplan, H., & Binici, H. (1995). Tras ve traslı çimentolar. *Pamukkale Üniversitesi Mühendislik Fakültesi Mühendislik Bilimleri Dergisi*, 1(2–3), 121–127.

Kappes, J. A. (1995). *Characterization of pozzolanic cement for radioactive waste disposal* [Unpublished Master's Thesis]. University of Nevada.

Kara, F. (n.d.). *Manganez (Mangan) Nedir? Nerelerde Kullanılır? Nasıl Üretilir? Özellikleri Nelerdir?*https://malzemebilimi.net/manganez-mangan-nedir-nerelerde-kullanilir-nasil-uretilir-ozellikleri-nelerdir.html

Karimi, E., & Ghorashi, S. M. B. (2017). Simulation of perovskite solar cell with P3HT hole-transporting materials. *Journal of Nanophotonics*, 11(3), 032510–032510. DOI: 10.1117/1.JNP.11.032510

Karna, S. K., & Sahai, R. (2012). An overview on Taguchi method. International journal of engineering and mathematical sciences, 1(1), 1-7.

Karthikeyan, O. P., Visvanathan, C., & Thayanithy, K. (2013). Management of urban biodegradable solid waste through biogas production: A review. *Renewable & Sustainable Energy Reviews*, 21, 978–988.

Kawamura, K., & Yasuda, K. (2016). Advances in silicon carbide (SiC) technology for electric vehicles. *Semiconductor Science and Technology*, 31(7), 073002. DOI: 10.1088/0268-1242/31/7/073002

Keerthi, N., Kori, S., (2018). Study on Improvement of Sub Grade Soil using Soil - Reinforcement Technique 13, 126–134.

Keklikcioglu, O., Dagdevir, T., & Ozceyhan, V. (2019, November). Heat transfer and pressure drop investigation of graphene nanoplatelet-water and titanium oxide-water nanofluids in a horizontal tube. *Applied Thermal Engineering*, 162(114256), 5.

Kenfack, A. K., Mashamba, D. R., Thantsha, N. M., & Msimanga, M. (2023). Prediction of band gap and optimum electrical parameters of a thin homojunction perovskite solar cell based on FA1-xCsxSnyPb1-yI3 through a combination of SCAPS-1D and Machine Learning based modelling. *Materials Today. Communications*, 37, 107318. DOI: 10.1016/j.mtcomm.2023.107318

Khanafer, K., & Vafai, K. (2011). A critical synthesis of thermophysical characteristics of Nanofluids. *International Journal of Heat and Mass Transfer*, 54(19-20), 4410–4428. DOI: 10.1016/j.ijheatmasstransfer.2011.04.048

Khattak, Y. H., Baig, F., Shuja, A., Beg, S., & Soucase, B. M. (2020). Numerical analysis guidelines for the design of efficient novel nip structures for perovskite solar cell. *Solar Energy*, 207, 579–591. DOI: 10.1016/j.solener.2020.07.012

Khodir, S. A., & Shibayanagi, T. (2007). Dissimilar friction stir welded joints between 2024-T3 aluminum alloy and AZ31 magnesium alloy. *Materials Transactions*, 48(9), 2501–2505. DOI: 10.2320/matertrans.MRA2007093

Kim, S., Lee, Y., & Jang, Y. (2020). Recycling of electric vehicle batteries: A review of current trends and future prospects. *Renewable & Sustainable Energy Reviews*, 117, 109477. DOI: 10.1016/j.rser.2019.109477

Kin Yuen Leonga. (2017, November 20). Idayu Razalia, K.Z. Ku Ahmada, Hwai Chyuan Ongb, M.J. Ghazalic, Mohd Rosdzimin Abdul Rahmana, "Thermal conductivity of an ethylene glycol/water-based nanofluid with copper-titanium dioxide nanoparticles: An experimental approach". *International Communications in Heat and Mass Transfer*, 90, 34–34.

Kitsopoulos, K. P., & Dunham, A. C. (1996). Heulandite and mordenite-rich tuffs from Greece: A potential source for pozzolanic materials. *Mineralium Deposita*, 31(6), 576–583. DOI: 10.1007/BF00196138

Kore Sudarshan, D., & Vyas, A. K. (2019). Impact of fire on mechanical properties of concrete containing marble waste. *Journal of King Saud University. Engineering Sciences*, 31(1), 42–51. DOI: 10.1016/j.jksues.2017.03.007

Kore, S. D., & Vyas, A. K. (2016). Impact of marble waste as coarse aggregate on properties of lean cement concrete. *Case Studies in Construction Materials*, 4, 85–92. DOI: 10.1016/j.cscm.2016.01.002

Kore, S. D., Vyas, A. K., & Syed, S. A. (2020). A brief review on sustainable utilisation of marble waste in concrete. *International Journal of Sustainable Engineering*, 13(4), 264–279. DOI: 10.1080/19397038.2019.1703151

Krishnan, M. M., Maniraj, J., Deepak, R., & Anganan, K. (2018). Prediction of optimum welding parameters for FSW of aluminium alloys AA6063 and A319 using RSM and ANN. *Materials Today: Proceedings*, 5(1), 716–723. DOI: 10.1016/j.matpr.2017.11.138

Küçükyildirim, E., & Uzal, B. (2014). Characteristics of calcined natural zeolites for use in high-performance pozzolan blended cements. *Construction & Building Materials*, 73, 229–234. DOI: 10.1016/j.conbuildmat.2014.09.081

Kumar, P., Hussain, S. S., Kumar, A., Srivastava, A. K., Hussain, M., & Singh, P. K. (2024). 10 Finite element method investigation on delamination of 3D printed hybrid composites during the drilling operation. 3D Printing Technologies: Digital Manufacturing, Artificial Intelligence, Industry 4.0, 223.

Kumar, R., & Kumar, S. K. (2018). " *Partial Replacement of Cement with Marble Dust Powder* " " *Partial Replacement of Cement with Marble Dust Powder* ." 5(August 2015), 106–114.

Kumar, A., Kumar, P., & Liu, Y. (2024). *Industry 4.0 Driven Manufacturing Technologies*. Springer Nature., DOI: 10.1007/978-3-031-68271-1

Kumar, A., Kumar, P., Sharma, N., & Srivastava, A. K. (Eds.). (2024). *3D Printing Technologies: Digital Manufacturing, Artificial Intelligence, Industry 4.0*. Walter de Gruyter GmbH & Co KG. DOI: 10.1515/9783111215112

Kumar, A., Kumar, P., Srivastava, A. K., & Goyat, V. (Eds.). (2023). *Modeling, Characterization, and Processing of Smart Materials*. IGI Global., DOI: 10.4018/978-1-6684-9224-6

Kumar, A., Liu, Y., & Kumar, R. (2024). *Handbook of Intelligent and Sustainable Manufacturing: Tools, Principles, and Strategies*. CRC Press. DOI: 10.1201/9781003405870

Kumar, A., Mittal, R. K., & Goel, R. (Eds.). (2023). *Waste Recovery and Management: An Approach Toward Sustainable Development Goals*. CRC Press., DOI: 10.1201/9781003359784

Kumar, A., Mittal, R. K., & Haleem, A. (Eds.). (2022). *Advances in Additive Manufacturing: Artificial Intelligence, Nature-Inspired, and Biomanufacturing*. Elsevier., DOI: 10.1016/C2020-0-03877-6

Kumar, A., Rajak, D. K., Kumar, P., & Kumar, A. (Eds.). (2024). *Advances in Sustainable Biomaterials: Bioprocessing 4.0, Characterizations, and Applications* (1st ed.). CRC Press., DOI: 10.1201/9781003434313

Kumar, A., Rajak, D. K., Kumar, P., & Rathee, S. (Eds.). (2024). *Smart Electric and Hybrid Vehicles: Design, Modeling, and Assessment by Industry 4.0 Approaches* (1st ed.). CRC Press., DOI: 10.1201/9781003502470

Kumar, A., Rajak, D. K., Kumar, P., & Rathee, S. (Eds.). (2024). *Smart Electric and Hybrid Vehicles: Fundamentals, Strategies and Applications* (1st ed.). CRC Press., DOI: 10.1201/9781003495574

Kumar, A., Singh, H., Kumar, P., & AlMangour, B. (Eds.). (2023). *Handbook of Smart Manufacturing: Forecasting the Future of Industry 4.0*. CRC Press., DOI: 10.1201/9781003333760

Kumari, N., Gohel, J. V., & Patel, S. R. (2018). Optimization of TiO2/ZnO bilayer electron transport layer to enhance efficiency of perovskite solar cell. *Materials Science in Semiconductor Processing*, 75, 149–156. DOI: 10.1016/j.mssp.2017.11.030

Kumar, L., & Sharma, R. K. (2020). Smart manufacturing and Industry 4.0: State-of-the-art review. In *Handbook of Smart Manufacturing* (pp. 1–28). CRC Press., DOI: 10.1201/9781003333760-1

Kumar, M., Das, A., & Ballav, R. (2023). Influence of the Zn interlayer on the mechanical strength, corrosion and microstructural behavior of friction stir-welded 6061-T6 aluminium alloy and AZ61 magnesium alloy dissimilar joints. *Materials Today. Communications*, 35, 105509. DOI: 10.1016/j.mtcomm.2023.105509

Kumar, M., Kumar, A., Raj, A., Sati, P. C., Sahni, M., & Anshul, A. (2022). Organic-inorganic perovskite-based solar cell designs for high conversion efficiency: A comparative study by SCAPS simulation. *Materials Today: Proceedings*, 49, 3081–3087. DOI: 10.1016/j.matpr.2020.11.035

Kumavat, S. R., Sonvane, Y., & Gupta, S. K. (2020). Structural, optical, transport, and solar cell properties of 2D halide perovskite MAZX3 (Z= Pb, Sn, and X= Cl, Br, I). *Journal of Applied Physics*, 128(11), 114304. DOI: 10.1063/5.0016331

Kumawat, V., Meena, R. V., Beniwal, A. S., & Gupta, G. (2023). *Materials Today : Proceedings '' Utilization of Kota stone slurry along with demolition and construction waste in concrete ". xxxx*. DOI: 10.1016/j.matpr.2023.06.103

Kurudirek, M., Özdemir, Y., Türkmen, I., & Levet, A. (2010). A study of chemical composition and radiation attenuation properties in clinoptilolite-rich natural zeolite from Turkey. *Radiation Physics and Chemistry*, 79(11), 1120–1126. DOI: 10.1016/j.radphyschem.2010.06.003

Laakri, M., Oudjit, M. N., & Saidani-Scott, H. (2019). Influence of natural pozzolan and blast furnace slag on hydration and physico-mechanical properties of cement. *International Journal of Sustainable Building Technology and Urban Development*, 10(3), 112–124. DOI: 10.22712/susb.20190012

Lakshmi, S. M., Gayathri, S., Susanna, K. D., Ammaiappan, M., & Manoharan, M. (2018). An Experimental Study On Soil Strength Enhancement Using Geosynthetics. *Int. J. Innov. Res. Explor.*, 5, 433–442.

Laturkar, K., & Laturkar, K. (2023). Environmental sustainability measures in the oil and gas industry. In *Multidisciplinary Approaches to Sustainable Human Development* (pp. 176–205). IGI Global. DOI: 10.4018/978-1-6684-8223-0.ch008

Le Pennec, J. L., Temel, A., Froger, J. L., Şen, S., & Gourgaud, A.; Boudier, J. L. (2005). Stratigraphy and Age of the Cappadocia Ignimbrites, Turkey: Reconciling Field Constraints with Paleontologic, Radiochronologic, Geochemical and Paleomagnetic Data. *Of Volcanology and Geothermal Research, 141*, 45–64.

Lee, M. M., Teuscher, J., Miyasaka, T., Murakami, T. N., & Snaith, H. J. (2012). Efficient hybrid solar cells based on meso-superstructured organometal halide perovskites. science, 338(6107), 643-647.

León, A., Reuquen, P., Garín, C., Segura, R., Vargas, P., Zapata, P., & Orihuela, P. A. (2017). FTIR and Raman characterization of TiO2 nanoparticles coated with polyethylene glycol as carrier for 2-methoxyestradiol. *Applied Sciences (Basel, Switzerland)*, 7(1), 49.

Lezzerini, M., Luti, L., Aquino, A., Gallello, G., & Pagnotta, S. (2022). Effect of Marble Waste Powder as a Binder Replacement on the Mechanical Resistance of Cement Mortars. *Applied Sciences (Basel, Switzerland)*, 12(9), 4481. Advance online publication. DOI: 10.3390/app12094481

Liguori, B., Iucolano, F., De Gennaro, B., Marroccoli, M., & Caputo, D. (2015). Zeolitized tuff in environmental friendly production of cementitious material: Chemical and mechanical characterization. *Construction & Building Materials*, 99, 272–278. DOI: 10.1016/j.conbuildmat.2015.09.035

Linton, V. M., & Ripley, M. I. (2008). Influence of time on residual stresses in friction stir welds in agehardenable 7xxx aluminium alloys. *Acta Materialia*, 56(16), 4319–4327. DOI: 10.1016/j.actamat.2008.04.059

Li, S., Dong, H., Shi, L., Li, P., & Ye, F. (2017). Corrosion behavior and mechanical properties of Al-Zn-Mg aluminum alloy weld. *Corrosion Science*, 123, 243–255. DOI: 10.1016/j.corsci.2017.05.007

Liu, K. V., Choi, U. S., & Kasza, K. E. (1988). *Measurements of pressure drop and heat transfer in turbulent pipe flows of particulate slurries* (No. ANL-88-15). Argonne National Lab.(ANL), Argonne, IL (United States).

Liu, T., Zhang, Y., Lin, L., & Wang, J. (2022). Trends and challenges in recycling of automotive lightweight materials. *Journal of Cleaner Production*, 346, 131161. DOI: 10.1016/j.jclepro.2022.131161

Li, Y., Fernández-Seara, J., Du, K., Pardiñas, Á. Á., Latas, L. L., & Jiang, W. (2016). Experimental investigation on heat transfer and pressure drop of ZnO/ethylene glycol-water nanofluids in transition flow. *Applied Thermal Engineering*, 93, 537–548.

Li, Y., Wang, B., Wang, Y., Yang, M., & Sun, R. (2015). Influence of pyrolysis temperature on properties of biochar obtained from pyrolysis of paper mill sludge. *Journal of Analytical and Applied Pyrolysis*, 111, 173–180.

Lua, A. C. (2008). Aromaticity and the degree of aromatic condensation of char. *Carbon*, 46(12), 1606–1612.

Lu, L., Han, X., Li, J., Hua, J., & Ouyang, M. (2013). A review on the key issues for lithium-ion battery management in electric vehicles. *Journal of Power Sources*, 226, 272–288. DOI: 10.1016/j.jpowsour.2012.10.060

Luo, W., Li, H., Shi, W., & Zhang, Y. (2021). Lightweight magnesium alloys for automotive applications: A review. *Materials Science and Engineering A*, 808, 140928. DOI: 10.1016/j.msea.2021.140928

Luxan, M. P., Madruga, F., & Saavedra, J. (1989). Rapid Evaluation of Pozzolanic Activity of Natural Products by Conductivity Measurement. *Cement and Concrete Research*, 19(1), 63–68. DOI: 10.1016/0008-8846(89)90066-5

Maage, M. (1984). Effect of micro silica on the durability of concrete structures, Report STF65 A 84019, FCB/SINTEF. *The Norwegian Institute of technology, Tronddheim, Norway.*

Madhu, P., Yadav, G., Bharath, S., Kumar, M. M., Reddy, M. N., & Reddy, G. C. (2018).. . *International Journal of Engineering Sciences & Research Technology Usage of Geogrids in Flexible Pavement Design*, 7, 144–153.

Maggenti, R., Knapp, C., & Fereira, S. (2013, July). Controlling Shrinkage Cracking. *Concrete International*, (July), 36–41.

Majid, M. (2020). Renewable energy for sustainable development in India: Current status, future prospects, challenges, employment, and investment opportunities. *Energy, Sustainability and Society*, 10(1), 1–36. DOI: 10.1186/s13705-020-00249-2

Malinowski, R., & Garfinkel, Y. (1991). Prehistory of concrete. *Concrete International*, 13(3), 62–68.

Mandadapu, U., Vedanayakam, S. V., Thyagarajan, K., & Babu, B. J. (2018). Optimisation of high efficiency tin halide perovskite solar cells using SCAPS-1D. International journal of simulation and process modelling, 13(3), 221-227.

Mandadapu, U., Vedanayakam, S. V., & Thyagarajan, K. (2017). Simulation and analysis of lead based perovskite solar cell using SCAPS-1D. *Indian Journal of Science and Technology*, 10(11), 65–72. DOI: 10.17485/ijst/2017/v10i11/110721

Manikandan, K., JafarAhamed, A., & Brahmanandhan, G. M. (2017). Synthesis, Structural and Optical Characterization of TiO2 Nanoparticles and its Assessment to Cytotoxicity Activity. *Journal of Environmental Nanotechnology*, 6(3), 94–102.

Manju, M., & K, A. N. (2016). Durability Studies on Concrete by Replacing Natural Sand with M-Sand-A Review. *International Journal of Emerging Technology and Advanced Engineering*, 6(3), 3–7.

Manjunatha, M., Preethi, S., Malingaraya, Mounika, H. G., Niveditha, K. N., & Ravi. (2021). Life cycle assessment (LCA) of concrete prepared with sustainable cement-based materials. *Materials Today: Proceedings, 47*(xxxx), 3637–3644. DOI: 10.1016/j.matpr.2021.01.248

Manzetti, S., & Mariasiu, F. (2015). Electric vehicle battery technologies: From present state to future systems. *Renewable & Sustainable Energy Reviews*, 51, 1004–1012. DOI: 10.1016/j.rser.2015.07.010

Mary chyne, J., Kiran Sepuri, H., & K, T. H. (2019). A Review on Recycled Asphalt Pavement in cement concrete. *International Journal of Latest Engineering Research and Applications (IJLERA), 04*(February), 9–18. www.ijlera.com

Mathias, V., Sedran, T., & Larrard, F. de. (2004). *Recycling reclaimed asphalt pavement in concrete roads. November 2004*, 66–75.

Maxwell, J. C. (1881). *A Treatise on Electricity and Magnetism*. Clarendon Press.

Mazzoni, G., Bocci, E., & Canestrari, F. (2018). Influence of rejuvenators on bitumen ageing in hot recycled asphalt mixtures. [English Edition]. *Journal of Traffic and Transportation Engineering*, 5(3), 157–168. DOI: 10.1016/j.jtte.2018.01.001

McCarthy, M. J., & Dyer, T. D. (2019). Pozzolanas and pozzolanic materials. In Hewlett, P. C. (Ed.), *Lea's Chemistry of Cement and Concrete* (5th ed., pp. 363–467). Elsevier., DOI: 10.1016/B978-0-08-100773-0.00009-5

Meena, R. V., Beniwal, A. S., & Gupta, G. (2023). Materials Today : Proceedings " Fresh characteristics and compressive strength evaluation of self- compacting concrete prepared with ceramic waste tile and fly ash. *Materials Today: Proceedings*, xxxx. Advance online publication. DOI: 10.1016/j.matpr.2023.05.687

Meena, R. V., Beniwal, A. S., Jain, A., Choudhary, R., & Mandolia, R. (2023). Evaluating resistance of ceramic waste tile self-compacting concrete to sulphuric acid attack. *Construction & Building Materials*, 393(March), 132042. DOI: 10.1016/j.conbuildmat.2023.132042

Meena, R. V., Jain, J. K., Chouhan, H. S., & Beniwal, A. S. (2021). Mechanical and durability performance of self-compacting concrete with waste ceramic tile as a replacement for natural river sand. *Materials Today: Proceedings*, xxxx. Advance online publication. DOI: 10.1016/j.matpr.2021.12.303

Memon, M. J., Jhatial, A. A., Rid, Z. A., Rind, T. A., & Sandhu, A. R. (2019). Marble Powder As Fine Aggregates in Concrete. *Engineering, Technology &. Applied Scientific Research*, 9(3), 4105–4107. DOI: 10.48084/etasr.2698

Mikulčić, H., Klemeš, J. J., Vujanović, M., Urbaniec, K., & Duić, N. (2016). Reducing greenhouse gasses emissions by fostering the deployment of alternative raw materials and energy sources in the cleaner cement manufacturing process. *Journal of Cleaner Production*, 136, 119–132. DOI: 10.1016/j.jclepro.2016.04.145

Minažek, K., Kopić, Ž., & Mulabdić, M. (2012). Simple Procedure of Geogrid-Soil Interaction Efficiency Assesment. *Tech. Gaz.*, 19, 399–405.

Mishra, R. K., Tripathi, R. K., & Dubey, V. (2016). Early age shrinkage pattern of concrete on replacement of fine aggregate with industrial by-product. *Journal of Radiation Research and Applied Sciences*, 9(4), 386–391. DOI: 10.1016/j.jrras.2016.05.003

Mittal, A., Kaisare, M. B., & Shetti, R. (2005). Experimental Study on use of fly ash in concrete. *Tarapur Atomic Power Project, 3.*

Mittal, A., & Shukla, S. (2019). Strength improvement of poor subgrade soil reinforced with polyester biaxial geogrid. *Jordan J. Civ. Eng.*, 13, 214–225.

Mittal, R. K. (2020). *Incremental sheet forming technologies: principles, merits, limitations, and applications.* CRC Press., DOI: 10.1201/9780429298905

Mittal, S., Korde, P., Shenbagalakshmi, S., & Sharma, K. (2023). *A systematic review on recent advancement in electric vehicle materials.* IRJEAS., DOI: 10.55083/irjeas.2023.v11i04006

Miyan, N., Omur, T., Amed, B., Özkan, H., Aydın, R., & Kabay, N. (2024). Recycled waste concrete and metakaolin based alkali-activated paste: Characterization, optimization, and life cycle assessment. *Construction & Building Materials*, 416(February), 135233. Advance online publication. DOI: 10.1016/j.conbuildmat.2024.135233

Modak, M., Srinivasan, S., Garg, K., Chougule, S. S., Agarwal, M. K., & Sahu, S. K. (2015). Experimental investigation of heat transfer characteristics of the hot surface using Al_2O_3 -water nanofluids. *Chemical Engineering and Processing*, 91, 104–113. Advance online publication. DOI: 10.1016/j.cep.2015.03.006

Mohan, D., Pittman, C. U.Jr, & Steele, P. H. (2006). Pyrolysis of wood/biomass for bio-oil: A critical review. *Energy & Fuels*, 20(3), 848–889. DOI: 10.1021/ef0502397

Moiz, S. A., Alahmadi, A. N. M., & Aljohani, A. J. (2021). Design of a novel lead-free perovskite solar cell for 17.83% efficiency. *IEEE Access : Practical Innovations, Open Solutions*, 9, 54254–54263. DOI: 10.1109/ACCESS.2021.3070112

Moropoulou, A., Cakmak, A., Labropoulos, K. C., Van Grieken, R., & Torfs, K. (2004). Accelerated microstructural evolution of a calcium-silicate-hydrate (C-S-H) phase in pozzolanic pastes using fine siliceous sources: Comparison with historic pozzolanic mortars. *Cement and Concrete Research*, 34(1), 1–6. DOI: 10.1016/S0008-8846(03)00187-X

Mostafa, N. Y., & Brown, P. W. (2005). Heat of hydration of high reactive pozzolans in blended cements: Isothermal conduction calorimetry. *Thermochimica Acta*, 435(2), 162–167. DOI: 10.1016/j.tca.2005.05.014

Nair, K.S., Deepthy, B.L., (2017). Improvement of Soft Subgrade Using Geogrid Reinforcement 181–183.

Nath, P., & Sarker, P. (2011). Effect of fly ash on the durability properties of high strength concrete. *Procedia Engineering*, 14, 1149–1156. DOI: 10.1016/j.proeng.2011.07.144

National Renewable Energy Laboratory. (2020). Best Research-Cell Efficiency Chart. Available online at: https://www.nrel.gov/pv/cell-efficiency.html (accessed April 6, 2020). [Ref list] https://www.ncbi.nlm.nih.gov/pmc/articles/PMC7399487/#B23

Ng, T. S., & Nyan, T. S. H. (2017). Structural Application of Steel Fibres Reinforced Concrete With and Without Conventional Reinforcement. *The New Zealand Concrete Industry, 3101*(2006), 2–15.

Nguyen, L., Moseson, A. J., Farnam, Y., & Spatari, S. (2018). Effects of composition and transportation logistics on environmental, energy and cost metrics for the production of alternative cementitious binders. *Journal of Cleaner Production*, 185, 628–645. DOI: 10.1016/j.jclepro.2018.02.247

Ng, Y. L., & Ng, D. K. S. (2011). A review on the utilization of fly ash. *Progress in Energy and Combustion Science*, 37(4), 512–537.

Nian, C. Y., Yang, W. H., & Tarng, Y. S. (1999). Optimization of turning operations with multiple performance characteristics. *Journal of Materials Processing Technology*, 95(1-3), 90–96. DOI: 10.1016/S0924-0136(99)00271-X

Nie, R., Lee, K. S., Hu, M., Paik, M. J., & Seok, S. I. (2020). Heteroleptic tin-antimony sulfoiodide for stable and lead-free solar cells. *Matter*, 3(5), 1701–1713. DOI: 10.1016/j.matt.2020.08.020

Ni, T., Xu, F., Gao, C., Yang, Y., Ma, W., & Kong, D. (2022). Chemical activation of pozzolanic activity of sludge incineration ash and application as row bonding materials for pervious ecological brick. *Construction & Building Materials*, 329, 127199. DOI: 10.1016/j.conbuildmat.2022.127199

Nogata, F. (1997). Learning about design concepts from natural functionally graded materials. *ASME International Mechanical Engineering Congress and Exposition*, 18343, 11–18. DOI: 10.1115/IMECE1997-0646

Ogundare, D.A., Familusi, A.O., Osunkunle, A.B., Olusami, J.O., (2019). American Journal of Engineering Research (AJER) Utilization Of Geotextile For Soil Stabilization 224–231.

Ogunkunbi, G., & Jimoh, Y. (2019). Design and Economic Analysis of a Flexible Pavement on a Geosynthetic Reinforced Subgrade. *Journal of Applied Science & Environmental Management*, 23(1), 121. DOI: 10.4314/jasem.v23i1.18

Olivetti, E. A., Ceder, G., Gaustad, G. G., & Fu, X. (2017). Lithium-ion battery supply chain considerations: Analysis of potential bottlenecks in critical metals. *Joule*, 1(2), 229–243. DOI: 10.1016/j.joule.2017.08.019

Onyango, M., Macha, I., & Busch, C. (2007). Use of naturally occurring pozzolans for road construction in Tanzania. *9th International Conference on Low-Volume Roads Proceedings, 2*, 1–17. DOI: 10.3141/1989-61

Onyango, M. (1989). Innocent, M., & Busch, C. (2007). se of Naturally Occurring Pozzolans for Road Construction in Tanzania. *Transportation Research Record: Journal of the Transportation Research Board*, (2), 169–177.

Ouslimane, T., Et-taya, L., Elmaimouni, L., & Benami, A. (2021). defect density, and operating temperature on the performance of MAPbI3 solar cells based on ZnO electron transporting material. Touria Ouslimane, Lhoussayne Et-taya, Lahoucine Elmaimouni, Abdellah Benami. *Heliyon*, 7(3), e06379. DOI: 10.1016/j.heliyon.2021.e06379 PMID: 33732928

Özata, Ş. (2015). *Kapadokya Bölgesi Kaya Oyma Yapı Sorunları ve Çözüm Öneri leri* . [Master's Thesis, Yildiz Technical University]. Institute of Natural and Applied Sciences https://tez.yok.gov.tr/UlusalTezMerkezi/tezSorguSonucYeni.jsp

Özata, Ş. (2018). Sustainability of Rock Carving Techniques in the World Heritage Site, Cappadocia. In A. C. Antunes & Marius Müller (Eds). *New Perspectives in Interdisciplinary Cultural Heritage Studies* (pp. 85-101.). Mazu Press.

Ozata, S., Akturk, B., & Yuzer, N. (2022). Utilization of waste Cappadocia earth as a natural pozzolan in alkali activation: A parametric study. *Construction & Building Materials*, 329, 127192. DOI: 10.1016/j.conbuildmat.2022.127192

Ozata, S., & Arun, E. G. (2018). Damage assessment for rock-cut dwellings and warehouses: A case study in Bahceli village, Cappadocia. *Masonry International*, 30(3), 81–90.

Özata, Ş., & Arun, E. G. (2018). Damage Assessment of Rock-Cut Ortahisar Castle in Cappadocia Region. *Gazi University Journal of Science*, 31(1), 1–11.

Özata, Ş., Yüzer, N., Ulukaya, S., Aktürk, B., & Kızılkanat, A. B. (2019). Alkali Activation and Pozzolanic Activity of Waste Material Excavated from Rock-Cut Buildings of Cappadocia. *10th International Concrete Congress Proceedings*, 42–50.

Ozcelik, B. (2011). Optimization of injection parameters for mechanical properties of specimens with weld line of polypropylene using Taguchi method. *International Communications in Heat and Mass Transfer*, 38(8), 1067–1072. DOI: 10.1016/j.icheatmasstransfer.2011.04.025

Pak, B. C., & Cho, Y. I. (1998). Hydrodynamic and heat transfer study of dispersed fluids with submicron metallic oxide particles. *Experimental Heat Transfer*, 11(2), 151–170. DOI: 10.1080/08916159808946559

Palladino, S., Minutolo, V., & Esposito, L. (2022). Hybrid semi-analytical calculation of the stress intensity factor for heterogeneous and functionally graded plates. *Engineering Fracture Mechanics*, 274, 108763. DOI: 10.1016/j.engfracmech.2022.108763

Park, M., Zhang, X., Chung, M., Less, G. B., & Sastry, A. M. (2010). A review of conduction phenomena in Li-ion batteries. *Journal of Power Sources*, 195(24), 7904–7929. DOI: 10.1016/j.jpowsour.2010.06.060

Pathak, C., & Pandey, S. K. (2020). Design, performance, and defect density analysis of efficient eco-friendly perovskite solar cell. *IEEE Transactions on Electron Devices*, 67(7), 2837–2843. DOI: 10.1109/TED.2020.2996570

Patil, R. (2017). Fuzzy based MPPT technique for PV system. In *2017 International Conference on Energy, Communication, Data Analytics and Soft Computing (ICECDS)* (pp. 120). IEEE. DOI: 10.1109/ICECDS.2017.8389686

Patil, R., & Anantwar, H. (2017). Comparative analysis of fuzzy based MPPT for buck and boost converter topologies for PV application. In *2017 International Conference on Smart Technologies for Smart Nation (SmartTechCon)* (pp. 1-6). IEEE. DOI: 10.1109/SmartTechCon.2017.8358610

Pekmezci, B. Y., & Akyüz, S. (2004). Optimum usage of a natural pozzolan for the maximum compressive strength of concrete. *Cement and Concrete Research*, 34(12), 2175–2179. DOI: 10.1016/j.cemconres.2004.02.008

Perumal, K., & Sundararajan, R. (2004, August). Effect of partial replacement of cement with silica fume on the strength and durability characteristics of high performance concrete. In *29th Conference on our world in Concrete & Structures* (pp. 25-26).

Pillai, V. K., & Sundaram, V. (2021). Recent advances in magnesium alloys for structural applications in electric vehicles. *Journal of Magnesium and Alloys*, 9(2), 432–452. DOI: 10.1016/j.jma.2020.08.011

Pusateri, A., & Risser, S. M. (2021). Silicon carbide inverters: A key technology for electric vehicle performance improvements. *IEEE Transactions on Power Electronics*, 36(5), 5845–5854. DOI: 10.1109/TPEL.2020.3048558

Qi, C., Liu, M., Luo, T., Pan, Y., & Rao, Z. (2018, December). Effects of twisted tape structures on thermo-hydraulic performances of nanofluids in a triangular tube. *International Journal of Heat and Mass Transfer*, 127(Part C), 146–159. DOI: 10.1016/j.ijheatmasstransfer.2018.08.017

Qi, C., Yang, L., Chen, T., & Rao, Z. (2018). Experimental study on thermo-hydraulic performances of TiO_2-H_2O nanofluids in a horizontal elliptical tube. *Applied Thermal Engineering*, 129, 1315–1324. DOI: 10.1016/j.applthermaleng.2017.10.137

Qiu, Y., Li, J., & Li, B. (2019). Thermal management in lithium-ion batteries for electric vehicles: Current and future trends. *Energy Conversion and Management*, 182, 262–282. DOI: 10.1016/j.enconman.2018.12.051

Qu, F., Su, Y., Lu, D., Li, N., Zeng, X., & Li, W. (2024). Anti-cracking and shrinkage performance of sustainable concrete incorporating high-volume natural pozzolans: A case design for high-speed railway concrete slab tracks. *Case Studies in Construction Materials*, 20(February), e03207. DOI: 10.1016/j.cscm.2024.e03207

Raei, B. (2021). Forced Convective Heat Transfer of MgO/Water Nanofluid under Constant Heat Flux: Experimental and Statistical Investigation. *Chall. Nano Micro Scale Sci. Tech.*, 9(1), 69–80.

Raghuwanshi, P., Parashar, R., Pathak, T., & Raghubanshi, A. S. (2023). *Integration of paper-based analytical tools with smartphones and IOMT Paper-Based Diagnostic Devices for Infectious Diseases*. IOP Publishing.

Rahman, M. T., Mohajerani, A., & Giustozzi, F. (2020). Recycling of waste materials for asphalt concrete and bitumen: A review. *Materials (Basel)*, 13(7), 1495. Advance online publication. DOI: 10.3390/ma13071495 PMID: 32218261

Rai, R. K. (2022). Fuzzy logic based control technique using MPPT for solar PV system. In *2022 First International Conference on Electrical, Electronics, Information and Communication Technologies (ICEEICT)* (pp. 56-65). IEEE. DOI: 10.1109/ICEEICT53079.2022.9768650

Rai, S., Pandey, B. K., & Dwivedi, D. K. (2020). Modeling of highly efficient and low cost CH3NH3Pb (I1-xClx) 3 based perovskite solar cell by numerical simulation. *Optical Materials*, 100, 109631. DOI: 10.1016/j.optmat.2019.109631

Raj, A., Kumar, M., Bherwani, H., Gupta, A., & Anshul, A. (2021). Evidence of improved power conversion efficiency in lead-free CsGeI3 based perovskite solar cell heterostructure via scaps simulation. *Journal of Vacuum Science & Technology B Microelectronics and Nanometer Structures*, 39(1).

Rajeev Singh Parihar. Anil Sharma, PrashantBaredar(2018) MSWMinBhopal city: a critical analysis and a roadmap for its sustainable management. Article in ICE Proceedings Municipal Engineer · February 2018 DOI: DOI: 10.1680/jmuen.17.00011

Rajesh, U., Sajja, S., & Chakravarthi, V. K. (2016). Studies on Engineering Performance of Geogrid Reinforced Soft Subgrade. *Transportation Research Procedia*, 17, 164–173. DOI: 10.1016/j.trpro.2016.11.072

Ramteke, N.B., (2014). A Review : Effect of Geo-grid reinforcement on soil ISSN : 2348 9510 Geo-grid : 1, 35–47.

Ramteke, P. M., & Panda, S. K. (2023). Computational Modelling and Experimental Challenges of Linear and Nonlinear Analysis of Porous Graded Structure: A Comprehensive Review. *Archives of Computational Methods in Engineering*, 30(5), 3437–3452. DOI: 10.1007/s11831-023-09908-x

Rangrazian, M., Madandoust, R., Mahjoub, R., & Raftari, M. (2023). Experimental investigations on the engineering properties of pozzolanic concrete reinforced with fibers and the reduction of CO2 emissions: A case study on local mineral pozzolan. *Materials Today Communications, 34*(October 2022), 105129. DOI: 10.1016/j.mtcomm.2022.105129

Raouf Dar, A., & Singh, E. S. (2020). Use of Recycled Materials in Pavement Construction. *IJSDR2008038 International Journal of Scientific Development and Research, 5*(8), 295–304. www.ijsdr.org

Rathod, S., Raghubanshi, A. S., Srivastava, Y., Kiragi, V. R., & Mondal, D. P. (2019). Sliding wear behaviour study of alumina-silicate short fiber reinforced AA2014-Al alloy composite (SFRAC). *Materials Research Express, 6*(10), 1065a1067. DOI: 10.1088/2053-1591/ab3d60

Ravisankar, R., Venkatachalapathy, V. S. K., & Alagumurthy, N. (2017). "Thermal performance improvement of tractor radiator using CuO/Water nanofluid", Heat Tran. *Asian Res.*, 46(1), 61–74.

Ren, X., Li, X., Wang, S., Ma, X., & Zhang, Y. (2019). Recent advances in biomass pyrolysis for biochar, bio-oil, and bioenergy production. *Critical Reviews in Environmental Science and Technology*, 49(1), 40–71.

Revathi, S., Suresh, D., & Anwar, S. T. (2023). Behaviour of concrete with PET bottles as fibers & silica fume as partial replacement of cement. *Materials Today: Proceedings*. Advance online publication. DOI: 10.1016/j.matpr.2023.03.580

Revilla-Cuesta, V., Evangelista, L., de Brito, J., Skaf, M., & Ortega-López, V. (2022). Mechanical performance and autogenous and drying shrinkage of MgO-based recycled aggregate high-performance concrete. *Construction and Building Materials, 314*(November 2021). DOI: 10.1016/j.conbuildmat.2021.125726

Rispoli, C., Bonis, A., & De, . (2019). The ancient pozzolanic mortars of the Thermal complex of Baia (Campi Flegrei, Italy). *Journal of Cultural Heritage*, 40, 143–154. DOI: 10.1016/j.culher.2019.05.010

Rozhko, A. A., Petrov, V. V., & Sayenko, A. V. (2021). Study of the effect of the thickness of the photosensitive layer of perovskite on its efficiency using SCAPS-1D software. []). IOP Publishing.]. *IOP Conference Series. Materials Science and Engineering*, 1035(1), 012032. DOI: 10.1088/1757-899X/1035/1/012032

Sahu, M., Paul, A., & Ganguly, S. (2021). Optimization of process parameters of friction stir welded joints of marine grade AA 5083. *Materials Today: Proceedings*, 44, 2957–2962. DOI: 10.1016/j.matpr.2021.01.938

Sai, P.N.V., Surendra, S., Arjun, N.V.K., Sai, M.S., Rao, C.H.H., (2020). Reinforcement of Black Cotton Soil by Using Geogrid Reinforcement of Black Cotton Soil by Using Geogrid 0193–4120.

Sakib, S., Noor, M. Y. M., Salim, M. R., Abdullah, A. S., Azmi, A. I., & Ibrahim, M. H. (2023). Effect of transport layer thickness in lead-based perovskite solar cell: A numerical simulation. *Materials Today: Proceedings*, 80, 1022–1026. DOI: 10.1016/j.matpr.2022.11.456

Sarkar, J. (2011). A critical review on convective heat transfer correlations of nanofluids. *Renewable & Sustainable Energy Reviews*, 15(6), 3271–3277.

Scrosati, B., Hassoun, J., & Sun, Y. K. (2011). Lithium-ion batteries. A look into the future. *Energy & Environmental Science*, 4(9), 3287–3295. DOI: 10.1039/c1ee01388b

Senhadji, Y., Escadeillas, G., Khelafi, H., Mouli, M., & Benosman, A. S. (2012). Evaluation of natural pozzolan for use as supplementary cementitious material. *European Journal of Environmental and Civil Engineering*, 16(1), 77–96. DOI: 10.1080/19648189.2012.667692

Sethuraman & Kala. (2013). Shrinkage Properties of HPC using Granite Powder as Fine Aggregate. *International Journal of Engineering and Advanced Technology*, 2(3), 637–643.

Shafiee, S., & Topal, E. (2009). When will fossil fuel reserves be diminished? *Energy Policy*, 37(1), 181–189. DOI: 10.1016/j.enpol.2008.08.016

Shah, R. K. (1975, December). Thermal entry length solutions for the circular tube and parallel plates. In *Proceedings of 3rd national heat and mass transfer conference* (Vol. 1, pp. 11-75). Delhi: Indian Institute of Technology Bombay.

Shanmuga Priya, T. (2017). Experimental investigation on high performance RC column with manufactured sand and silica fume. *IOP Conference Series: Materials Science and Engineering, 263*(3), 0–8. DOI: 10.1088/1757-899X/263/3/032021

Sharma, K., & Kumar, D. (2016). Thermo-mechanical analysis of Functionally Graded Material plate. *Advanced Science Letters*, 22(11), 3813–3816. DOI: 10.1166/asl.2016.8046

Sharma, K., & Kumar, D. (2017). Elastoplastic stability and failure analysis of FGM plate with temperature dependent material properties under thermomechanical loading. *Latin American Journal of Solids and Structures*, 14(7), 1361–1386. DOI: 10.1590/1679-78253747

Shirayama, M., Kadowaki, H., Miyadera, T., Sugita, T., Tamakoshi, M., Kato, M., Fujiseki, T., Murata, D., Hara, S., Murakami, T. N., Fujimoto, S., Chikamatsu, M., & Fujiwara, H. (2016). Optical transitions in hybrid perovskite solar cells: Ellipsometry, density functional theory, and quantum efficiency analyses for CH 3 NH 3 PbI 3. *Physical Review Applied*, 5(1), 014012. DOI: 10.1103/PhysRevApplied.5.014012

Šídlová, M., Šulc, R., Škvára, F., Pulcová, K., Formáček, P., & Snop, R. (2023). Pozzolanic activity of stockpile ash: Comparison of test methods. *Case Studies in Construction Materials*, 19(June), 1–11. DOI: 10.1016/j.cscm.2023.e02396

Sigurdsson Houghton, B. M. S. R. H. y S. J. H., & Wedge, F. M. (2000). No Title. *Encyclopedia of Volcanoes., 3*, 662.

Silica Fume User's Manual. (2015). Silica Fume Association.

Silva, R. V., de Brito, J., & Dhir, R. K. (2019). Use of recycled aggregates arising from construction and demolition waste in new construction applications. *Journal of Cleaner Production*, 236, 117629. DOI: 10.1016/j.jclepro.2019.117629

singh, J., & S Bansal, E. R. (2015). Partial Replacement of Cement With Waste Marble Powder With M25 Grade. *International Journal of Technical Research and Applications, 3*(2), 202–205. www.ijtra.com

Singh, V., Kumar, A., Kumar, K., & Gupta, M. (2021). Synthesis and Stability of Al 2 O 3/Water Nanofluids. In Advances in Materials and Mechanical Engineering: Select Proceedings of ICFTMME 2020 (pp. 3-11). Springer Singapore.

Singh, V., Kumar, A., Kumar, K., & Gupta, M. (2021). Synthesis and Stability of Al_2O_3/Water Nanofluids. In Advances in Materials and Mechanical Engineering: Select Proceedings of ICFTMME 2020 (pp. 3-11). Springer Singapore.

Singh, A., Srivastava, A. K., Kumar, A., Bajaj, R., & Singh, G. K. (2023). Design and development of plentiful fly ash-based glass powder-reinforced plastic composite bricks for low water absorption and high compressive and flexural strength. [IJIDeM]. *International Journal on Interactive Design and Manufacturing*, •••, 1–10. DOI: 10.1007/s12008-023-01580-6

Singhal, V., Nagar, R., & Agrawal, V. (2020). Sustainable use of fly ash and waste marble slurry powder in concrete. *Materials Today: Proceedings, 32*(xxxx), 975–981. DOI: 10.1016/j.matpr.2020.05.650

Singh, M., Choudhary, K., Srivastava, A., Singh Sangwan, K., & Bhunia, D. (2017). A study on environmental and economic impacts of using waste marble powder in concrete. *Journal of Building Engineering*, 13, 87–95. DOI: 10.1016/j.jobe.2017.07.009

Singh, M., Srivastava, A., & Bhunia, D. (2019). Analytical and Experimental Investigations on Using Waste Marble Powder in Concrete. *Journal of Materials in Civil Engineering*, 31(4), 1–13. DOI: 10.1061/(ASCE)MT.1943-5533.0002631

Singh, N., Agarwal, A., & Agarwal, M. (2020). Numerical simulation of highly efficient lead-free all-perovskite tandem solar cell. *Solar Energy*, 208, 399–410. DOI: 10.1016/j.solener.2020.08.003

Singh, N., & Balasubramanian, R. (2017). A comprehensive review on the pyrolysis of lignocellulosic biomass. *Renewable & Sustainable Energy Reviews*, 73, 68–86.

Singh, R. K., & Pant, D. (2012). Kinetics of thermal decomposition of municipal solid waste components. *Waste Management (New York, N.Y.)*, 32(6), 1120–1132.

Singh, T., & Miyasaka, T. (2018). Stabilizing the efficiency beyond 20% with a mixed cation perovskite solar cell fabricated in ambient air under controlled humidity. *Advanced Energy Materials*, 8(3), 1700677. DOI: 10.1002/aenm.201700677

Slami, A., Bouchaour, M., & Merad, L. (2019). Numerical study of based perovskite solar cells by SCAPS-1D. *International Journal of Energy and Environment*, 3, 17–21.

Snellings, R., Elsen, J., & Mertens, G. (2012). Supplementary Cementitious Materials Supplementary Cementitious Materials. *Reviews in Mineralogy & Geochemistry, 74*(December 2013), 211–278. DOI: 10.2138/rmg.2012.74.6

Song, Z., Abate, A., Watthage, S. C., Liyanage, G. K., Phillips, A. B., Steiner, U., Graetzel, M., & Heben, M. J. (2016). Perovskite solar cell stability in humid air: Partially reversible phase transitions in the PbI2-CH3NH3I-H2O system. *Advanced Energy Materials*, 6(19), 1600846. DOI: 10.1002/aenm.201600846

Song, Z., & Li, X. (2015). Carbon fiber composite materials: New advancements for electric vehicle structural applications. *Composite Structures*, 131, 111–121. DOI: 10.1016/j.compstruct.2015.04.025

Sonne, C., González, M. R., Solevic-Knudsen, T., Wenzel, H., Mosbæk, H., & Astrup, T. F. (2015). Life cycle assessment of thermal waste-to-energy technologies: Review and recommendations. *Waste Management (New York, N.Y.)*, 37, 104–115. DOI: 10.1016/j.wasman.2014.06.011 PMID: 25052337

Srivastava, A. K., Kumar, A., Kumar, P., Gautam, P., & Dogra, N. (2023). Research Progress in metal additive manufacturing: Challenges and Opportunities. [IJIDeM]. *International Journal on Interactive Design and Manufacturing*, •••, 1–17. DOI: 10.1007/s12008-023-01661-6

Srivastava, A. K., Tiwari, S., Pachauri, P., Gupta, N., Sunil, B., & Kumar, A. (2024). Bonding strength and microstructural features of Al5083-AZ31B alloys laminated sheet through friction stir additive manufacturing. *Journal of Adhesion Science and Technology*, 38(4), 583–596. DOI: 10.1080/01694243.2023.2240637

Standard, I. (2016). *Coarse and Fine Aggregate for Concrete — Specification. 19*(January).

Stanislao, C., Rispoli, C., Vola, G., Cappelletti, P., Morra, V., & De Gennaro, M. (2011). Contribution to the knowledge of ancient Roman seawater concretes: Phlegrean pozzolan adopted in the construction of the harbour at Soli-Pompeiopolis (Mersin, Turkey). *Periodico di Mineralogia*, 80(3), 471–488. DOI: 10.2451/2011PM0031

Stark, N. M., & Rowlands, R. E. (2017). Pyrolysis as a recycling technique for plastics. *Current Opinion in Green and Sustainable Chemistry*, 6, 12–16.

Stefanidou, M. (2016). Use of natural pozzolans with lime for producing repair mortars. *Environmental Earth Sciences*, 75(9), 1–8. DOI: 10.1007/s12665-016-5444-5

Steuwer, A., Dumont, M., Peel, M., Preuss, M., & Withers, P. J. (2007). The variation of the unstrained lattice parameter in an AA7010 friction stir weld. *Acta Materialia*, 55(12), 4111–4120. DOI: 10.1016/j.actamat.2007.01.024

Strezov, V., Evans, T. J., & Hayman, C. (2011). Thermal processing of sewage sludge to recover valuable metals: A critical review. *Renewable & Sustainable Energy Reviews*, 15(1), 471–481.

Sudhakar, C., & Linga, P. (2016). A review on thermal and catalytic pyrolysis of plastic solid waste (PSW). *Environmental Progress & Sustainable Energy*, 35(3), 779–799.

Sunarjono, S., & Hidayati, N. (2019). Physical properties of reclaimed asphalt pavement. *IOP Conference Series. Materials Science and Engineering*, 674(1), 012029. Advance online publication. DOI: 10.1088/1757-899X/674/1/012029

Suriya, D., Chandar, S. P., & Ravichandran, P. T. (2023). Impact of M-Sand on Rheological, Mechanical, and Microstructural Properties of Self-Compacting Concrete. *Buildings*, 13(5), 1126. Advance online publication. DOI: 10.3390/buildings13051126

Tagliaferro, A., Freda, C., & Sánchez-Silva, L. (2019). A critical review on the recycling of liquid crystal displays. *Journal of Cleaner Production*, 236, 117596.

Tang, Y., & Niu, M. (2019). The characteristics and environmental effects of plastics additives. *Environmental Chemistry Letters*, 17(2), 541–558.

Tarascon, J. M., & Armand, M. (2001). Issues and challenges facing rechargeable lithium batteries. *Nature*, 414(6861), 359–367. DOI: 10.1038/35104644 PMID: 11713543

Tege, S., & Kumar, P. Intelligent and Sustainable Manufacturing Applications in the Automotive Industry. In Handbook of Intelligent and Sustainable Manufacturing (pp. 305-335). CRC Press. https://doi.org/DOI: 10.1201/9781003405870-17

Thakur, P., Anand, P., Kumar, A., & Vijay, V. K. (2016). Impact of feedstock characteristics on bio-oil yield and product distribution during fast pyrolysis. *Energy Conversion and Management*, 108, 373–382.

Thangaraj, P., Janakiraman, G.G., Paramaguru, P., Pratheep, T., Scholar, P.G., (2019). Stabilization of Subgrade Using Geosynthetics. Artic. J. Adv. Res. Dyn. Control Syst. 12.

Thomas, M. D. A. (2007). *Optimizing the use of fly ash in concrete* (Vol. 5420). Portland Cement Association.

Tian, N., Deng, S., Zhang, S., Wang, H., Li, Z., & Zhang, Y. (2019). Biochar made from hydrothermal liquefaction of biomass: A review. *Renewable & Sustainable Energy Reviews*, 108, 268–279.

Tironi, A., Trezza, M. A., Scian, A. N., & Irassar, E. F. (2013). Assessment of pozzolanic activity of different calcined clays. *Cement and Concrete Composites*, 37(1), 319–327. DOI: 10.1016/j.cemconcomp.2013.01.002

Topçu, I. B., Bilir, T., & Uyguno lu, T. (2009). Effect of waste marble dust content as filler on properties of self-compacting concrete. *Construction & Building Materials*, 23(5), 1947–1953. DOI: 10.1016/j.conbuildmat.2008.09.007

Topçu, I. B., Karakurt, C., & Saridemir, M. (2008). Predicting the strength development of cements produced with different pozzolans by neural network and fuzzy logic. *Materials & Design*, 29(10), 1986–1991. DOI: 10.1016/j.matdes.2008.04.005

Toprak, V. (1998). Vent distribution and its relation to regional tectonics, Cappadocian Volcanics, Turkey. *Journal of Volcanology and Geothermal Research*, 85(1-4), 55–67. https://pascal-francis.inist.fr/vibad/index.php?action=getRecordDetail&idt=1640484. DOI: 10.1016/S0377-0273(98)00049-3

Toshniwal, A., Jariwala, A., Kheraj, V., Opanasyuk, A. S., & Panchal, C. J. (2017). Numerical simulation of tin based perovskite solar cell: Effects of absorber parameters and hole transport materials. *Journal of Nano-and Electronic Physics*, 9(3), 03038-1, 03038-4. DOI: 10.21272/jnep.9(3).03038

Tripathi, A., Manohar, S., & Ramasubramanian, K. (2023). Decoding of lime-based binder materials of ancient vernacular Chettinad constructions in Southern India through microanalytical characterization. *Journal of Building Engineering*, 79(September), 107844. DOI: 10.1016/j.jobe.2023.107844

Tural, H. G., Ozarisoy, B., Derogar, S., & Ince, C. (2024). Investigating the governing factors influencing the pozzolanic activity through a database approach for the development of sustainable cementitious materials. *Construction and Building Materials, 411*(November 2023), 134253. DOI: 10.1016/j.conbuildmat.2023.134253

Turanli, L., Uzal, B., & Bektas, F. (2004). Effect of material characteristics on the properties of blended cements containing high volumes of natural pozzolans. *Cement and Concrete Research*, 34(12), 2277–2282. DOI: 10.1016/j.cemconres.2004.04.011

Turkish Standards Institute. (2008). *Doğal Puzolan (Tras) - Çimento Ve Betonda Kullanılan - Tarifler, Gerekler ve Uygunluk Kriterleri* (TS 25).

Turkish Standards Institute. (2012). *Çimento Deney Yöntemleri - Bölüm 5: Puzolanik Çimentolarda Puzolanik Özellik Deneyi*, (TS EN 196-5).

Türkmeno lu, A. G., & Tankut, A. (2002). Use of tuffs from central Turkey as admixture in pozzolanic cements: Assessment of their petrographical properties. *Cement and Concrete Research*, 32(4), 629–637. DOI: 10.1016/S0008-8846(01)00734-7

Tyagi, A., Pandey, S. M., Walia, R. S., Murtaza, Q., & Kumar, A. (2021). Effect of Temperature on the Sliding Wear Composite Behavior Coating of HVOF Sprayed Al2O3. Advances in Materials and Mechanical Engineering: Select Proceedings of ICFTMME 2020, 23. DOI: 10.1007/978-981-16-0673-1_3

Ulukaya, S., & Yüzer, N. (2016). Assessment of Pozzolanicity of Clay Bricks Fired at Different Temperatures for Use in Repair Mortar. *J. Mater. Civ. Eng., 04016052–1*, 1–9. .DOI: 10.1061/(ASCE)MT.1943-5533.0001560

Ulukaya, S., Yoruç, A. B. H., Yüzer, N., & Oktay, D. (2017). Material characterization of byzantine period brick masonry walls revealed in Istanbul (Turkey). *Periodica Polytechnica. Civil Engineering*, 61(2), 209–215. DOI: 10.3311/PPci.8868

Umamaheswaran, V., Sudha, C., Ravichandran, P. T., & Kannan Rajkumar, P. R. (2015). Use of M Sand in High Strength and High Performance Concrete. *Indian Journal of Science and Technology*, 8(28). Advance online publication. DOI: 10.17485/ijst/2015/v8i28/84018

United Nations Environment Programme & Yale Center for Ecosystems + Architecture. (2023). *Building Materials And The Climate: Constructing A New Future.* https://wedocs.unep.org/20.500.11822/43293

Unni et al., (2017), 2017. Efficacy of Geosynthetics in Subgrade Stabilization-A Comparative Study in Laterite Soil 6709–6716. https://doi.org/DOI: 10.15680/IJIRSET.2017.0604087

Vaidevi, C. (2013). Study on marble dust as partial replacement of cement in concrete. *Indian Journal of Engineering*, 4(9), 14–16.

Vakili, M., Mohebbi, A., & Hashemipour, H. (2013). Experimental study on convective heat transfer of TiO 2 nanofluids. *Heat and Mass Transfer*, 49, 1159–1165.

Vardhan, K., Siddique, R., & Goyal, S. (2019). Strength, permeation and microstructural characteristics of concrete incorporating waste marble. *Construction & Building Materials*, 203, 45–55. DOI: 10.1016/j.conbuildmat.2019.01.079

Vasile, C., Stan, E. L., & Popa, M. V. (2016). Biochar—A sustainable approach for improving plant growth and soil properties. *Journal of Soils and Sediments*, 16(3), 883–897.

Vigneshpandian, G. V., Shruthi, E. A., Venkatasubramanian, C., & Muthu, D. (2017). Utilisation of Waste Marble Dust as Fine Aggregate in Concrete. *IOP Conference Series. Earth and Environmental Science*, 80(1), 012007. Advance online publication. DOI: 10.1088/1755-1315/80/1/012007

Vijaya, B., & Senthil Selvan, S. (2015). Comparative study on the strength and durability properties of concrete with manufactured sand. *Indian Journal of Science and Technology*, 8(36). Advance online publication. DOI: 10.17485/ijst/2015/v8i36/88614

Vijayan, V., Jayakesh, K., & Anand, K. B. (2022). Mechanical properties of recycled aggregates concrete with sisal fiber and silica fume. *Materials Today: Proceedings*, 65, 1887–1894. DOI: 10.1016/j.matpr.2022.05.055

Vilas Meena, R., Kumar Jain, J., & Singh Chouhan, H., mandolia, R., & Singh Beniwal, A. (. (2021). Impact of waste ceramic tile on resistance to fire and abrasion of self-compacting concrete. *Materials Today: Proceedings*, xxxx. Advance online publication. DOI: 10.1016/j.matpr.2021.12.287

Vilas Meena, R., Singh Chouhan, H., Kumar Jain, J., & Kumar Satankar, R. (2021). Construction and Demolition Waste as an Alternative of Rigid Pavement Material: A Review. *IOP Conference Series. Earth and Environmental Science*, 795(1), 012019. DOI: 10.1088/1755-1315/795/1/012019

Vilas, R., Kumar, J., & Singh, A. (2022). " *Sustainable self-compacting concrete containing waste ceramic tile aggregates : Fresh, mechanical, durability, and microstructural properties ." 57*(March).

Vilas, R., Kumar, J., Singh, H., & Singh, A. (2022). Use of waste ceramics to produce sustainable concrete : A review. *Cleaner Materials*, 100085, 100085. Advance online publication. DOI: 10.1016/j.clema.2022.100085

Vitruvius, M. (2017). *Mimarlık Üzerine (De Arquitectura)* (Yayıncılık, A., Trans.). Çiğdem Dürüşken.

Wagih, A., Junaedi, H., Mahmoud, H. A., Lubineau, G., Kumar, A., & Sebaey, T. A. (2024). Enhanced damage tolerance and fracture toughness of lightweight carbon-Kevlar fiber hybrid laminate. *Journal of Composite Materials*, 58(9), 00219983241235853. DOI: 10.1177/00219983241235853

Wahedy, M. N., Sharbatdar, M. K., & Rezaifar, O. (2023). Mechanical, environmental, and economic assessment of sustainable cement mortar using Afghan natural pozzolan as a partial replacement for cement. *Construction & Building Materials*, 386(May), 131574. DOI: 10.1016/j.conbuildmat.2023.131574

Wang, H., Liu, X., & Zhang, Z. (2023). Pozzolanic activity evaluation methods of solid waste: A review. *Journal of Cleaner Production*, 402, 136783. DOI: 10.1016/j.jclepro.2023.136783

Wang, J. Y., Zhang, J., Bai, X. M., & Luo, Y. (2016). Catalytic pyrolysis of lignin for the production of phenolic-rich bio-oil. *Bioresource Technology*, 216, 71–76.

Wang, T. Y., & Huang, C. Y. (2007). Improving forecasting performance by employing the Taguchi method. *European Journal of Operational Research*, 176(2), 1052–1065. DOI: 10.1016/j.ejor.2005.08.020

Wenting, S., Sun, Z., Ding, R., & Wang, X. (2023). Graphene-based materials for the separator functionality in lithium-ion batteries. *Materials (Basel)*, 16(12), 4449. DOI: 10.3390/ma16124449 PMID: 37374632

West, P., Shunmugasamy, V. C., Usman, C. A., Karaman, I., & Mansoor, B. (2021). Part I.: Friction stir welding of equiatomic nickel titanium shape memory alloy – microstructure, mechanical and corrosion behavior. *Journal of Advanced Joining Processes*, 4, 100071. DOI: 10.1016/j.jajp.2021.100071

Wood, E., Alexander, M., & Bradley, T. (2011). Investigation of battery end-of-life conditions for plug-in hybrid electric vehicles. *Journal of Power Sources*, 196(11), 5147–5154. DOI: 10.1016/j.jpowsour.2011.02.025

Wu, W., Zhu, H., & Xie, L. (2021). Recent developments in magnesium-based materials for electric vehicle applications. *Journal of Alloys and Compounds*, 867, 159074. DOI: 10.1016/j.jallcom.2021.159074

Xu, J., Zhang, X., Zhou, Y., & Jiang, G. (2015). Fast pyrolysis of biomass for bio-oil production: A review. *Journal of Engineering for Gas Turbines and Power*, 138(10), 101801.

Xu, W. F., Liu, J. H., Chen, D. L., Luan, G. H., & Yao, J. S. (2012). Improvements of strength and ductility in aluminum alloy joints via rapid cooling during friction stir welding. *Materials Science and Engineering A*, 548, 89–98. DOI: 10.1016/j.msea.2012.03.094

Xu, W., Li, W., & Ji, Y. (2021). Mechanical behavior investigation of reclaimed asphalt aggregate concrete in a cold region. *Materials (Basel)*, 14(15), 4101. Advance online publication. DOI: 10.3390/ma14154101 PMID: 34361295

Yang, W. S., Park, B. W., Jung, E. H., Jeon, N. J., Kim, Y. C., Lee, D. U., Shin, S. S., Seo, J., Kim, E. K., Noh, J. H., & Seok, S. I. (2017). Iodide management in formamidinium-lead-halide–based perovskite layers for efficient solar cells. *Science*, 356(6345), 1376–1379. DOI: 10.1126/science.aan2301 PMID: 28663498

Yang, Z., Zhang, J., Kintner-Meyer, M. C., Lu, X., Choi, D., Lemmon, J. P., & Liu, J. (2011). Electrochemical energy storage for green grid. *Chemical Reviews*, 111(5), 3577–3613. DOI: 10.1021/cr100290v PMID: 21375330

Yao, G., Wang, Z., Yao, J., Cong, X., Anning, C., & Lyu, X. (2021). Pozzolanic activity and hydration properties of feldspar after mechanical activation. *Powder Technology*, 383, 167–174. DOI: 10.1016/j.powtec.2021.01.042

Yoro, K. O., & Daramola, M. O. (2022). *Engineering a Sustainable Carbon Future.*

Yuzer, N., Ozata, S., & Akturk, B. (2023). *Atık Toprak Kapadokya Doğal Puzolanı ile Geopolimer Malzeme Üretimi ve Hazırlama Yöntemi* (Patent No. 2021/019771). Türk Patent. https://www.turkpatent.gov.tr/bultenler

Zaid, O., Mukhtar, F. M., M-García, R., El Sherbiny, M. G., & Mohamed, A. M. (2021, November). M-García, R., El Sherbiny, M. G., & Mohamed, A. M. (2022). Characteristics of high-performance steel fiber reinforced recycled aggregate concrete utilizing mineral filler. *Case Studies in Construction Materials*, 16, e00939. DOI: 10.1016/j.cscm.2022.e00939

ZairaZaman Chowdhury. Sagadevan Suresh (2017) Pyrolysis: A Sustainable Way to Generate Energy from Waste. , pp 1-36DOI: 10.5772/intechopen.69036

Zaumanis, M., Mallick, R. B., & Frank, R. (2016). 100% Hot Mix Asphalt Recycling: Challenges and Benefits. *Transportation Research Procedia*, 14(June), 3493–3502. DOI: 10.1016/j.trpro.2016.05.315

Zeeshan, K., Mirza, B. A., Student, M., Khan, Z., & Baig, M. A. (2017).. . *Utilization of Demolish Brick and Marble Waste in Concrete For Partial Replacement of Coarse Aggregate.*, 3, 1702–1707.

Zhang, H., Wang, M., Zhang, X., & Yang, G. (2015). Microstructural characteristics and mechanical properties of bobbin tool friction stir welded 2A14-T6 aluminum alloy. *Materials & Design (1980-2015), 65*, 559-566.

Zhang, H., Huang, Y., & Liu, Z. (2020). Current progress in advanced materials for high-performance sodium-ion batteries. *Energy Storage Materials*, 24, 1–22. DOI: 10.1016/j.ensm.2019.10.006

Zhang, S. S. (2006). A review on the separators of liquid electrolyte Li-ion batteries. *Journal of Power Sources*, 164(1), 351–364. DOI: 10.1016/j.jpowsour.2006.10.065

Zhao, X., & Park, N. G. (2015, November). Stability issues on perovskite solar cells. In Photonics (Vol. 2, No. 4, pp. 1139-1151). MDPI. DOI: 10.3390/photonics2041139

Zhao, H., Gao, X., & Hu, J. (2020). A review on next-generation lithium batteries for electric vehicles: Advances in cathode materials. *Energy Storage Materials*, 24, 329–353. DOI: 10.1016/j.ensm.2019.07.017

Zhao, S., Li, C., Zhao, M., & Zhang, X. (2016). Experimental Study on Autogenous and Drying Shrinkage of Steel Fiber Reinforced Lightweight-Aggregate Concrete. *Advances in Materials Science and Engineering*, 2016, 1–9. Advance online publication. DOI: 10.1155/2016/2589383

Zornberg, J. G. (2017). Functions and Applications of Geosynthetics in Roadways. *Procedia Engineering*, 189, 298–306. DOI: 10.1016/j.proeng.2017.05.048

Zumrawi, M. M. E., & Mansour, N. (2016). Laboratory Evaluation of Geogrid Used forStabilising Soft Subgrade. *Int. J. Civ. Environ. Eng.*, 10, 418–421.

About the Contributors

Ajay Kumar is currently serving as a Professor in School of Engineering and Technology, JECRC University, Jaipur, Rajasthan, India. He received his Ph.D. in the field of Advanced Manufacturing from Guru Jambheshwar University of Science & Technology, Hisar, India after B.Tech. (Hons.) in mechanical engineering and M.Tech. (Distinction) in manufacturing and automation. His areas of research include Incremental Sheet Forming, Artificial Intelligence, Sustainable Materials, Additive Manufacturing, Mechatronics, Smart Manufacturing, Industry 4.0, Waste Management, and Optimization Techniques. He has over 120 publications in international journals of repute including SCOPUS, Web of Science and SCI indexed database and refereed international conferences. He has also co-authored and co-edited many books with ELSEVIER, CRC Press, WILEY, De Gruyter and IGI Global and conference proceedings with IOP. He has organized various national and international events including an International conference on Mechatronics and Artificial Intelligence (ICMAI-2021), International conference on Artificial Intelligence, Advanced Materials, and Mechatronics Systems (AIAMMS-2023) as conference chair. He has more than 20 national and international patents to his credit. He has supervised more than 8 M.Tech, Ph.D scholars and numerous undergraduate projects/ thesis. He has a total of 15 years of experience in teaching and research. He is Guest Editors and Review Editor of the reputed journals including Frontiers in Sustainability. He has contributed to many international conferences/ symposiums as a session chair, expert speaker, and member of the editorial board. He has won several proficiency awards during the course of his career, including merit awards, best teacher awards, and so on. He is adviser of QCFI, Delhi Chapter student cell at JECRC University and has also authored many in-house course notes, lab manuals, monographs and invited chapters in books. He has organized a series of Faculty Development Programs, International Conferences, workshops, and seminars for researchers, PhD, UG and PG level students. He is associated with many research, academic, and professional societies in various capacities.

Parveen Kumar is currently serving as an Assistant Professor and Head in Department of Mechanical Engineering at Rawal Institute of Engineering and Technology, Faridabad, Haryana, India. He is pursuing Ph.D. from National Institute of Technology, Kurukshetra, Haryana, India. He completed his B.Tech. (Hons.) from Kurukshetra University, Kurukshetra, India and M.Tech. (Distinction) in Manufacturing and Automation from Maharshi Dayanand University, Rohtak, India. His areas of research include Intelligent Manufacturing Systems, Materials; Die less Forming, Design of Automotive systems, Additive Manufacturing, CAD/CAM and Artificial Intelligence, machine learning and Internet of Things in manufacturing, multi objective optimization techniques. He has over 50 publications in international journals of repute including SCOPUS, Web of Science and SCI indexed database and refereed international conferences. He has 8 national and international patents in his credit. He has supervised 2 M.Tech. scholars and numerous undergraduate projects/ theses. He has a total of 13 years of experience in teaching and research. He has co-authored/co-edited several books with ELSEVIER, CRC Press, WILEY, De Gruyter and IGI Global. He is also book series editor of "Industrial and Manufacturing Systems and Technologies: Sustainable and Intelligent Perspectives" in CRC Press, Taylor and Francis. He has organized a series of Faculty Development Programmes, workshops, and seminar for researcher, UG level students.

Tamer Sebaey graduated with honors in Industrial and Production Engineering in 2002, becoming a best-in-class student. Tamer commenced a career in academia as a Teaching Assistant at Zagazizg University and pursued a Master's degree with a specialization in lightweight structures. In 2009, he secured a scholarship from the Catalan Government, leading to an academic stint at the University of Girona. He completed his PhD which yielded nine research papers published in renowned Q1 journals, and received the coveted "Best Thesis Award" for the year. He contributed to institutions such as Qatar University, the University of Limerick in Ireland, and Prince Sultan University in Saudi Arabia. Currently, Tamer is a Full-Professor of Manufacturing and Design at Prince Sultan University. He has published over 100 papers, predominantly in Q1 journals and participated actively in more than 30 international academic events. His h-index is nearly 30and he successfully secured external funding amounting to approximately 1.5 million USD. He was granted external funding from Qatar foundation, Spanish Government, Government of Catalonia, and European Union. Tamer is a co-founder and leader in the domain of composite materials within the Structures and Materials Research Lab at Prince Sultan University. He served as the quality assurance chair for the College of Engineering at PSU as well as an external accreditation reviewer. He was recognized several times with the best paper award at international events and was recently acknowledged for exceptional contributions to the field.

Victor Gambhir is presently working as President/Vice Chancellor and Professor in the Mechanical Engineering Department at JECRC University, Jaipur, Rajasthan (since January 2021). He worked as Vice Chancellor and Professor in the Mechanical Engineering Department at SGT University Gurugram, Haryana from May 2020 to December 2020 and at Maharishi Markandeshwar (deemed to be University), Mullana, Ambala, Haryana from November 2016 to May 2020. He also worked as President/Vice Chancellor and Professor in the Mechanical Engineering Department at JECRC University, Jaipur, Rajasthan from June 2014 to November 2016. Prior to that, he worked as Pro-Vice Chancellor/Director Planning & Coordination and Professor in the Mechanical Engineering Department at Manav Rachna International Institute of Research and Studies from September 2006 to May 2014. He also worked as Workshop Superintendent at YMCA Institute of Engineering (now J. C. Bose University Science & Technology), Faridabad, Haryana from May 2005 to September 2006. Earlier, he worked as Joint Director, Technical Education Department, Government of Haryana from April 1998 to May 2005 and Deputy Director, Technical Education Department, Government of Haryana from April 1991 to June 1994. He has also worked as Head of the Department, Senior Lecturer, Workshop Superintendent and Lecturer in Government Polytechnics under the Department of Technical Education, Government of Haryana from December 1985 to March 1991 and July 1994 to April 1998. Prior to that, he worked in industry from June 1982 to May 1985 in the area of Engineering Projects and Pipelines. He obtained his Bachelor degree in Mechanical Engineering from the Thapar Institute of Engineering & Technology, Patiala (Punjabi University Patiala) in 1982 and Master degree in Mechanical Engineering with specialization in Rotodynamic Machines from Punjab Engineering College, Chandigarh (Punjab University Chandigarh) in 1998. He obtained his Doctoral Degree (Ph.D.) in Quality in Engineering Education from Manav Rachna International University (now Manav Rachna International Institute of Research and Studies, deemed to be University), Faridabad. He obtained his Diploma in Management from IGNOU, New Delhi in 1989 and attended a Foreign Fellowship Program sponsored by the World Bank at Northern Melbourne Institute of TAFE, Victoria, Australia from January 1998 to April 1998. He has published five textbooks for polytechnic students in the field of Applied Mechanics, Thermodynamics, Hydraulics and Hydraulic Machines, and Engineering Graphics. He has guided two Ph.D. dissertations. He has published nine research papers in international journals (six SCOPUS indexed) and six research papers in international/national conferences held by IISC Bangalore, IIT Delhi, and YMCA Institute of Engineering, Faridabad. He has been awarded two patents as co-inventor. He is a Member of the American Society for Mechanical Engineers, Life Member of the Indian Society for Technical Education, and Fellow Member of the Institution of Engineers (India).

***

Sarita Baghel is currently working as Assistant professor in Department of Applied Physics, Delhi Technological University, Delhi-110042, India. Her research interest is the design, synthesis, optimization and characterization of different types of solar cells.

Ankit Beniwal is an Assistant Professor at JECRC University, Jaipur, India M. Tech. - Transportation and Infrastructure Engineering, IISc Bangalore, India B.Tech. - Civil Engineering, MNIT Jaipur, India

Sonia Chahar is aresearch student currently pursuing her PhD at Chitkara University. Her primary research interest is in materials science and engineering, particularly in the fields of perovskite and two-dimensional (2D) materials. Sonia's PhD research revolves around the development and optimization of perovskite and 2D materials for their application in optoelectronic devices and solar cells. Her work aims to enhance the efficiency, stability, and overall performance of these materials, which hold great promise for the next generation of renewable energy technologies.

Ravpreet Kaur has a doctorate in Civil Engineering and a gold medalist in Structural Engineering. She has an over 8 years of academic experience and more than 15 research papers.

Rahul Kundara is a senior research fellow in Department of Applied Physics, Delhi Technological University, Delhi-110042, India. His research interests focus on synthesis, characterization and optimization of lead-free perovskite solar cell for the application in solar industry.

Kasturi Laturkar is a validation engineer for Validation Associates LLC. She graduated with a M.S. in Chemical Engineering from Syracuse University and a B.Tech in Chemical Engineering from Guru Gobind Singh Indraprastha University, Delhi, India. She has more than 5 years of experience working in commissioning, qualification and validation of upstream and downstream bioprocessing equipment and critical utilities.

Ram Meena is an Assistant Professor at JECRC University Jaipur, India

Krishna Mishra holds a Ph.D. Degree in Solid State Electronics and Master Degree in Physics with specialisation in Electronics. Dr. Mishra has 18 years of experience in teaching and research with more than 30 research papers to his credit published in various journals and conferences at national and international level. He has been a reviewer on various reputed journals and a member of various governing bodies and professional societies during his tenure. Currently, he is a Professor of Physics/Electronics and Pro Vice Chancellor (Quality Assurance) at Chitkara University, Punjab. He has visited UK, Indonesia, Kuwait, Philippines, Hong Kong, Macau, Singapore, Malaysia and UAE for presenting his research and as a speaker in various events of reputed ranking agencies.

Sivakumar N completed his B. E. EEE at CIT, Coimbatore and M. E Power Electronics at Anna university. In 2018 he completed his Ph.D in the area of Multilevel Inverters and Fuzzy Logic Applications. His area of intrest include Solar PV power harvesting technology, Power Electronics, Electric Vehicles, and Application of Artificial intelligence in power system operation etc. He has more than 14 Years of Industrial Experience and 16 years of Academic with 10 years in Research included. He published 23 papers in reputed International Journals and attended 6 international conferences.

Serife Ozata is an Assistant Professor in the Department Architecture at Samsun University, Faculty of Design and Architecture. She earned both her Bachelor's and Master's degrees from Yildiz Technical University (YTU), and subsequently completed her PhD at YTU in 2022. With a profound expertise in in-situ and laboratory testing, Dr. Ozata specializes in the documentation, diagnosis, conservation, and restoration of built and cultural heritage. Over the past decade, she has imparted knowledge on material-oriented and structural systems of concrete, steel, wooden, and masonry buildings as a dedicated teaching assistant. Her primary research interests encompass heritage preservation, structural analyses, material analyses and characterization, waste valorisation, alkali activation, geopolymer, compatible and sustainable materials for historical buildings, diagnosis, restoration, on-site application of new materials, in-site NDT, destructive tests, and the application of artificial intelligence in architecture. Dr. Ozata has made significant contributions to multidisciplinary project groups supported by the Scientific and Technological Research Council of Türkiye, and has filed two patent applications related to her PhD thesis and interdisciplinary projects. She is deeply committed to advancing the field of architecture and building materials through her exemplary research and teaching endeavours.

Gourisankar Roymahapatra is currently working as an Associate Professor of Chemistry at the School of Applied Science and Humanities, Haldia Institute of Technology (HIT) Haldia, India. He completed his Ph.D. from Jadavpur University, Kolkata (2014) with N-Heterocyclic Carbene Chemistry as a major field under the mentoring of Dr. Joydev Dinda of HIT Haldia, and Prof. Ambikesh Mahapatra of Jadavpur University. Before joining to Global Institute of Science and Technology (GIST Haldia), as a Lecturer in Chemistry in the year 2011, he worked at MCC PTA

Chem. Corp. Pvt. Ltd India (MCPI), Haldia, (2003 ~ 2011) as a Senior Chemist. He joined HIT Haldia in 2015 as an Assistant Professor of Chemistry. His area of research interest includes NHC complexes, organometallics, catalysis, antibiotics, anti-carcinogenic, gas adsorption, super alkali, computational chemistry and DFT. For his contribution in research, he got the 'Distinguished Young Scientist Award in Chemistry -2014'; 'Bharat Gaurabh Award – 2018'; and 'Indian Chemical Society Research Excellent Award 2021'. Dr. Roymahapatra is a Life Fellow of the Indian Chemical Society (FICS), and a Life Fellow of the Institutes of Chemistry

Nishant Saxena is a PhD Research Scholar in the Department of Mechanical Engineering, UIT RGPV Bhopal. His research work is related to the treatment of municipal solid waste materials.

Ranjodh Singh has a doctorate in Civil Engineering. He has an over 10 years of academic experience and authored more than 15 research papers.

Serhan Ulukaya graduated from Yildiz Technical University, Faculty of Civil Engineering, Department of Civil Engineering in 2005. He achieved his M Sc and Ph D degrees at Istanbul Technical University and Yildiz Technical University in 2008 and 2016, respectively. He has been working in Yildiz Technical University since 2010. His main study areas are cementitous materials, concrete technology and historic construction materials.

Nabi Yüzer graduated from Yıldız Technical University of İstanbul, Faculty of Civil Engineering, Department of Civil Engineering in 1989. He worked in a construction company in 1990. In 1991, he worked as a Research Assistant at Yıldız University. He completed his master's degree at Yıldız Technical University, Institute of Science and Technology in 1992. He earned his PhD degree in 1998. He became an associate professor in 2005, and Professor in 2011. He works on the durability of concrete such as high temperature, reinforcement corrosion, chloride and sulfate effects, and material properties in historical buildings. He has been serving as the Vice Dean of YTU Faculty of Civil Engineering since 2020.

Index

www.ingramcontent.com/pod-product-compliance
Lightning Source LLC
LaVergne TN
LVHW061540290926
849162LV00004B/859